THE ART OF STRUCTURAL DESIGN

A SWISS LEGACY

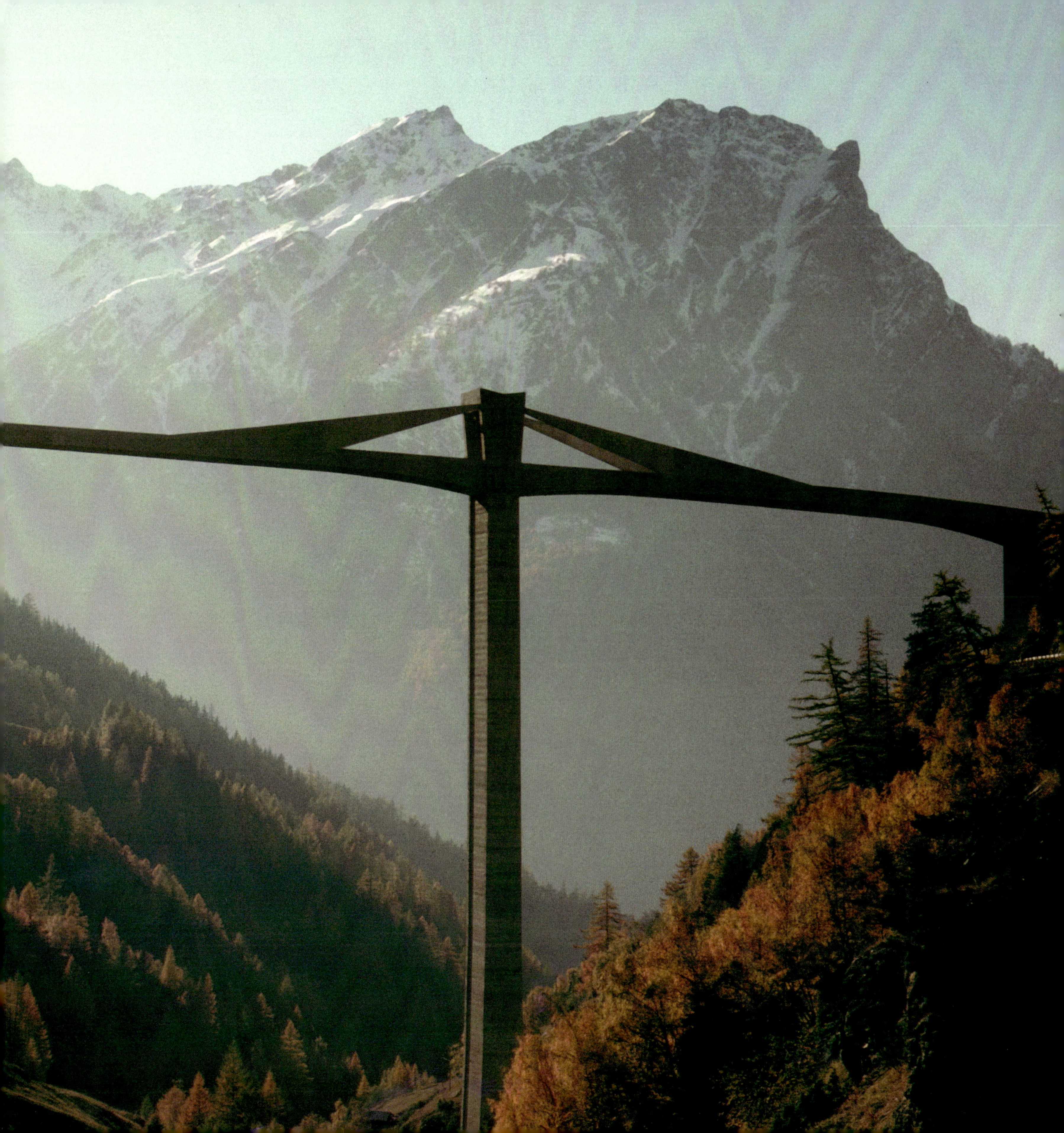

THE ART OF STRUCTURAL DESIGN

A SWISS LEGACY

David P. Billington
With a contribution by Jameson W. Doig

Princeton University Art Museum

Distributed by Yale University Press
New Haven and London

The Art of Structural Design: A Swiss Legacy
is published in cloth and paper editions
by the Princeton University Art Museum
Princeton, New Jersey 08544-1018

The cloth edition is distributed by Yale University Press,
New Haven and London

The book is published on the occasion of the exhibition
"The Art of Structural Design: A Swiss Legacy"
Princeton University Art Museum
March 8 – June 15, 2003

Managing Editor: Jill Guthrie

Assistant Editor: Nicola Knipe

Editorial Assistants: Nathan T. Arrington and Mary Cason

Copy Editor: Sharon Herson

Design: CoDe. New York Inc.,
Jenny 8 del Corte Hirschfeld, Mischa Leiner

Indexer: Kathleen M. Friello

Printer: Meridian Printing, East Greenwich, Rhode Island

Library of Congress Control Number: 2002116510
ISBN 0-300-09786-7 (cloth); 943812-38-4 (paper)

The book was set in Caslon, Berthold Helvetica,
on Vintage Velvet.

Jacket illustrations (from front flap):
Bronx-Whitestone Bridge (Ammann), Sicli Garden Center (Isler), Ganter Bridge (Menn), George Washington Bridge (Ammann), Heimberg Tennis Center (Isler), Vessy Bridge (Maillart), Sunniberg Bridge (Menn), George Washington Bridge (Ammann), Felsenau Bridge (Menn), Salginatobel Bridge (Maillart), Bayonne Bridge (Ammann), Grötzingen Outdoor Theater (Isler), Chiasso Shed (Maillart), Bürgi Garden Center (Isler), Leonard P. Zakim Bunker Hill Bridge (Menn)

Frontispiece: Ganter Bridge (Menn)

Printed and bound in America

The publication was made possible by:
Preston H. Haskell, Class of 1960, Ammann, Whitney, Frances E. and Elias Wolf Fund, Swiss-American Cultural Exchange Council, Pro Helvetia Arts Council of Switzerland, Princeton University School of Engineering and Applied Science, Princeton University Department of Civil and Environmental Engineering, the Friends of the Princeton University Art Museum, and the Partners of the Princeton University Art Museum.

This book, and everything else, is dedicated to Phyllis.

CONTENTS

FOREWORD

In a recent letter, Christian Menn, who is widely regarded as the greatest bridge designer practicing today, wrote to David Billington: "Over the course of many years, you did much for me. I learned a lot from you, and I am well aware that I owe to you the most important part of my professional career." Menn's esteem is matched in the judgments of present and past students alike. A recent alumnus, now a writer, told me without hesitation that the most memorable course he had ever taken was one taught by David Billington. He went on to explain the reason: it had transformed the way he looked at and experienced manmade structures – especially bridges.

David P. Billington, Gordon Y. S. Wu Professor of Engineering and director of the Program in Architecture and Engineering, is a renowned scholar, beloved teacher, respected colleague, and generous mentor. His enthusiasm and passion for his work infuse every aspect of his career. A leading educator in the field of structural design, Professor Billington has challenged the profession to teach not only the principles of structural engineering but also the importance of aesthetics and beauty as they relate to function. In so doing, he has created a discipline in which efficiency, economy, and elegance are on equal footing. Using the case-study method, he has pioneered an educational approach that deliberately and formally links engineering to the liberal arts. His placement of engineering within the context of design, politics, and aesthetics has resulted in a curriculum that is recognized internationally as an innovative approach to the study of structural design. The Princeton University Art Museum is honored to join with Professor Billington in presenting *The Art of Structural Design: A Swiss Legacy.* This exhibition and the associated publication, the culmination of forty-five years of research and teaching, give visual form to his principles and celebrate his achievements.

This is not the museum's first collaboration with Professor Billington. Five other exhibitions over a thirty-year period welcomed new audiences to the art museum, broadening their understanding of the built environment and of the roles design and aesthetics played in relation to structural considerations. The current exhibition is the first to focus on the group of Swiss engineers associated with the Federal Institute of Technology in Zurich. Robert Maillart and Othmar Ammann studied there with Wilhelm Ritter; Heinz Isler and Christian Menn worked under Pierre Lardy. The pupils, widely recognized as the most innovative structural designers of the twentieth century, developed their approach and practice as a result of the educational philosophy of their teachers, for whom aesthetic considerations weighed as heavily as those of structure. By examining the work of the four engineers as artists and the philosophies of the two educators who were their mentors, the exhibition exemplifies a vital aspect of the museum's mission – to extend the boundaries of the definition of art and art education.

I would like to acknowledge Princeton alumnus and University Trustee Preston H. Haskell, Class of 1960, who introduced me to David Billington and encouraged this new undertaking. His generosity extended well beyond an introduction to include significant financial support. The Francis E. and Elias Wolf Fund and a challenge grant from the National Endowment for the Arts, a federal agency, have also helped to make this initiative possible. In addition, the museum would like to thank the Friends of the Princeton University Art Museum, the Partners of the Princeton University Art Museum, the Swiss-American Cultural Exchange Council, Pro Helvetia Arts Council of Switzerland, the Princeton University Department of Civil and Environmental Engineering, Princeton University School of Engineering and Applied Science, and Ammann & Whitney for their generous contributions to the exhibition and catalogue.

I join David Billington in recognizing the important contributions of Jameson W. Doig, professor of Politics and Public Affairs at Princeton, as co-author of the chapter on Othmar Ammann. My gratitude extends to Rebecca Sender, associate director, and Jill Guthrie, managing editor, for shepherding this complex endeavor through all its phases. They, along with the entire museum staff, deserve special recognition. In addition, a talented group of students worked with lab supervisor Joseph Vocaturo during the summer of 2002 to design and build models for the exhibition: Courtney Clark, Class of 2003; Elizabeth Grau, Class of 2003; Maria Janaro, graduate student; and Joshua White, Class of 2004.

The museum and the Department of Civil and Environmental Engineering, together with generations of engineering students and an international group of teachers and designers, celebrate David Billington's accomplished career, recognizing that he, like the structural artists about whom he writes, stands in the tradition of Wilhelm Ritter and Pierre Lardy.

Susan M. Taylor,
Director

This book is the culmination of more than thirty years of research in Switzerland, which began formally when my son David and I entered the Zurich apartment of Marie-Claire Blumer-Maillart in midsummer 1970. Extending her friendship to me, Marie-Claire graciously shared her recollections and her documents, and encouraged my colleague Robert Mark and me to hold a symposium celebrating the centennial of the birth of her father, Robert Maillart. John Abel, then at Princeton, played a major role in organizing that celebration and in editing the conference proceedings. David Steadman, acting director of the Princeton University Art Museum, embraced the idea of an exhibition concurrent with the symposium. The resulting "Bridges and Sculpture" was devoted primarily to Maillart but included works of the sculptor Max Bill, who had written the first book on the engineer.

During succeeding trips to Switzerland, I worked closely with Marie-Claire and her husband, Edouard Blumer, to collect more documents, visit works designed and built by Maillart, and establish contact with his friends and colleagues as well as other family members. Through our symposium in 1972, I came to know Christian Menn, who was then a professor at the Federal Institute of Technology (ETH) in Zurich. He opened numerous doors for me to local archives and provided introductions to canton engineers, as well as to colleagues and students at the ETH. He understood the structures of Robert Maillart better than anyone else, and better still, he introduced me firsthand to the ideas of structural art when we visited his own bridges along with those of Maillart. Thus began my education and early attempts to teach and write about the new art of structural engineering.

With my colleague Robert Mark, a brilliant experimentalist, I began a scholarly program aimed at understanding the greatest works of structure. He focused on structures before the Industrial Revolution and I concentrated on those after the late eighteenth century. Through the invaluable guidance and support of Princeton president Robert F. Goheen, this fruitful collaboration gained financial support in the late 1960s from the newly formed National Endowment for the Humanities, and especially from its sympathetic program officer, Herbert MacArthur. Our chairman of Civil Engineering, Norman Sollenburger, who was responsible for bringing me onto the Princeton faculty, strongly encouraged the new scholarship on historical structures.

Following the 1972 exhibition and symposium, I began to work on a biography of Robert Maillart, and in 1974 Marshall Cloggett agreed to sponsor me as a visitor to the Institute for Advanced Study. In that same year, with the backing of our then chairman Ahmet Cakmak, I introduced the course "Structures and the Urban Environment," which focused on structures as an art form. Also in 1974 we held our second exhibition in the museum, this one on the 1874 Eads Bridge, thanks to the fine cooperation of Peter C. Bunnell, director of the museum. He also endorsed a second exhibition on the works of Maillart in 1976, the year in which "Structures and the Urban Environment" received during the spring term the highest student rating of all large courses (over fifty students) in the entire university. In 1978, a fourth exhibition was held in the museum, "The Bridges of Christian Menn," and in 1980, a related one, entitled "Heinz Isler as a Structural Artist"; for each of these shows, the respective museum directors, Fred Licht and Allen Rosenbaum, were exemplary collaborators. I also over the years have received invaluable assistance on the scholarly works of the Swiss engineers from the archivists at the ETH–Alvin Jaeggli, Beat Glaus, and more recently, Flavia Janini and Christine Bärtsch. My work in Switzerland was materially furthered by a grant from the Swiss Society of Cement, Chalk, and Gypsum Manufacturers, whose executive officer, Dr. H. Eichenberger, provided not only funds but considerable aid in documentation and insight into Swiss culture.

I met Heinz Isler at a conference in the Netherlands in 1961, but I did not fully appreciate his work until a second meeting at Morgantown, Virginia, in 1978, when his lecture there stunned the audience and stimulated me to visit him in Switzerland to see his thin shells and to plan the art museum exhibition of 1980. When we first met, I was writing my text book on thin shell concrete structures; after the second meeting I reworked that text, influenced considerably by Isler's innovations. Although I had done research, writing, and consulting on thin shells, Heinz Isler changed my views drastically and taught me in the same way as did Christian Menn. Both Isler and Menn have generously given of their time, and both have spent many weeks at Princeton presenting teaching seminars. I am also grateful for the help of Dorothy Candela for giving me many materials on her husband, Felix Candela, and on Edward Torroja, and for her friendship.

In 1977 I wrote a paper on Othmar Ammann and the George Washington Bridge, which received severe criticism from leaders in the profession and caused me to study more deeply that great engineer's ideas and works. Thanks to Hans Hauri and his colleague Tom Peters, I became involved with the Ammann Centennial of 1979, wrote a paper on Wilhelm Ritter, and began more serious study of Ammann's New York bridges. In that same year, my first book on Maillart appeared with the help of my editor, Edward Tenner. In 1981, while I was struggling to write a full biography of Robert Maillart, Martin Kessler, president of Basic Books, appeared in my office at the suggestion of my brother, James H. Billington, to ask me to write a book on the new course. Kessler was an ideal editor, as was the copy editor, Sheila Friedling. The book, *The Tower and the Bridge: The New Art of Structural Engineering* (title invented by my wife, Phyllis), was published in 1983 and incorporated Ammann, Isler, and Menn as well as Maillart into the central argument for structure as a new art form. Shortly after *The Tower and the Bridge* appeared, Victoria Newhouse and Edgar Kaufman Jr. came to my archive at the suggestion of George Collins, art historian at Columbia University, who had written a seminal article on Maillart for the 1972 symposium. The result of that visit was another book, *Robert Maillart and the Art of Reinforced Concrete*, published in 1990 by Victoria Newhouse's Architecture History Foundation, and underwritten by Mr. Kaufman. Mrs. Newhouse also accepted my Maillart biography and did a marvelous job of editing it before deciding to close her publishing house and giving my manuscript to Cambridge University Press, where Beatrice Rehl did fine work in getting it published in 1997.

Beginning in the late 1980s, my long-time colleague Jameson Doig and I began talking about Ammann. Jim was working on his history of the Port Authority of New York and New Jersey and, in going through my archive on Ammann, he discovered correspondence between the engineer and George Silzer, governor of New Jersey from 1923 to 1926. This immediately set Doig on the track of that relationship. He soon discovered an archival gold mine in Trenton at the New Jersey State Archives on "Ammann's Bridge," which the Trenton archivest did not recognize as the George Washington Bridge. This treasure revealed an important and completely unstudied side of Ammann's career. It led us to collaborate on a scholarly paper in 1994, and also on a seminar given in 1992 and 1999 in the American Studies Program. I owe a great deal to Jim Doig's political perspective on public works, an essential context for most of the works discussed in the present volume. We have taken parts of our previous writing, reworked them, and added to them to form chapter 3 in this book. We are deeply indebted to Dr. Margot Amman-Durrer for her continuing help and for providing documents for all our works on her father.

Beginning in 1983, I had occasion to think more deeply about teaching, thanks in part to a program begun in that year by the Alfred P. Sloan Foundation called "The New Liberal Arts," which focused on bringing technology into the liberal arts curriculum. Foundation president Albert Rees and program officer Samuel Goldberg were instrumental in stimulating me to organize the structures materials and disseminate them to more than thirty institutions of higher learning. The Sloan Foundation program also forced me to think more carefully about the impact of teaching, both on engineers and on liberal arts students. Gradually I came to see the need to study not just pedagogy in the abstract, but also—and more importantly—to study the ideas and practices of exemplary teachers.

In 1951–52, while on a Fulbright renewal fellowship in Ghent, Belgium, I was profoundly influenced by Gustave Magnel, and in 1966 on a National Science Foundation fellowship to Delft, the Netherlands, I came under the powerful influence of Warner Koiter. In addition to the paper on Ritter, written in 1979, I presented a retirement celebration paper in 1990 on the teaching of Alexander Scordelis, an important educational figure at the University of California at Berkeley. In many discussions with Isler and Menn, I had come to recognize the exemplary teaching of Pierre Lardy. Then in 1999, Viva Hardigg interviewed me and published an article in *The Journal of Engineering Education* on ideas about education—experiences that further convinced me of the need to probe beneath the surface into the nature of engineering education. These most recent events helped turn my attention to a more consistent study of great teaching and provided the impetus for the present book.

In the fall of 2001, Peter R. Jaffé, chairman of the Department of Civil and Environmental Engineering, to whom I am deeply indebted, and I began to discuss the possibility of an exhibition in the art museum to summarize my work. It seemed natural to focus on teaching but in a scholarly sense and to center on the great Swiss tradition that I had been studying since the late 1960s. Earlier, Preston H. Haskell, Class of 1960, had introduced me to Susan Taylor, the new director of the Princeton University Art Museum, and so Peter and I went to see her about an exhibition. She received the idea with considerable enthusiasm, and the project was soon underway.

My long-time colleague J. Wayman Williams prepared an elegant book of photographs to give an overview of the exhibition and the book. In addition, he has made a photographic record of the students' works in designing and building models for the exhibition. Early in 2002, I began to work closely with the museum's managing editor, Jill Guthrie, on the publication, which under her careful direction was completed by early fall. Her sympathetic understanding and exacting standards are responsible for the high quality of this book. Moreover, it has been a personal pleasure and an extraordinary privilege to work with her. I am indebted to Professor David G. De Long of the University of Pennsylvania for his thoughtful review of the manuscript, which led to many important revisions. I also thank my colleague Mary

Sansalone for her excellent review and strong encouragement over many years for the type of teaching described in this volume. The highly professional art museum staff, working with Rebecca Sender, associate director, who has been in overall charge, has provided the expertise needed at crucial times: Caroline Cassells, curator of Education and Academic Programs; Frances Yuan, manager of community relations; Mike Jacobs, chief preparator; Nicola Knipe, assistant editor; Mary Cason, editorial assistant; and Nathan Arrington, museum intern. All have been enthusiastic from the start and it has truly been a joy to work with them. I would also like to thank Tee Williams, associate director of Corporate and Foundation Relations; Jane Maggard, assistant dean for Development, School of Engineering and Applied Science; and Michele Minter, senior associate director, Leadership Gifts, for their efforts in securing funding for this project.

The quality of the book is in great measure is due to the superb photographs by Bruce M. White. Frank Döring, who traveled to Switzerland and Germany, also provided effective photographs for both the book and exhibition. The ETH archive under Flavia Lanini has been particularly responsive to our photographic requests, and many other images have come from Marie-Claire Blumer-Maillart, Margot Ammann-Durrer, Heinz and Maria Isler, and Christian and Hedy Menn. Hans Hauri sent me his recollections of Lardy, as did Hans von Gunten. The manuscript was meticulously and sensitively copyedited by Sharon Herson. Mischa Leiner and Jenny 8 del Corte Hirschfeld of CoDe, New York, are responsible for the beautiful and imaginative book and exhibition design. The excellent quality of the printing is the work of Meridian Printing, under the direction of Daniel Frank.

The entire undertaking—book and exhibition—was conceived as an educational event and therefore relied heavily on students at Princeton, especially Maria Janaro, a graduate student working on Isler; Elizabeth Grau, a senior studying Maillart; Joshua White, a junior focusing on Ammann; and Courtney Clark, a senior concentrating on Menn. These students did research, helped collect photographs, and designed models for the exhibition. Along with technical supervisor Joseph Vocaturo, they built models for the exhibition, aided later by Theodore Griffith. William Cooch, a senior, prepared drawings for the book and helped in the research on Lardy. Ryan Woodward, a graduate student, traveled to Switzerland to see Menn and visit archives at the ETH. Michael Barelli, Class of 2002, did additional research on Ammann and collected photographs. And Rick Ellis, a graduate student, helped Maria Janaro in her work on Isler. Chelsea Honigmann, a former graduate student, began the organization of the archival materials and other documents, and after she left, museum intern Nathan T. Arrington, Class of 2002, completed that very detailed task. My secretary, Kathy Posnett, has typed the original manuscript, organized my files, as well as anyone can, and has been for many years a splendid and faithful coworker.

Many former and present students have contributed to this work; they include Daniel Chung, Peter Cole, Nicholas Edwards, Paul Gauvreau, Wendy Gottshall, James Guest, Roger Haight, Greg Hasbrouck, Neil Hauck, Michael Hein, Mark Herron, Eric Hines, Scott Hunter, Nicolas Janberg, Gayle Katzman, David Lamb, Nicole Leo, Abbie Liel, Ellen Lieng, Susan Lyons, Sinead Mac Namara, John Matteo, Karen Mielich, John Ochsendorf, Karena Ostrem, Angela Ovecka, Christopher Peck, Caroline Pennington, Rosemary Secoda, Robert Shulock, Kent Smith, Michael Starc, Michael Tantala, Moira Treacy, David Wagner, and Ronald Wakefield. I am indebted to James Chiu, who voluntarily organized the Geneva Maillart Archive and thus founded the Princeton Maillart Archive, so essential to the research for this book.

For the research on Ammann, I am grateful for the help provided by the Metropolitan Transit Authority of New York, the Port Authority of New York and New Jersey, the Museum of the City of New York, the archives at Case-Western Reserve University and Syracuse University; special thanks to Edward Wladas, engineering librarian at Princeton University.

To a group of my closest colleagues—my family—I extend heartfelt thanks and deepest gratitude. My historian brother, James H. Billington, has been mentoring me for forty years on historical writing; the best kind of critic, insightful and sympathetic, he always raises exciting ideas. My eldest son, David, also an historian, has read all my historical works in manuscript and insisted on crucial changes; he learned early that disagreeing with his father was the proper road to higher ground. My eldest daughter, Elizabeth, a musician with an artist's eye for photography, has also edited some of my early work. My youngest daughter, Sarah, a professor of structural engineering at Stanford University and a distinguished structural engineer in her own right, is another independent critic. My other children, Jane, Philip, and Stephen, have all traveled with me to study structures, serve as scale factors, and respond to my ideas as the good liberal arts students that they are. Finally, above all, is my wife, Phyllis, to whom all my work is dedicated.

David P. Billington

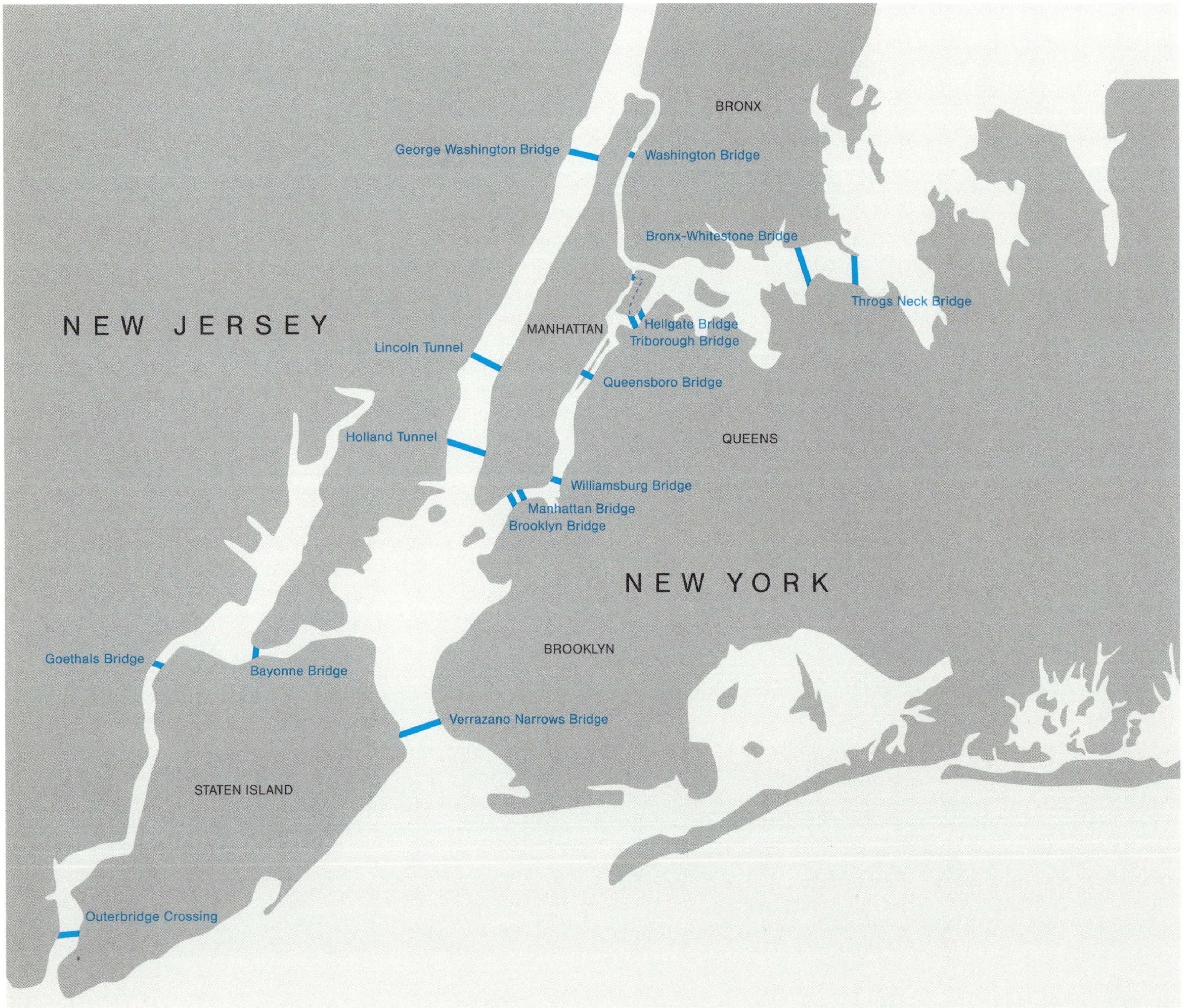
BRONX
George Washington Bridge
Washington Bridge
Bronx-Whitestone Bridge
Throgs Neck Bridge
NEW JERSEY
MANHATTAN
Hellgate Bridge
Triborough Bridge
Lincoln Tunnel
Queensboro Bridge
Holland Tunnel
QUEENS
Williamsburg Bridge
Manhattan Bridge
Brooklyn Bridge
NEW YORK
BROOKLYN
Goethals Bridge
Bayonne Bridge
Verrazano Narrows Bridge
STATEN ISLAND
Outerbridge Crossing

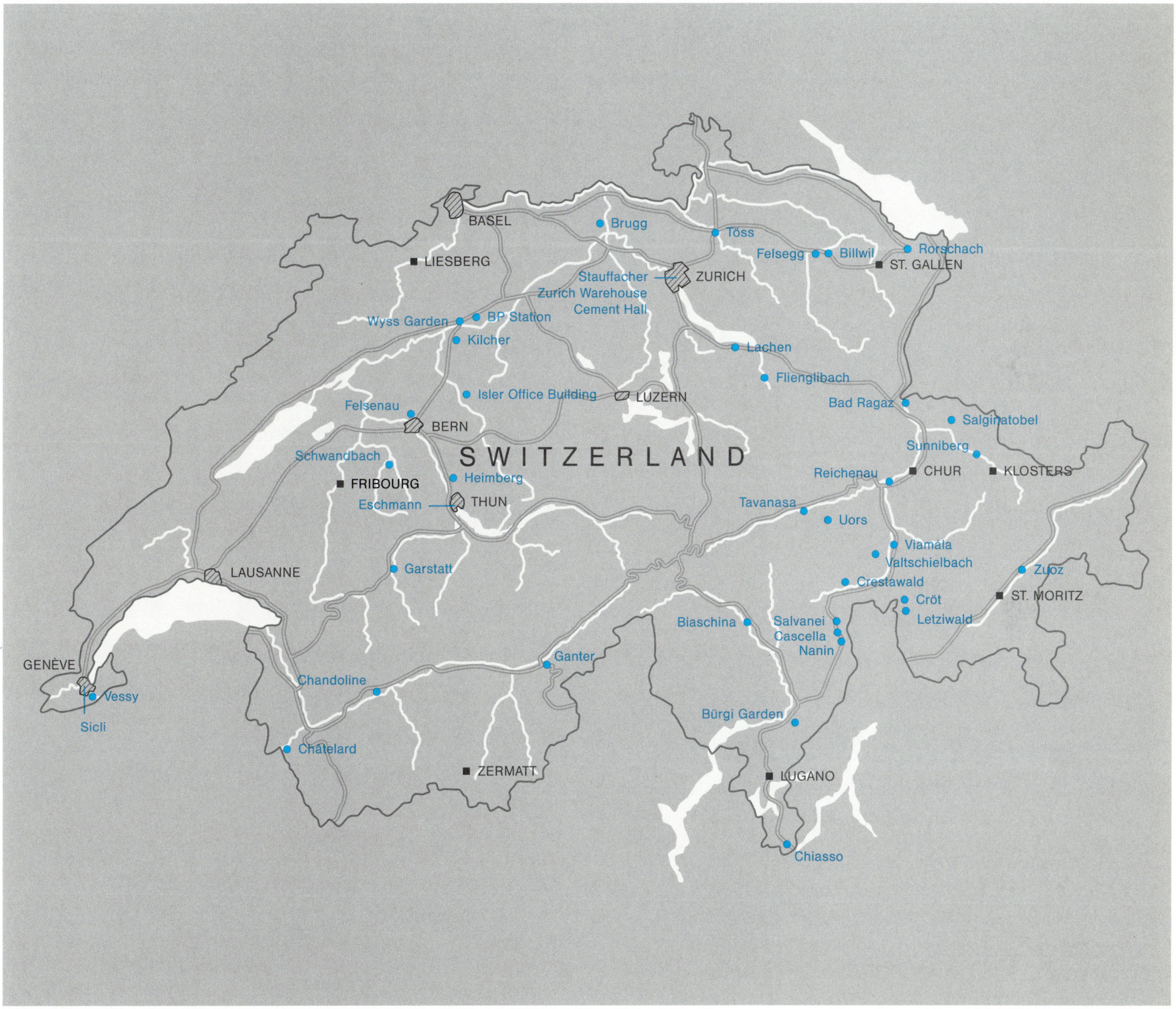
SWITZERLAND
BASEL
LIESBERG
Brugg
Töss
Felsegg
Billwil
Rorschach
ST. GALLEN
ZURICH
Stauffacher
Zurich Warehouse
Cement Hall
Wyss Garden
BP Station
Kilcher
Lachen
Flienglibach
Isler Office Building
LUZERN
Bad Ragaz
Salginatobel
Felsenau
BERN
Sunniberg
Schwandbach
FRIBOURG
Heimberg
Reichenau
CHUR
KLOSTERS
Eschmann
THUN
Tavanasa
Uors
Viamala
Valtschielbach
Zuoz
Garstatt
LAUSANNE
Crestawald
Cröt
ST. MORITZ
Letziwald
Biaschina
Salvanei
Cascella
Nanin
Ganter
GENÈVE
Chandoline
Vessy
Sicli
Bürgi Garden
Châtelard
ZERMATT
LUGANO
Chiasso

Rarely does a new art form emerge to challenge old ideas about artistic boundaries. This has happened in our present technological age, which has given birth to the new art of structural engineering.[1] My goal here is twofold: to illustrate how this new art, as a powerful expression of our culture, is fully accessible to the general public; and to demonstrate why it is an essential part of the education of engineers. The focus is on four of the greatest structural artists of the twentieth century—Robert Maillart, Othmar Ammann, Heinz Isler, and Christian Menn—and on their unique engineering education at the Swiss Federal Institute of Technology in Zurich.

Certainly, great buildings existed long before the Industrial Revolution. Many included major structural elements—for example, the Pantheon and the High Gothic cathedrals—but these are clearly works of architecture.[2] Their engineering was unscientific and also subordinate to the larger goals of spatial organization and direct human use: unscientific because designers did not use structural mechanics, the formal study of which did not exist until the nineteenth century; and subordinate because the leading designers were architects whose ideals included decoration and the creation of spaces in which people would congregate. The designers learned the technical part of their profession from field observation and from codified writings like those of Vitruvius and Alberti.[3] Some of their works approached pure structural form—for example, the aqueduct at Segovia—but the scale was still small enough that empirical rules could ensure a safe structure, the material was usually heavy enough to avoid the dangers of buckling, and the live loading was slight enough to permit forms that were appropriate to the weight of the structure alone. The introduction of industrialized iron would change all that.

Before the nineteenth century, nearly all public construction was in the hands of architects. Even bridges were regularly called works of architecture, especially when they were built out of stone. This ancient building material had been beautifully refined in France during the eighteenth century, and construction techniques had advanced as well. Thus, when iron began to dominate in the nineteenth century, a tradition of building already existed. Much of that tradition has had to be abandoned or radically reconceptualized. This effort did not fit within the ancient canons of architecture and soon required a new type of education. By the mid-nineteenth century, many engineering schools had been founded and the modern engineering profession established. Thus, there was not a split between architecture and engineering; rather, a new profession grew up alongside an older one. Even in the early twenty-first century, the general public still seems to believe that bridges and all buildings are part of architecture. Undoubtedly, many buildings in which structure is subordinate are properly works of architecture, but some buildings are not, as the work of Heinz Isler demonstrates. Sadly, many recently constructed bridges have forms imagined by architects, but as these chapters will show, the best bridges are purely the work of engineers, illustrated by the structures of Robert Maillart, Othmar Ammann, and Christian Menn.

It is thus that most structural engineers work without any relationship to architects and that engineering students need to study the two hundred-year history of structural art to recognize its significance for the future. Likewise, some structural engineers work closely with architects and make important contributions to edifices that are nevertheless properly called architecture. There is some analogy here between structure and music. Architects often speak of their art as that of the conductor who leads a variety of specialists to bring a great symphony to life. Structural engineers, as one specialty for a complex building, are also like the soloists who perform a complex work on their own without other instruments or even a conductor. Here I focus on the engineer as a soloist who conceives a new form, visualizes its final appearance, defines it by calculations, and develops a means of building it. This does not mean that the designer alone can bring the structure into being. Other engineers may do the detailed calculations necessary to dimension the elements, construction specialists will organize the building site and arrange the elements, and experienced architects may provide a helpful critique of the design. But the work of structural art remains the product of one person's imagination. And that imagination can operate only after substantial formal education and practical experience. It is the interrelationship of education, experience, and expression that this text seeks to explore.

Therefore, one principal theme of this book is that structural engineering at its best is a new art form, beginning with the Industrial Revolution of the late eighteenth century. It is parallel to but independent of architecture in the same way that photography, that other new art of the nineteenth century, is parallel to but independent of painting. Furthermore, structural art has flourished most impressively in Switzerland, in no small part because of the engineering educational tradition at the Federal Institute of Technology in Zurich. This book presents that tradition through two exemplary teachers, Wilhelm Ritter (1847–1906) and Pierre Lardy (1903–1958) and four of their students: Robert Maillart (1872–1940) and Othmar Ammann (1879–1965) studied with Ritter, and Heinz Isler (born 1926) and Christian Menn (born 1927) with Lardy.

Imagine studying twentieth-century architecture without mentioning Frank Lloyd Wright, Le Corbusier, Walter Gropius, and Mies Van der Rohe. Imagine studying American literature of the last century without knowing about—or reading—the works of Ernest Hemingway, F. Scott Fitzgerald, William Faulkner, and T. S. Elliot. Education in structural engineering—indeed, in all engineering—has deprived students and the general public of the humanistic values inherent in engineering and so central to architecture and literature. One goal of this book is to help overcome that truncated education by describing one major part of the grand tradition of structural engineering, stressing the pivotal role of educators in the process.

For the civil engineer, the great innovation of the Industrial Revolution was the introduction of industrialized iron—a process that improved upon an old material by producing a higher quality more economically than had previously been possible. From the eighteenth century on, this new iron enabled erection of large-scale structures made out of metal instead of wood or stone. The best nineteenth-century structural engineers came to see how this new material could make possible unprecedented forms for bridges, towers, and vaults.

But the new possibility required a new type of education because iron, being still expensive, could be used economically only in small quantities. With such light structures, safety became critical and the means to analyze these new forms required a more scientific study than had been usual for structures in wood or stone. The need for such study produced a new education for the modern engineer, typified by the curricula at the École Polytechnique in Paris (established in 1794) and the early-nineteenth-century Technische Hochschulen in Germany. Thus, a large intellectual gap opened between classically educated architects and scientifically educated engineers.

In Western culture, the pre-Industrial Revolution tradition of structural design as the province of architects was challenged in the nineteenth century by a few gifted engineers in this new scientific tradition, most prominently Thomas Telford (1757–1834) in Britain, Gustave Eiffel (1832–1923) in France, and John Roebling (1806–1869) in the United States. Each of these engineers believed that the best designs for iron and steel bridges, towers, and vaults could be made only by those educated in the new discipline of structural engineering. They based their ideas not on an abandonment of aesthetic sensitivity, but rather on the development of an aesthetic appropriate to the new industrial democratic societies that began to flourish after the three great eighteenth-century revolutions: the British (Industrial), the American, and the French. In different ways each of the great nineteenth-century artists symbolized in his best works the revolution that changed his own nation.

Telford's dramatic iron designs for arch and suspension bridges still stand today in full service as exemplars of the British revolution. Eiffel's Parisian iron tower of 1889 literally commemorates the French Revolution, and his great railroad bridges in France's Massif Central demonstrate the restructuring of his country in the modern world. At the same time, Roebling's steel and stone Brooklyn Bridge showed an amazed Europe the growing dominance of the United States following the founding of its democratic government. In each case, the structural forms came exclusively from the engineering imagination, and for each designer, aesthetic sensitivity was central to his thinking.

But this sensitivity arose within the new disciplines of structural mechanics and political accountability. Wasting materials and money was an anathema to these early structural artists. Their careers were formed by the new disciplines, and their search for elegance stayed within their boundaries. Indeed, those disciplines served as stimuli to art because structural engineering is not, as many people suppose, a mere application of science or a search for "the one best way." As these artists recognized, there are many possible designs for a given function, all of which would be roughly equivalent in efficiency (minimum materials) and economy (competitive costs). While respecting this discipline, structural artists seek to make designs that appeal most to their own aesthetic sensitivities.

Telford first described this basis for structural art in an article in 1812; Eiffel expressed the same ideals in his defense of the tower against the high-art world of late-nineteenth-century Paris; and Roebling expressed an equivalent vision in the preliminary report for his great bridge.[4] These arguments and the works that express them in metal are the bases upon which the great structural artists of the twentieth century would create these structures, but two competing ideologies have delayed and deflected public acceptance of this art. The first expresses the belief that all structure is merely a part of architecture and that the engineer's work is purely technical while the architect determines the aesthetics. The second holds that engineering is applied science and hence any aesthetic arises from the laws of nature, which dictate an optimum for each case. The first requires an architect to be in charge, whereas the second claims that architecture is irrelevant, as is any aesthetic choice by the engineer. Both ideologies are particularly damaging to the wider understanding of structural art in bridges because the first often implies that beauty requires great cost and thus little economic discipline, while the second suggests (erroneously) that beauty naturally occurs because efficiency automatically leads to elegance.

Nevertheless, while both positions have something to add to structural art, they lack its crucial element, the centrality of cost. Engineering educators can learn much about aesthetic sensitivity from architects, while the engineering students must learn the discipline of engineering mechanics to guide their future ideas about design. The key to structural art, however, is the way in which mechanics and aesthetics are linked through practice, which means through continual study of both construction processes and the behavior of structures in service.

While the practice of structural art had an impact on engineers and the general public during the nineteenth century, the education of engineers by the end of that century had begun to turn away from the practice and toward the mechanics. This in part explains why the new art did not flourish as widely as one might have anticipated in Britain, France, and the United States after the three pioneers died. Rather, the most important tradition of engineer-artists began to appear in Switzerland. To a large extent, this small, land-locked confederation adopted the ideals of the nineteenth-century leaders and went on to create the greatest works of structural art in the new century.

Notes

1. For a detailed presentation of structural art, see D. P. Billington, *The Tower and the Bridge: The New Art of Structural Engineering* (Princeton, 1985).

2. For the clearest structural study of these buildings, see R. Mark, *Light, Wind and Structure* (New York, 1989).

3. Vitruvius, *Ten Books on Architecture*, trans. M. H. Morgan (1914, reprinted New York, 1960). For a general discussion of Alberti's works in context, see Rudolf Wittkower, *Architectural Principles in the Age of Humanism* (New York, 1962).

4. Billington, *The Tower and the Bridge*, chap. 1.

CHAPTER 1

WILHELM RITTER: THE ORIGINS OF SWISS STRUCTURAL ART

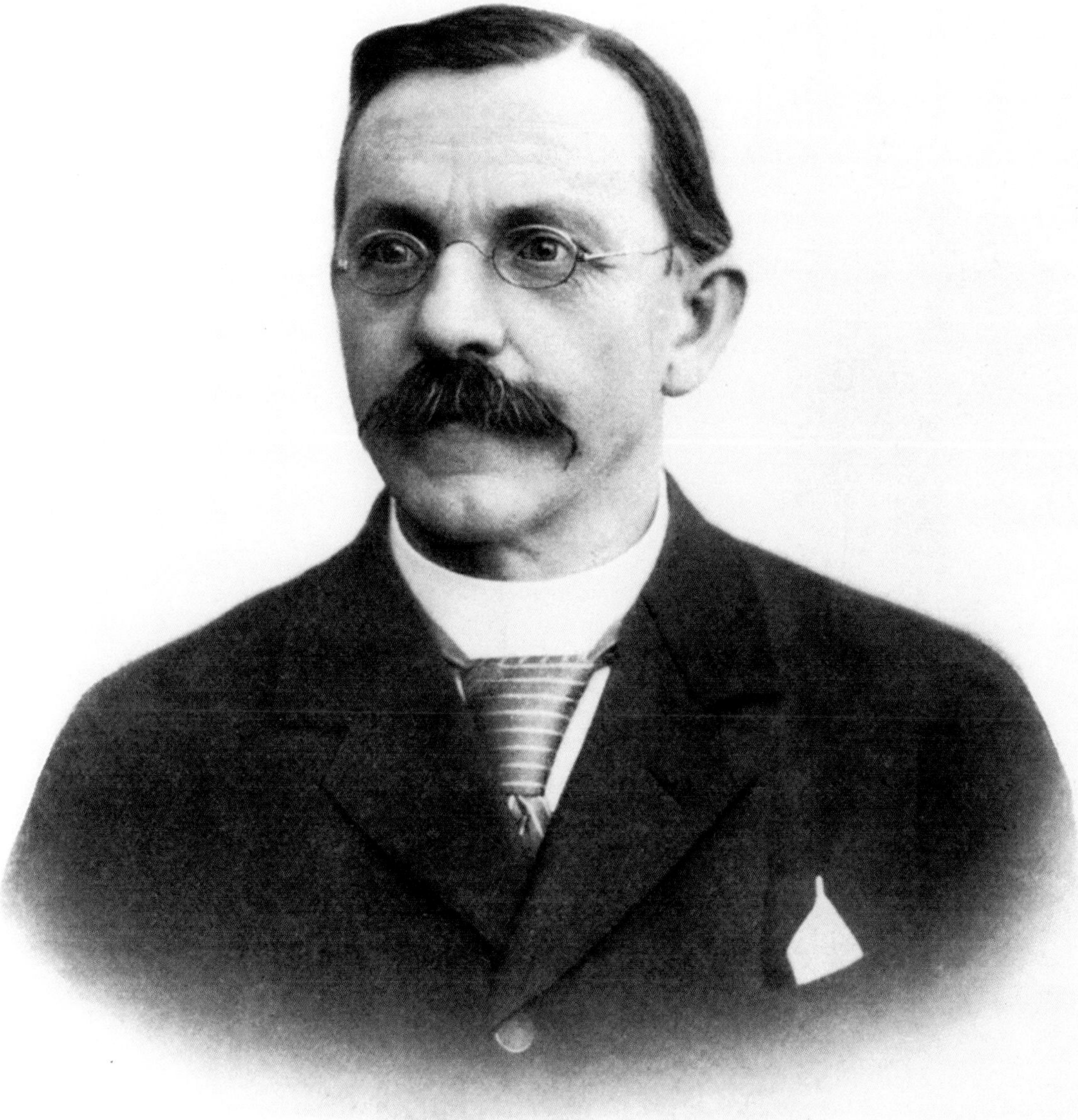

Figure 1 (previous page)
Karl Wilhelm Ritter (1847–1906).

Figure 2
Federal Institute of Technology, Zurich, Switzerland, ca. 1900.

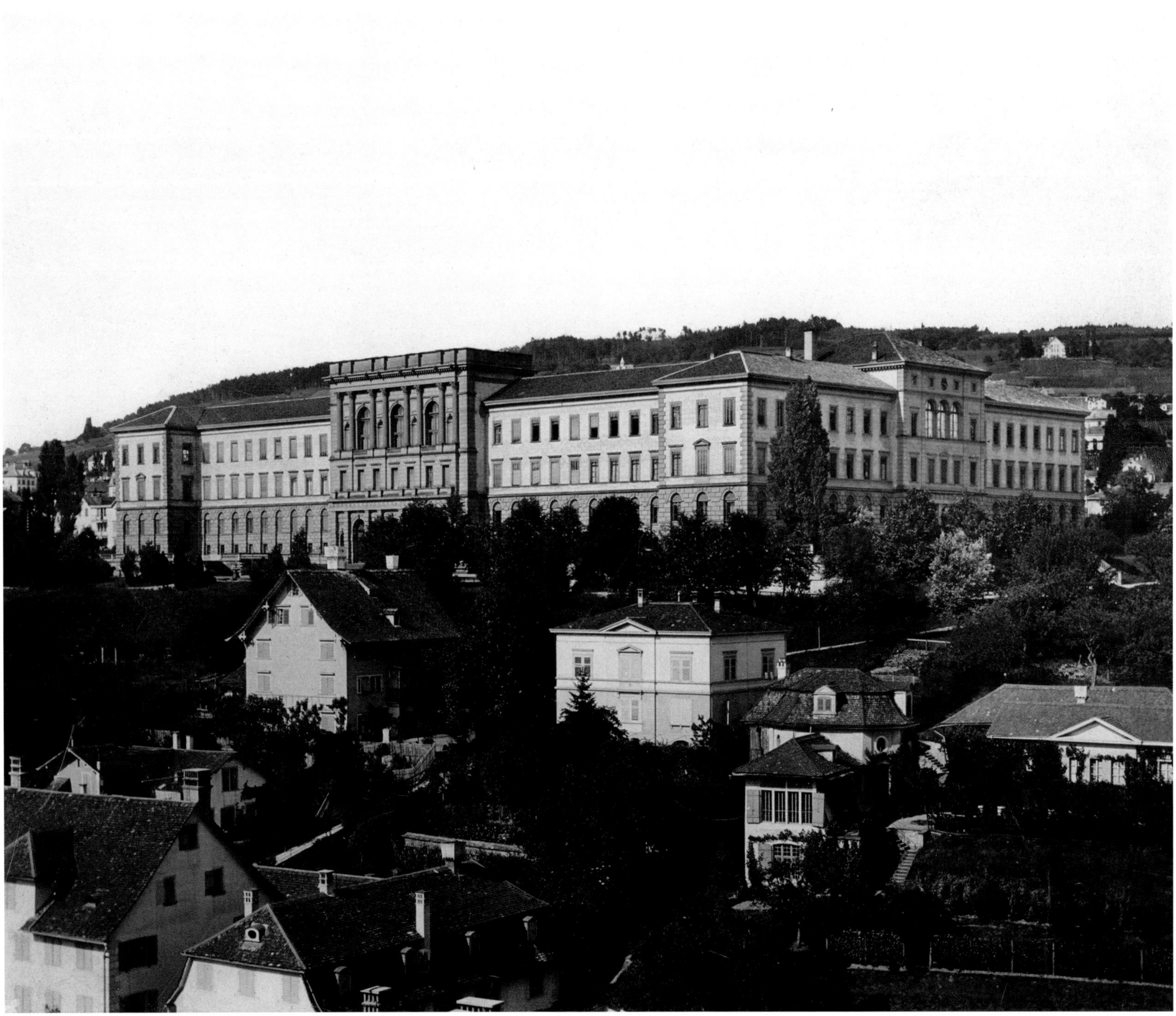

The revolutionary fervor that swept through Europe in 1848 played a major role in Swiss history by providing a setting in which the Swiss could form a centralized government based on a new constitution. Although a Swiss confederation had existed since 1291, it did not function as a modern nation until after 1848. A very brief civil war in late 1847 actually precipitated the founding of modern Switzerland. Born in that year, Wilhelm Ritter (fig. 1) would become Switzerland's first teacher of structural art. Although to this day Switzerland is the most decentralized of the major European nations, its impressive economic strength owes much to the unity achieved in 1848.

A primary goal of the new government in 1848 was to establish two federal institutions of higher learning in Switzerland: a federal university in Zurich and a federal technological institute in Lausanne.[1] It was assumed that education, like the postal service, the railways, and the customs agency, would benefit by a more centralized presence within the reconstructed confederation.

Two distinct traditions of engineering education predominated in the Europe of 1848. The French method, founded during the revolution, standardized and controlled all education from Paris; indeed, most engineering was taught in the capital city. The other practice was the German, in which educational institutions were founded by individual states, each having different local characteristics; thus, engineering was taught in many different places without one school being dominant. Common to both traditions was the principle that engineering required an institution completely separated from the university, where classical studies and pure sciences were taught. This clear division originated at least in part with the ideas that engineering was a new subject and that its content was barely, if at all, connected to subjects taught in universities. The field was seen as closely linked to the growing prosperity of an emerging industrial economy and thus symbolic of something new, different, and useful. To a mid-nineteenth-century government, the technical school seemed to stand as an emblem of the nation's future.

Thus, when the new federal government in Bern tried to establish two institutions, it found two different responses. There was no major objection to an engineering school, but to a new federal university there arose strenuous dissent from the cantons in which universities already existed. The result was that a federal university never appeared, but a Federal Institute of Technology did, in Zurich in 1855 (fig. 2).[2]

There were at least three good reasons why conditions were excellent for the founding of a new Swiss engineering school in 1855: it came late; it was undertaken as a national enterprise; it was built after substantial and high-quality planning and debate. Its lateness with respect to the French and German schools meant that its founders could benefit from a study of numerous working examples and could choose well-trained people for the first faculty. The first proposal, drawn up in 1851 by French-Swiss, closely followed French examples, especially those in the Parisian École Centrale des Arts et Manufactures, but a proposal later in the same year, guided by the German- Swiss Alfred Escher (1819–1882), adhered more to the German example of Karlsruhe. One significant difference was in the organization of the faculty: the French system had individual professors and central administrators, while the German had groups of professors in departments (*Fachschulen*) with fewer central personnel. Eventually, the Federal Assembly chose the Karlsruhe system.[3]

With the selection of the Karlsruhe system, it was only natural that the search for the founding faculty would turn in that direction as well. In 1854, nearly two hundred individuals were contacted as potential professors; of these 113 were German, 57 Swiss, 11 French, 3 Belgian, 3 English, and 2 Italian.[4] The first group to be appointed was impressive. At least 4 of the original 31 professors were men whose international stature was so high that each could have been called the best academic in his field anywhere: architect Gotfried Semper (1803–1879), who designed the institute building, civil engineer Carl Culmann (1821–1881), physicist Rudolf Clausius (1822–1888), and historian Jacob Burckhardt (1818–1897).[5] Only Burckhardt was Swiss, the other three were German, and of those Semper and Culmann were the most important to the education of Robert Maillart and Othmar Ammann.

In addition to having the advantage of a comparatively late beginning, the new school had the even greater benefit of being a national institute, the only one of its kind in Switzerland, and thus a unifying symbol. At the opening ceremonies on October 15, 1855, the first president of the board of regents, Senator Johann Konrad Kern (1808–1888), spoke about how this type of institute could reconcile Switzerland's different traditions of locality, nationality, and faith.[6] Recalling the strife that led to the 1848 constitution, Kern saw the new school as a catalyst for uniting Protestants and Roman Catholics, French-Swiss and German-Swiss, those from the canton of the city of Basel (density of 830 people per square kilometer in 1850) and those from the Graubünden (density of 12.4 people per square kilometer). That this ideal would be made possible through an engineering school was characteristic of one mid-century view of technology.

Figure 3
Carl Culmann (1821–1881).

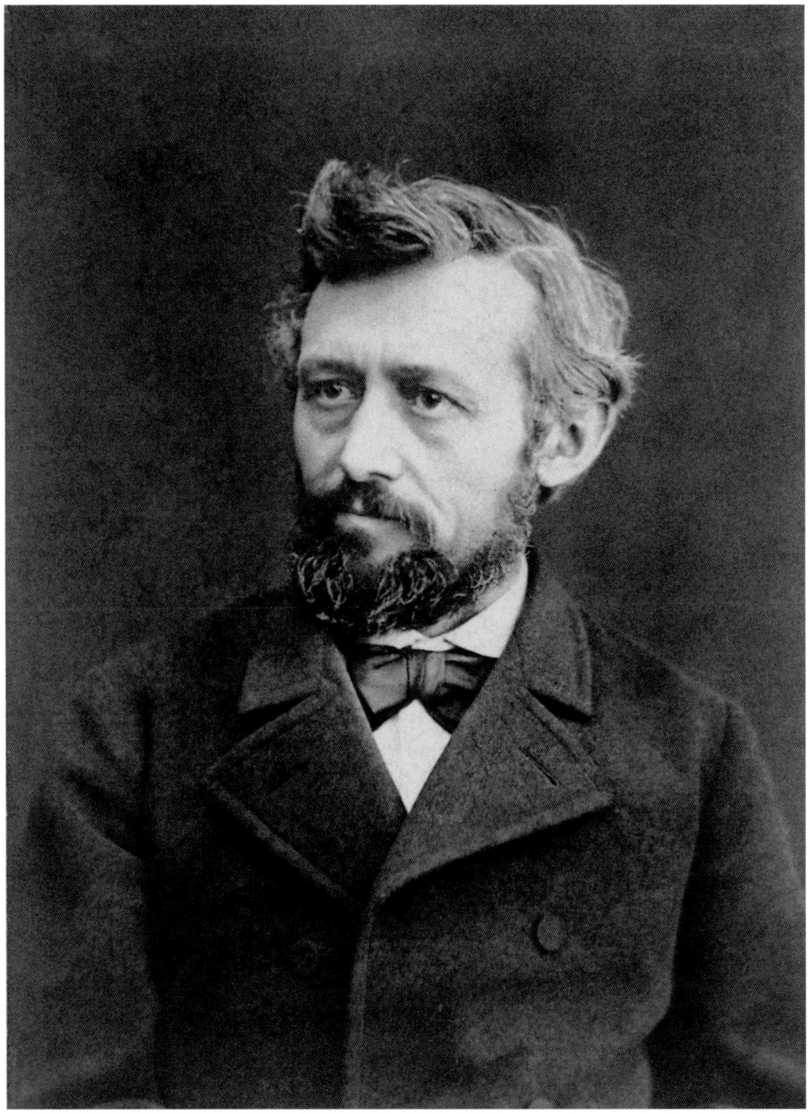

But perhaps most important of all to the success of the new school was the quality of the debate and planning that led to the federal law establishing the school (February 7, 1854), and then to the regulations passed by the Federal Assembly (July 31, 1854). The two most important figures in the debate, planning, and successful launching of the school were Alfred Escher from Zurich and Johann Kern from the Thurgau. Both were presidents of the Federal Assembly after 1848 and remained leading national figures until they died. They succeeded largely because of the people they chose—the most important for civil engineering was Carl Culmann.

Following the designation of Gotfried Semper as head of the Department of Architecture, Culmann was Kern's second major appointment (fig. 3). Born in Bavaria, Culmann received an engineering diploma from Karlsruhe in 1841, after which he worked for the Bavarian state railways until called to Zurich in 1855.[7] In addition to having firsthand field experience with railroad structures built during the early days of the railroad boom, Culmann was strongly motivated to study recently completed structures elsewhere and to systematize structural engineering. These goals led him in 1849 to make a two-year study trip to Britain and the United States to learn about bridge building and railway construction; the result was a widely read report that was published in 1851–52.[8] Next, he began detailed studies of structural analysis. Early in his teaching career, he began to systemize these studies, and in 1866 he published them in *Die Graphische Statik* [Graphic Statics], probably the most influential book on structural analysis of the time.[9] The basic idea behind his work was to demonstrate structural behavior through geometric diagrams rather than through algebraic formulas. "Drawing is the language of the Engineers," he used to say, "because the geometric way of thinking is a view of the thing itself and is therefore the most natural way; while with an analytic method, as elegant as that may also be, the subject hides itself behind unfamiliar symbols."[10]

In addition to teaching and writing, Culmann served as a consultant on practically every important Swiss bridge built during his tenure at Zurich. Moreover, he studied stream flooding in Switzerland as well as retaining-wall problems, and published a famous study in 1865 on similarities between stresses in the human hipbone and a loaded building crane. A revised and expanded version of *Die Graphische Statik* appeared in 1875, and in 1879 it was translated into French. At his death in 1881, Culmann's work was known and used throughout Europe, but its value was not uncontested. As his contrast (quoted above) between geometric and algebraic methods implies, there was another school of thought, one that believed more in abstract analyses and formulas. But the Culmann tradition was more than just geometry over algebra; it set the tone for civil engineering in the Zurich school for half a century. The bases for his influence were his intensive experience in the field from 1841 to 1855; his extensive travel for the study of foreign public works; and his individual research that led to the development of visual methods for structural analysis.

To become a tradition, the Culmann way required more than just a gifted founding professor; it demanded a successor to carry forward teaching in the same vein. It is not too much to say that the second individual is crucial to the establishment of a tradition, as opposed to the enshrinement of a master. Culmann, Robert Maillart, and Othmar Ammann were fortunate that the professor chosen in 1882 to succeed the founder was Karl Wilhelm Ritter (1847–1906).[11]

In 1865 Karl Wilhelm Ritter (he dropped his first name very early) entered the Federal Institute of Technology in Zurich, graduating first in a class of twenty civil engineers in 1868. After working for a year on a railroad line in Hungary, he was invited by Culmann to be one of his assistants. After another year he qualified as a *Privatdozent*, which allowed him to run classes on structural mechanics for architects and on practical geometry for students preparing to enter the school. He also began to help students who were having difficulty—as many did—understanding Culmann's lectures.

In 1873, Ritter was called to Riga to become a regular professor, an unusual honor for such a young engineer. The polytechnic school in Riga had opened in 1862 and had quickly become a major institute in Russia, thanks mainly to its German-language base and its location so near to other Western European centers of industry and education. On the strength of Culmann's international reputation—he was renowned even by 1862—the new school invited one of his former assistants, Henri Bessard (1837–1873), to be its first professor of civil engineering. In the spring of 1873, Bessard fell from a bridge on an inspection trip and died. The Riga school sent back to Culmann for another assistant, and he warmly recommended Ritter.[12]

Following Culmann's death, Ritter was called back to Switzerland and immediately began to think about completing Culmann's great project of writing the second volume of *Die Graphische Statik*. The more he looked into the project, the more he became convinced that such an effort could not be done as Culmann had imagined it. Ritter finally decided to write his *Anwendungen der Graphischen Statik* [Applications of Graphic Statics] as five separate volumes (only four ever appeared) in which he would think out on his own the organization and content. Clearly, he was continuing the Culmann tradition, but the new books were to be his own. Ritter explained the relationship in his foreword: "Evidently Culmann had the idea to keep the second volume as close as possible to the first. The writer [Ritter] had a feeling of reverence leading him that same way; on the other hand, by following Culmann's way, the integrity of the treatment would have been prejudiced; and especially so when one looks at the recent developments… Thus, it seemed more advisable for a fresh and independent work to appear."[13] Not only did Culmann's published approaches need substantial reworking, but also some of his general methods of presentation seemed awkward to Ritter. For example, in his second edition, Culmann had presented both geometric and algebraic methods of analysis. Ritter thought that this was redundant and that the algebraic treatment did not belong in a work entitled *Graphic Statics*. Thus, Ritter, at age thirty-five, began his tenure in Zurich within the spirit of Culmann's independent thinking and critical judgment; ironically, this involved respectful criticism of Culmann himself.

In a brief biographical sketch in 1903 for a German biographical dictionary, Ritter, near the end of his own career, reflected critically on Culmann's influence: "Culmann's method of teaching was not easy to follow. Because of his lively temperament, his thoughts often ran ahead of his words. Also, his books in many ways lacked the desirable clarity and thoroughness."[14] But he followed this rather severe judgment with an image that is more credible and suitable—perhaps because it followed a negative critique: "He was like an eagle who draws his circle high over the heads of his students. What his verbal presentation and his written publications lacked was compensated for by the personal inspiration with which he enlivened his lectures and the warm personal interest which he took in each of his students."[15] Ritter's words give a better sense of Culmann than mere superlatives ever could. For the student, it was more important to have discovered his teacher's inspiration and personality than clarity and thoroughness in his work. Ritter's own clear critique of the work makes us that much more sympathetic toward his final eulogy to the man: "The gifted mastery of his material, the astounding ease with which he answered the most difficult questions, and the underlying bases of his character—kindness and modesty—gained him the undivided respect and honor of all who sat at his feet."[16] This is not the uncritical adulation of a loyal assistant, but the human response of one great man to another. Indeed, it might be suggested

Figure 4
Cover of Ritter's book *Der Brückenbau in den Vereinigten Staaten Amerikas* [The Bridges in the United States] (Zurich, 1895), based on a three-month trip to the United States.

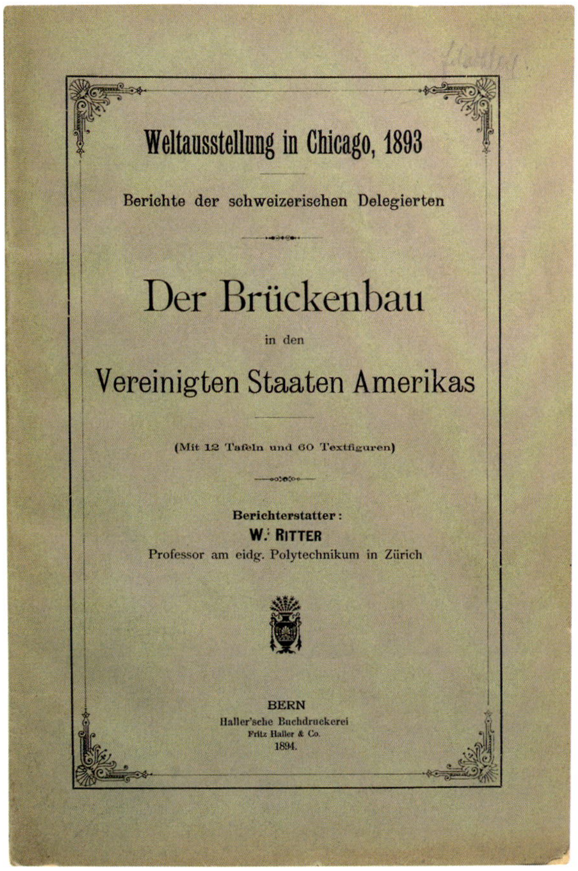

that Ritter, very near the end of his own life, was unconsciously writing his own epitaph and, in a way, the epitaph of a nineteenth-century engineering education that was already on the wane.

Between 1873 and 1882 Ritter matured in Riga as a major European engineering academic, and when the call came from Zurich he was ready—not as an assistant is ready to move into his master's office, but as Culmann himself had been ready in 1855—an independent mind prepared to assume major responsibilities in what was by then a well-developed school. Ritter would later reflect that his first years back in Zurich were the best of his life: "The scholarly activity of my colleagues spurred me on to a great extent. Scientific study was highly respected here and found its way quickly into practice, while I was looking into questions of structural analysis more and more each year." [17]

Culmann's field of teaching had so broadened by 1881 that he was actually replaced by two professors: Ritter taught Graphic Statics and Bridges, and Eduard Gerlich (1836–1904), who had been chief engineer for the construction of the Gotthard line, instructed in the fields of Railroad Engineering and Management.[18] Ritter's work therefore focused on developing applications for Culmann's graphic statics, on studying structural problems in bridge design, and on answering a wide range of practical questions for which his advice was sought. His earliest major books were the first two parts of *Anwendungen der Graphischen Statik* [Applications of Graphic Statics]: *Die Inneren eines Balkens Wirkeneden Kräfte* [Internal Stresses in Beams] appeared in 1888 and *Das Fachwerk* [Trusses] in 1890. The next volumes, *Der Kontinuierliche Balken* [Continuous Beams] and *Der Bogen* [Arches], were published in 1900 and 1906, respectively. Since he was too sick to complete the final volume, his son, engineer Hugo Ritter, finished it for him during the last year of his life.[19]

Another important piece of writing was a long article that appeared in the first six issues of the new *Schweizerische Bauzeitung* [Swiss Building Journal] of 1883. Its subject, calculations for stiffened arches and stiffened suspension bridges, emphasized the wide compass of Ritter's research and especially his deep inquiry into modern bridge forms. This investigation later included a three-month trip to the United States in 1893 to visit the Chicago World's Fair and to study American bridges. His lectures in 1893–94 were greatly enlivened as a result of this tour, and in 1895 *Der Brückenbau in den Vereinigten Staaten Amerikas* [The Bridges of the United States] was published (fig. 4).[20]

Ritter had also been serving as a consultant on a wide variety of projects and had studied the full-scale load testing of completed bridges. In 1885, he became a member of the city of Zurich's building council. By the late 1880s, he was so well known beyond Switzerland that in 1889 he received an offer from the prestigious Technical University in Munich, which he declined "out of love for the polytechnic school and for his country."[21] The Federal Council, out of gratitude, granted him a lifetime appointment, and the city of Zurich gave him and his family citizenship in that city as a token of their thanks. These were unusual honors for an engineering professor, and they seemed to express the nation's pride in their new school and the place it now held in European circles.

Ritter made a deep personal impression on his students and associates. He was a family man, known to his students for his hospitality. He and his wife, an American from Boston whom he met on a visit home from Riga in the summer of 1874, had five children. He played an important role in their early education, including personally teaching his sons how to make engineering drawings—graphics even in the home. He was also a fine pianist. Deeply religious and a devoted member of the Methodist church, he understood and respected the beliefs of others. He was noted for his fairness and for his unwillingness to speak against anyone behind his back. He liked to say that he worked for knowledge and not for money.

In 1887, he was made director of the Federal Institute of Technology, just as Culmann had been fifteen years before, and he remained in this position until 1891. Between 1896 and 1898, he was president of the Zurich Natural Science Research Organization, and in 1898 the University of Zurich awarded him an honorary doctoral degree. Ritter was much honored and had great talents, but his susceptibility to deep feelings that were almost depressions seems to have been connected with the onset of a nervous agitation that began to make work difficult for him after 1900. In the spring of 1902, the illness forced him to give up his work, and this, in turn, plunged him into deep melancholy. He traveled to Locarno, Albisbrunn, and Spiez for rest. Finally in the spring and summer of 1904, he taught again, but the effort was too demanding. He entered a sanatorium in Kussnacht until March 1905. He then moved to the Asyl Remismuhle, where he died on October 18, 1906.

Ritter's death concluded a half-century tradition of structural engineering education in Zurich, which was to produce over the next half century at least two men who can be called the outstanding bridge designers during the first part of the twentieth century; one worked in concrete and the other in steel. Their general interest in bridges and much of their later work with special forms can be easily traced to the Zurich tradition, and especially to the ideas of Wilhelm Ritter.

Ritter's ideas, like those of Culmann, followed from local field experience, international study and travel, and individual research that emphasized visual methods of analysis. These three components permitted, in more general terms, a teaching that included firsthand experience with structures in the natural environment, a wide variety of structural images drawn from the results of construction in different social settings, and scholarship in which the teacher alone worked out a new approach to older ideas. This new approach needed to be one that arose directly from design practice but was systematic and general enough to be of basic value to students throughout their entire careers.

On April 9, 1892, there appeared in the *Schweizerische Bauzeitung* a short note arguing that full-scale load tests on steel bridges were not only worthless but possibly even misleading.[22] A lively debate followed, ending on July 9 with a brief report from a distinguished German professor from Karlsruhe, Franz Engesser, who "protested against the situation in which load tests even on little structures with a span of 2 meters will be widely made, because such experiences are not only useless but also are actually detrimental since through them time, energy, and money are squandered and operations are blocked and endangered."[23]

In the next issue, Ritter wrote a detailed defense of what had become the common Swiss practice of full-scale load testing.[24] In effect, he presented a Swiss position at odds with a German one. In a broader sense, he reflected a pragmatic and balanced attitude toward understanding structural behavior against a more theoretical and dogmatic approach that emphasized analysis. It would be wrong to overgeneralize this distinction; but insofar as one can characterize national attitudes, the Swiss tended to rely less on the emerging mathematical theories in engineering and to be more open to the need for visual demonstrations of performance. The difference between Engesser and Ritter on this bridge question was primarily a difference in attitude and philosophy.

What Ritter presented was the viewpoint that public works set in a difficult environment are always built with uncertainty. There was no way in the late nineteenth century to predict mathematically the full response of a public structure; and despite many new mathematical theories, detailed textbooks, and immense computing power, the same condition exists in the early twenty-first century. The validity of any work rests, as Ritter emphasized near the end of his article, with "the probing expert" who must give "a reliable judgment." In short, the final test relies on the judgment of a person and not on the solution to an equation.

Thus, for Ritter the firsthand field experience so useful to the engineer was to be gained partly through full-scale load tests. Unlike Culmann, he did not have extensive field experience himself; his early brilliance led him too soon into a professor's chair for that. But by way of a satisfactory replacement, he placed high value on the insight obtained from those load tests. Ritter's defense of such field experiences, against the German objections, allowed structures to be built for which complex and supposedly more rigorous analyses would have obscured the design potentials.

Another aspect of Ritter's ideas, widening the engineer's horizons, was illustrated by his trip to the United States, as reported in *Der Brückenbau* and in

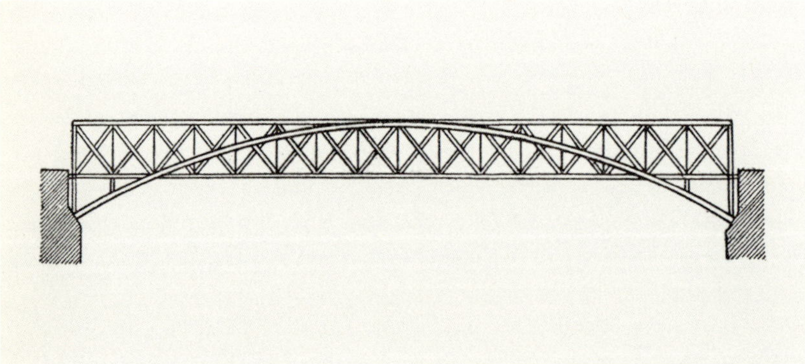

Figure 5
Ritter's image of a thin arch stiffened by a deep truss.

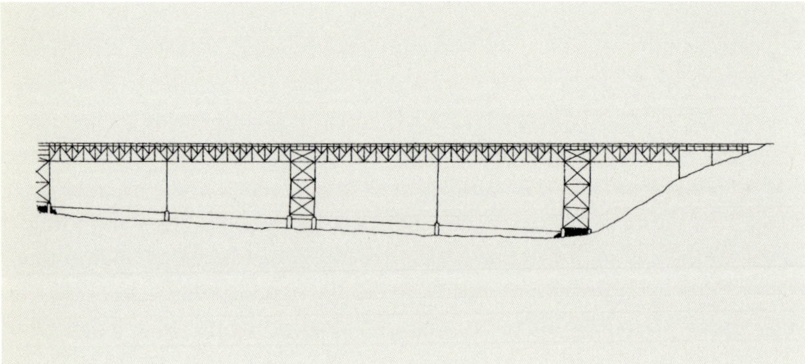

Figure 6
Drawing accompanying Ritter's criticism of the high bridge built in 1988-89 over the Mississippi River at St. Paul, Minnesota.

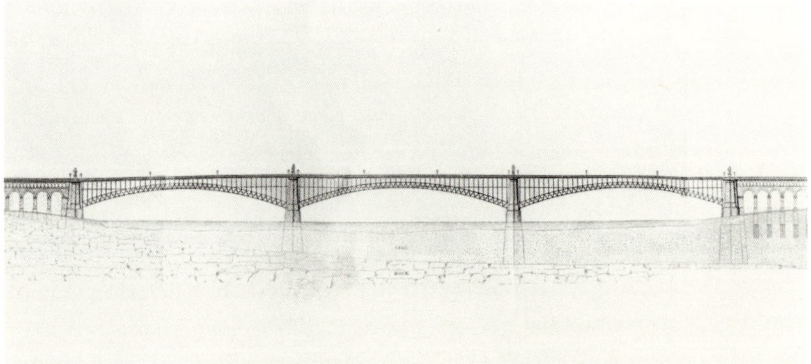

Figure 7
Eads Bridge, built in 1874 over the Mississippi River at St. Louis.

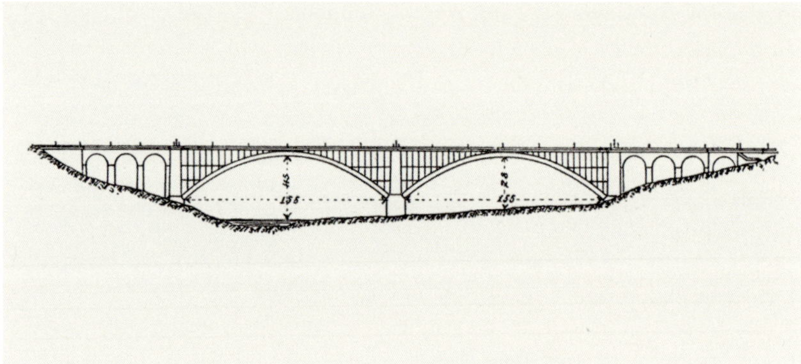

Figure 8
Ritter's depiction of the Washington Bridge spanning the Harlem River, New York, accompanying his written approval.

numerous articles in the *Bauzeitung*.[25] These articles showed Ritter's breadth of view. They ranged from sketches of the Chicago Columbian Exposition, to reports on Chicago's bascule bridges, to a study on engineering education in the United States, and finally to a report on bridges throughout the country. The book was devoted to wood and metal bridges; the heading "Eiserne Brücken" includes both iron and steel bridges even though the proper translation of *Eiserne* is iron and the word for steel is *Stahl*. Early in the text Ritter showed a thin arch stiffened by a deep truss typical of American wooden covered bridges (fig. 5).

Ritter was surprised that so little attention was paid to bridges exhibited at the Chicago World's Fair. He reported: "With all its expanse and plentifulness, the Fair offered relatively little information on bridge structures in the United States."[26] Knowing the spectacular bridge works completed in the United States since the Civil War, Ritter had assumed that such world leadership would be proudly displayed. Instead, the exposition represented the American romance with the machine that so startled Henry Adams (1838–1918) and set off his contemplation of its potentially destructive force.[27]

For Ritter, structures were the main consideration. He wrote little about his travels but painted a vivid picture of what interested him: the wide variety of overall forms, the broad use of certain non-European details, and, in elegant plates, a complete technical picture of a few selected bridges. This combination of insights is what most intrigues the design-oriented student, who would be genuinely excited to see the array of visual forms that have already been used to solve what is essentially the same set of problems. Ritter did not hesitate to introduce aesthetic judgments, as, for example, in describing the high bridge built in 1888–89 over the Mississippi River at St. Paul (fig. 6): "The structure, in spite of its extraordinary size, makes a rather insipid impression; it leads us to realize the aesthetic ideas must have been fully withdrawn in favor of some utilitarian principles."[28]

He went on to note that American tendencies to avoid complex analyses and to think only of utility rather than beauty had worked against the frequent use of arches, a point he illustrated with the Eads Bridge of 1874 and its three-arch spans across the Mississippi at St. Louis (fig. 7). Ritter noted that many

Figure 9

Full view of the Washington Bridge (Harlem River), New York.

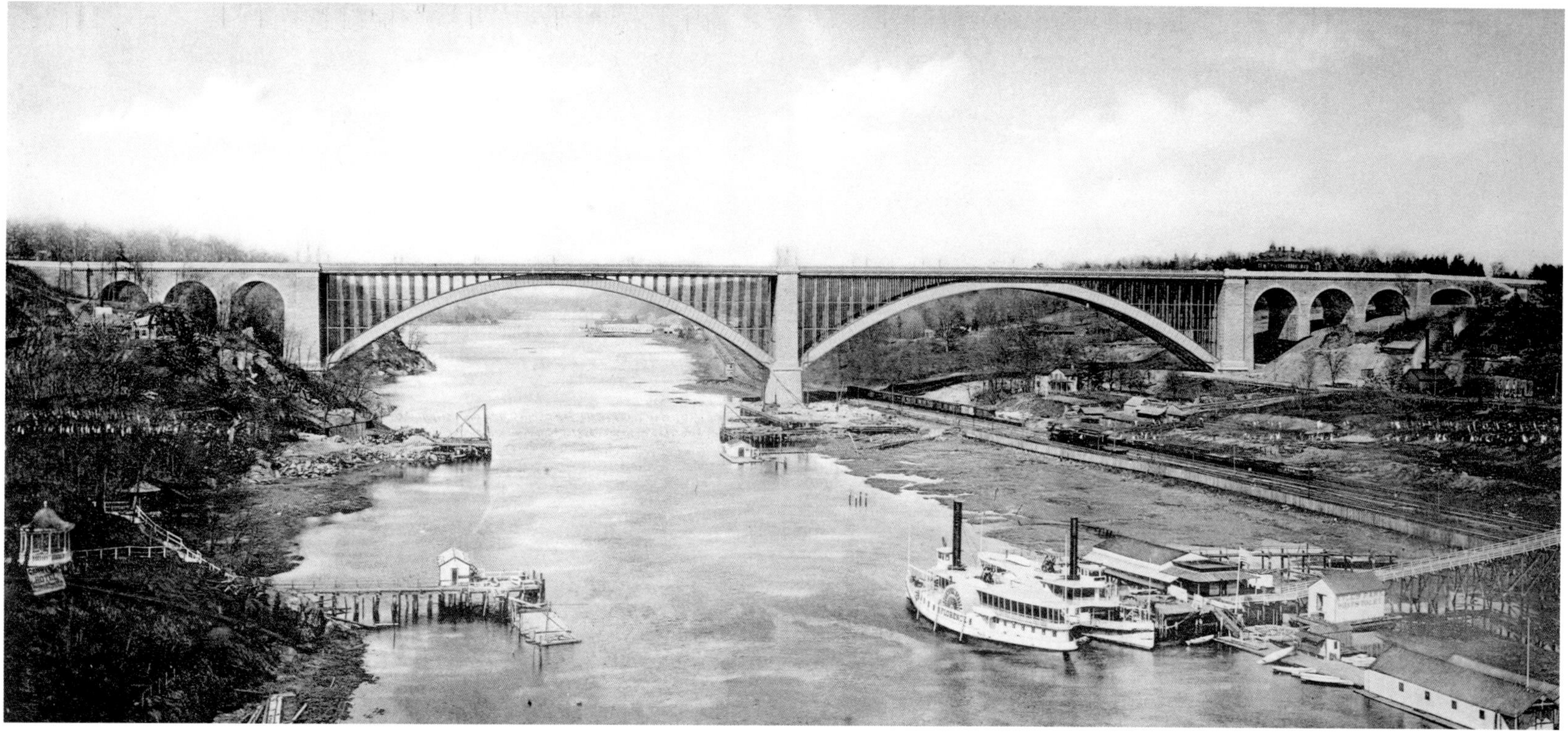

difficulties, including very high cost, discouraged Americans from using such arches on bridges, but he did show the much less costly and still striking Washington Bridge built in 1889 over the Harlem River in New York with its two-arch spans (figs. 8–9). Also included was a concise summary of American suspension bridges, including the great Brooklyn Bridge of 1883 (fig. 10) and ending with a sketch of Gustav Lindenthal's proposal of 1888 (fig. 11) for a Hudson River crossing.[29] This is the final text figure and a fittingly prophetic one, since one of Ritter's last students, Othmar Ammann, would go to the United States and begin his spectacular bridge career by designing the George Washington Bridge that would, thirty-six years later, cross that very river.

As every bridge designer knows, however, overall form means nothing if the details are not well done. All the pieces must fit together, and none can be structurally weak. To the watchmaking Swiss, details are an aesthetic part of design because they require great care. Ritter proceeded from overall form to detail, emphasizing both. A meticulous review of joints, connections, eyebars, and rivets (fig. 12) occupied approximately one-third of the text and illustrations in *Der Brückenbau.* Many of these drawings are elegant, and Ritter criticized other drawings shown in the book as not elegant. The final part of the book consisted of twelve plates, each providing the full details for individual bridges: there were three wooden bridges and nine metal bridges, of which one was a plate-girder bridge, one a suspension bridge, one a cantilever, and the others were truss bridges.

Der Brückenbau was a unique work in the Culmann tradition; Ritter's other writings were also far more significant than the preceding examples might suggest. A defense of load testing and a report on a study trip are very different from the type of solitary research needed for the numerous papers and books that Ritter published during his lifetime. One example of such research illustrates the high quality of his mind as well as his taste for simplicity and elegance in calculations. In 1877 he had first written about stiffened suspension bridges. Six years later, in 1883, he published a major article that expanded the subject and introduced advanced ideas on analysis for suspension bridges and arch bridges.[30] This article represents as clear a study of suspension-bridge

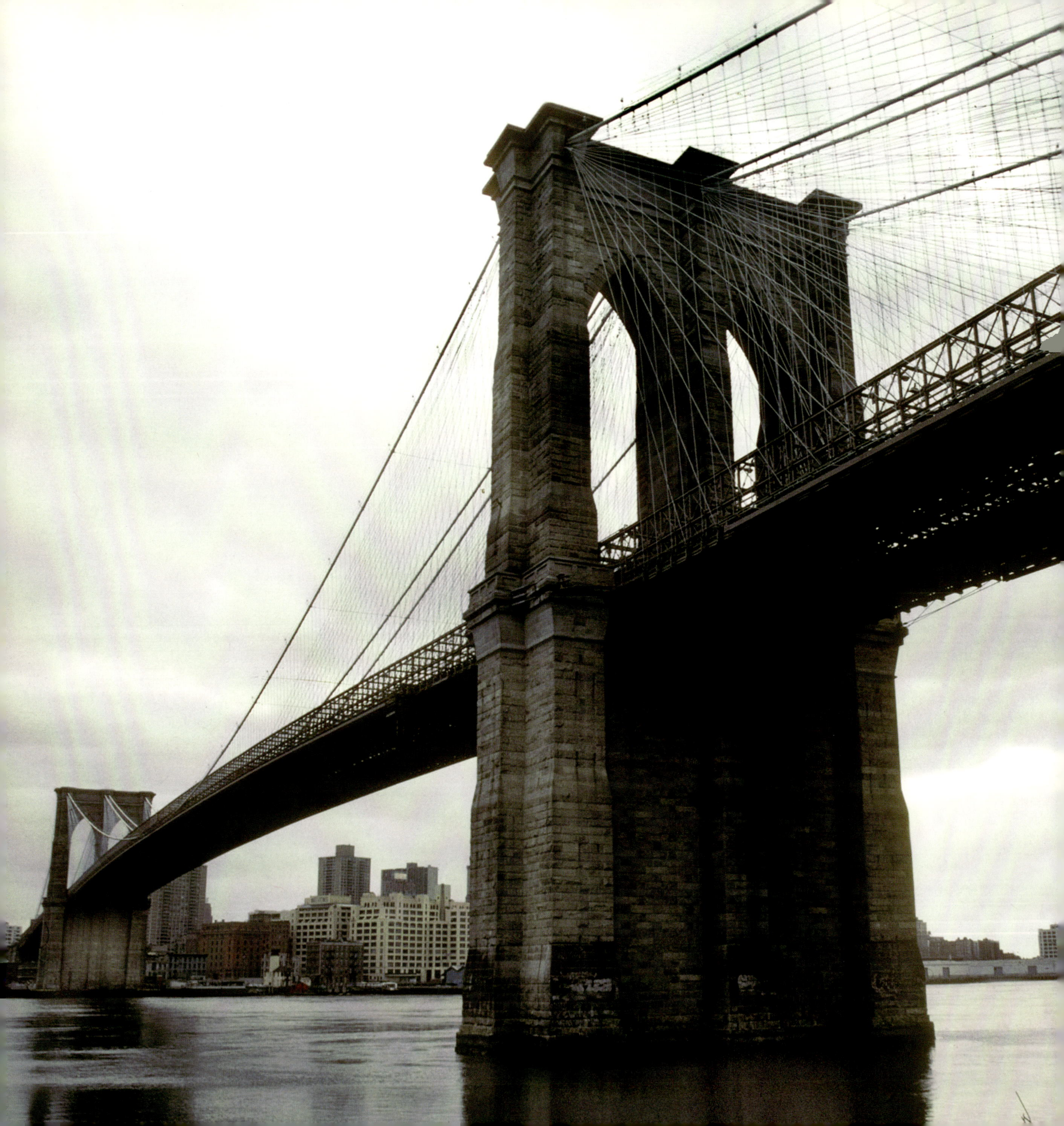

Figure 10
Brooklyn Bridge, opened in 1883.

Figure 11
Ritter's sketch of Gustav Lindenthal's 1888 suspension bridge proposal for a Hudson River crossing.

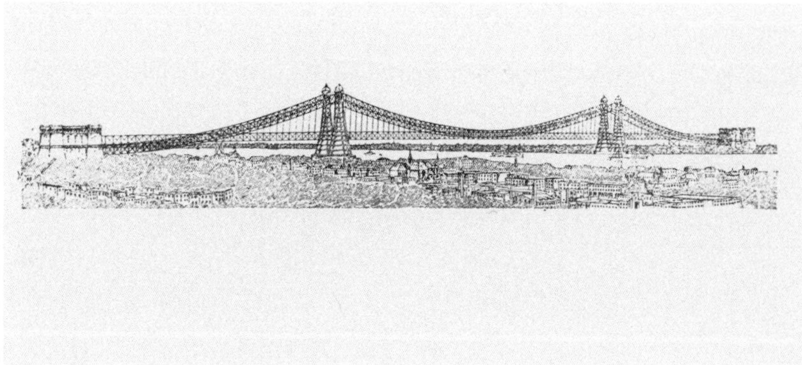

design as any in the nineteenth century. It did not develop the deflection theory, a mathematical theory credited to Josef Melan and first published five years after Ritter's article.[31] Its historical interest is not for its presentation of a new analysis but rather for its elegant simplicity, its design ideas, and its essential correctness: even today, Ritter's method of analysis would be a reasonable basis for the conceptual design of a bridge on the scale of the one illustrated in his paper; deflection theory became important only for longer spans.

In the late twentieth century, it became common practice to read only the most recent works on any technical subject, the earlier ones presumed to be outdated and useless. But like some old bridges, a few early articles on structural analysis and design retain their importance because they reflect the mind of a masterful teacher. One such is Hardy Cross's article on the simplified analysis of frame buildings, published in 1932;[32] Ritter's article on suspension bridges is another.

The paper begins with a clear statement of intention in which Ritter referred to his paper of 1877, identified its omissions, and stated that "the following development has the goal of making up for the earlier omissions; at the same time, the essential thing will be to present that earlier work in as concise a way as possible for the reader who is unfamiliar with it."[33] The paper is thus self-contained and written for a broad audience rather than a specialized one. He emphasized that since the theory of elasticity must be used, complicated formulations result, and thus "one is compelled to make simplifying assumptions."[34] In fact, his whole presentation avoids even calculus and yet sets forth a sound analysis. He then gave a simple, elegant, and practical theory of the beam-stiffened arch. Not only did Ritter develop a simplified analytic method, he also considered its design implications: "The stiffer the beam and the more flexible the arch, the greater the bending moment in the beam and the less in the arch."[35] Taking such simple ideas, he applied them to the suspension bridge, where the ratio of the horizontal truss stiffness to the cable stiffness determines how much bending goes into the truss. Ritter developed the analysis by setting the deflections of the truss equal to those of the cable because the suspenders tie the two parts together, and thus they determine the load carried by the truss and that carried by the cable. Ritter's characteristic modesty and concern for clarifying and simplifying is exemplified in a footnote in which he observed that Professor A. Ritter (no relationship) in a recent book had published essentially the same theory about the suspension bridge theory but that it was not presented clearly enough to be easily useful.

Ritter illustrated his article with diagrams using graphic statics, and his equations were always simple. Where they involved more than a few terms, he placed nondimensional elements in tables for easier use. Toward the end of the article, he furnished a fully worked-out design for an actual suspension footbridge in Switzerland with a span of fifty-seven meters. A section on approximations follows that provides a reasonable basis for preliminary design as long as the final design is analyzed by the more exact procedure he has already outlined. At the end, he reviews practical problems, such as construction sequence and trusses of variable section, and how they relate to the design.

In summary, Ritter was doing research and presenting his results for the designer, not for other researchers. He sought to clarify the behavior of structures rather than to develop methods of analysis. Finally, he used examples of actual structures in order to illustrate his ideas numerically, and he considered construction questions along with those of analysis.

To describe Ritter's influence, an old metaphor may be useful. In his approach to his students' education, he stands very much as a bridge does in its environment. As an educator, carrying students for only a small fraction of their lives, he was an essential link between their past inborn talents and their relatively long future careers. As a teacher, he communicated the recent traditions of his country and profession through the specific objects about which he lectured. Finally, as a writer, he took scientific formulations and shaped them with as little complexity as possible into simple clear ideas so that his students would see the design potentials more easily. He was not a designer, although he taught from designs; he was not a public official, although he taught public works; he was not a natural scientist, although he worked with science to reduce general formulations to specific applications. He thus stood at junctions

Figure 12
Ritter's drawing of a connection, showing his emphasis on details.

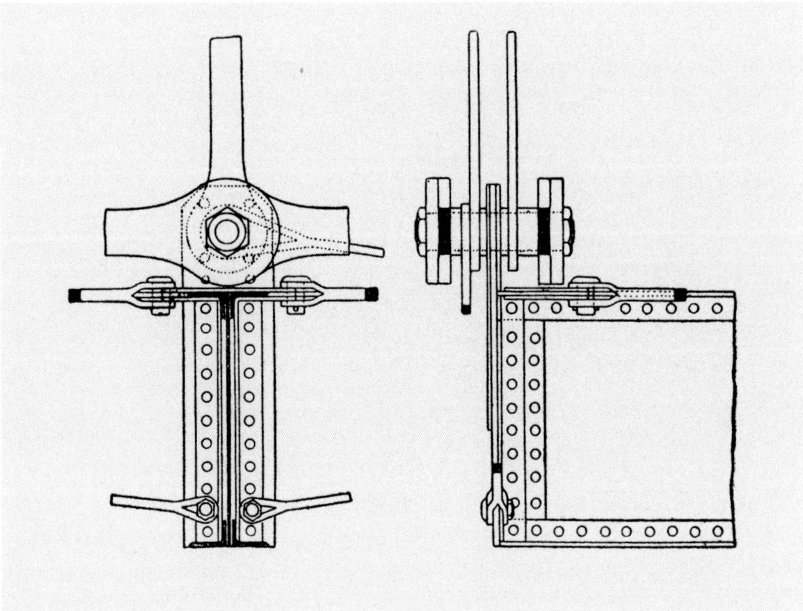

and let the students pass on. He did not become either their competitor by having a design office or their master by having a design ideology. Standing as the essential link between the awkward and sometimes unclear theories of Culmann and the elegant clarity of the late works of Maillart and Ammann, he interpreted technical events: to students through his lectures, to the profession through his writings, and to public officials through his detailed consulting studies that led to Swiss codes for both metal structures and works of reinforced concrete.[36]

Ritter's name has been nearly lost, and his style of teaching and research has been out of fashion for years, but he touched students, of whom Maillart and Ammann were only the most outstanding examples. Thus, the study of great designers like Maillart and Ammann has now led to the question of how education influences practice. It seems clear that Wilhelm Ritter played a significant role with these two men. More generally, the example of Ritter may suggest the need to rethink structural engineering so as to include teaching and research that encompasses field experience and the visual study of structures, wide knowledge of international structures, and solitary individual writing that interprets new ideas for the profession in a practical and clear manner. The Culmann-Ritter tradition cannot be duplicated, but it can stimulate education for the future.[37]

Notes

1. W. Oechsli, *History of Switzerland, 1499–1914*, trans. E. Paul and C. Paul (Cambridge, 1922), 402. From 1855 to 1911, the school was called the Eidgenössische Polytechnische Schule (Federal Polytechnical Institute). In 1911, the name was officially changed to the Eidgenössische Technische Hochschule, or ETH (Federal Institute of Technology).

2. W. Oechsli, *Geschichte der Gründung des Eidg. Polytechnikums mit einer Ubersicht seiner Entwicklung, 1855–1905* (Frauenfeld, 1905), 35–121. See especially the role of Alfred Escher in bringing the school to Zurich, 65-66.

3. Ibid., 70, 75–76.

4. Ibid., 175.

5. Ibid., 221.

6. Ibid., 234.

7. See R. Goldschmidt, *Die Stadt Karlsruhe* (Karlsruhe, 1915), 300–306. Karlsruhe was not designated a polytechnic school until 1885. From its founding in 1825 until the mid-1830s, it was also a secondary school; by the late 1830s, there was an engineering department. By 1860, Zurich's polytechnic school was probably more fully developed than the one in Karlsruhe.

8. C. Culmann, "Der Bau der holzernen Brücken in den Vereinigten Staaten von Nordamerika," *Allgemeine Bauzeitung* 16 (1851): 69–129 and "Der Bau der eisernen Brücken in England und Amerika," *Allgemeine Bauzeitung* 17 (1852): 163–222.

9. C. Culmann, *Die Graphische Statik* (Zurich, 1866).

10. W. Ritter, "Karl Culmann," in *Allgemeine Deutsche Biographie*, König. Akademie der Wissenschaft (1903; reprint, Berlin, 1971), vol. 47, 571–74.

11. For Ritter's influence on Maillart, see D. P. Billington, *Robert Maillart's Bridges: The Art of Engineering* (Princeton, 1979), 7–8, 11–12, 16, 22, 24; for his influence on Ammann, see F. Stüssi, *Othmar H. Ammann: Sein Beitrag zur Entwicklung des Brückenbaus* (Basel, 1974), 36–38, 85.

12. For Bessard, see *Dictionnaire Historique et Biographique de la Suisse*, ed. M. Godet, H. Türler, and V. Altinger, 7 vols. (Neuchâtel, 1924), vol. 2, 151. His death is given as 1872, but the story makes more sense if we take it as 1873, the date in the article by G. Thurnheer, "Wilhelm Ritter," *Vierteljahrsschrift, Zürcherischen Naturforschenden Gesellschaft* 1 (1906): 510, 513. For the Riga part of the article, Thurnheer quotes a colleague of Ritter's, Alexander Beck from Schaffhausen, a professor of mathematics at Riga and later at Zurich.

13. W. Ritter, *Anwendungen der Graphischen Statik nach Professor Dr. C. Culmann*, Part 1, *Die Inneren eines Balkens Wirkeneden Kräfte* [Internal Stresses in Beams] (Zurich, 1888), v (foreword).

14. Ritter, "Karl Culmann," 573.

15. Ibid.

16. Ibid.

17. Thurnheer, "Wilhelm Ritter," 510, 513.

18. E. Meister, "Professor Dr. W. Ritter," memorial in the *Schweizerische Bauzeitung* [hereafter *SBZ*] 48, no. 17 (Oct. 27, 1906): 206–8.

19. W. Ritter, *Anwendungen der Graphischen Statik nach Professor Dr. C. Culmann*: Part I, *Die Inneren eines Balkens Wirkeneden Kräfte* (Zurich, 1888), 148 pages; Part 2, *Das Fachwerk* (Zurich, 1890), 229 pages; Part 3, *Der Kontinuierliche Balken* (Zurich, 1900), 270 pages; Part 4, *Der Bogen* (Zurich, 1906), 269 pages.

20. W. Ritter, *Der Brückenbau in den Vereinigten Staaten Amerikas* (Zurich, 1895), 66 pages.

21. Thurnheer, "Wilhelm Ritter," 516.

22. "Über den Werth der Belastungsproben eiserner Brücken," *SBZ* 19, no. 15 (April 9, 1892): 106.

23. "Über den Werth der Belastungsproben eiserner Brücken," *SBZ* 20, no. 2 (July 9, 1892): 12.

24. W. Ritter, "Über den Werth der Belastungsproben eisernen Brücken," *SBZ* 20, no. 3 (July 16, 1892): 14–18.

25. The following articles are by Wilhelm Ritter, "Skizzen von der Chicagoer Ausstellung," *SBZ* 22, no. 11 (Sept. 23, 1893): 79–82; no. 12 (Sept. 30, 1893): 86; no. 13 (Oct. 7, 1893): 91; no. 14 (Oct. 14, 1893): 100–101; no. 15 (Oct. 21, 1893): 116; "Die Neue Falten-Zugbrücke in Chicago," *SBZ* 22, no. 12 (Sept. 30, 1893): 86; "Das Technische Unterrichtswesen der Vereinigten Staaten Amerikas," *SBZ* 24, no. 2 (July 14): 10–11 and no. 3 (July 21, 1894): 21–23. For the book, published in 1895, see note 20 above; the following report is a summary, but it appears not to have been written by Ritter: "Der Brückenbau in den Vereinigten Staaten Americkas," *SBZ* 24, no. 24 (Dec. 14, 1894): 165–70.

26. Ritter, *Der Brückenbau*, 3.

27. H. Adams, *The Education of Henry Adams* (1918; reprint, New York, 1931).

28. Ritter, *Der Brückenbau*, 52.

29. Ibid., 63.

30. W. Ritter, "Statische Berechnung der Versteifungsfachwerke der Hängebrücken," *SBZ* 1, no. 1 (Jan. 6, 1883): 6–7; no. 2 (Jan. 13, 1883): 14; no. 3 (Jan. 20, 1883): 19–21; no. 4 (Jan. 27, 1883): 23–25; no. 5 (Feb. 3, 1883): 31–33; no. 6 (Feb. 10, 1883): 36–38.

31. D. B. Steinman, *A Practical Treatise on Suspension Bridges: Their Design, Construction, and Erection*, 2d ed. (New York, 1929), 246.

32. H. Cross, "Analysis of Continuous Frames by Distributing Fixed-End Moments," *Transactions of the American Society of Civil Engineers* 96 (1932): 1–10, with 147 pages of discussion.

33. Ritter, "Hängebrücken," 6. Ritter's article of 1877 is in the *Zeitschrift für Bauwesen*, 189.

34. Ibid.

35. Ibid.

36. W. Ritter and L. Tetmajer, "Der Bericht der Eidgenössischen Experten über die Mönchensteiner Brücken-Katastrophe," *SBZ* 18, no. 18 (Oct. 31, 1891): 114–15. This introductory report and a series of later additions to the main report led to the federal law of August 19, 1892, concerning the calculations and testing for metal bridges and roofs for the railway; see Meister, "Professor Dr. W. Ritter," Memorial in *SBZ* 18, no. 17 (Oct. 27, 1906): 206–8. Ritter co-authored reports in 1901 and 1902 on the failure of a concrete building in Basel; these publications led directly to the formation of a commission to prepare a Swiss building code for reinforced concrete.

37. Much of the discussion in this chapter originally appeared in D. P. Billington, "Wilhelm Ritter: Teacher of Maillart and Ammann," *Journal of the Structural Division of the American Society of Civil Engineers* 106, no. ST5 (May 1980): 1103–16; reprinted in *Schweizer Ingenieur und Architekt* 87, no. 7 (Mar. 26, 1987): 94–100.

CHAPTER 2

ROBERT MAILLART: MASTER OF CONCRETE FORMS

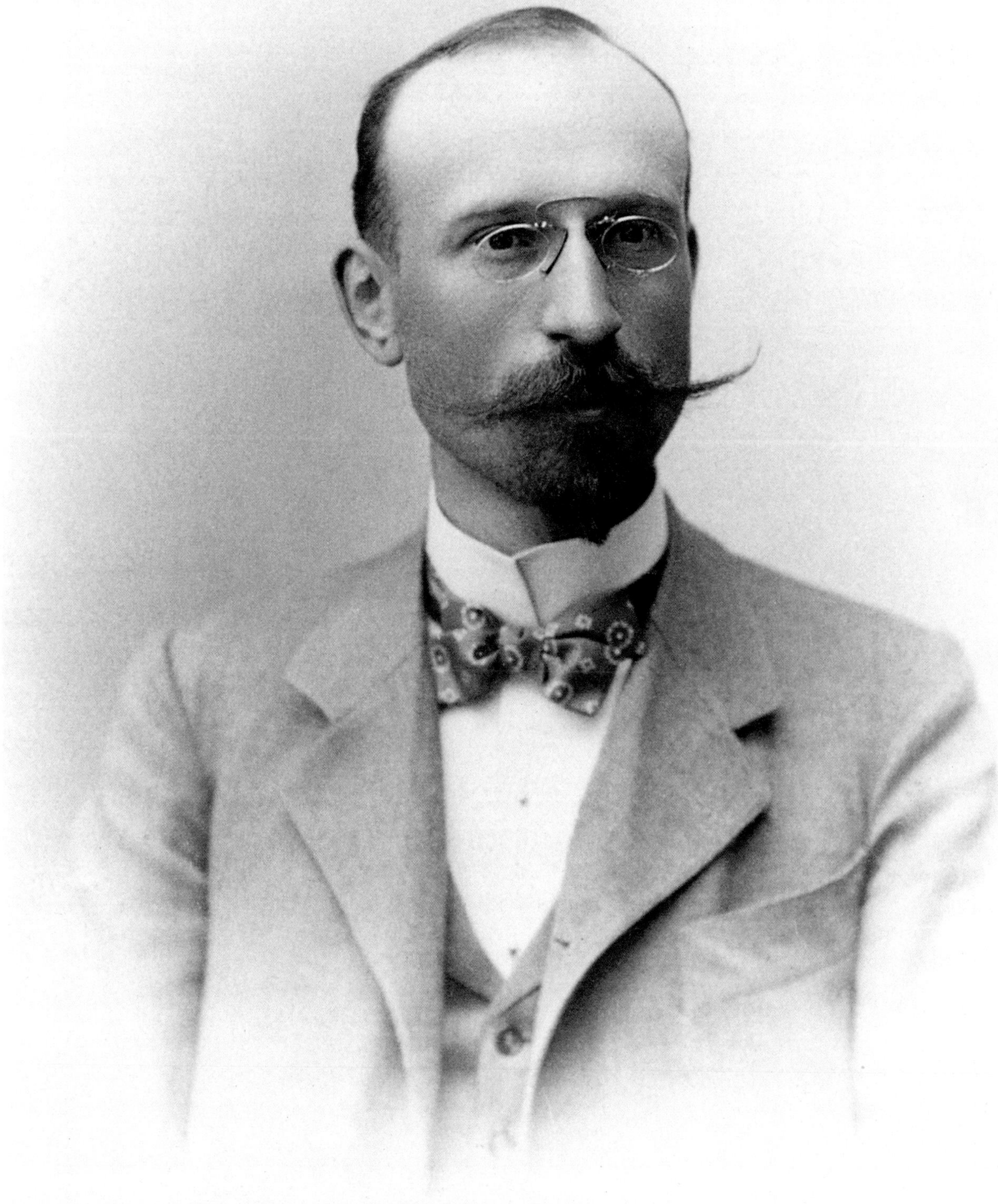

Figure 1 (previous page)
Robert Maillart (1872–1940), ca. 1901.

Wilhelm Ritter's prestige as Switzerland's leading consultant on bridge design was crucial to the early design career of Robert Maillart (1872–1940), and his support enabled Maillart to develop his own early fundamental ideas on design. Ritter exercised influence on Maillart (fig. 1) both through his teaching of structural engineering and through his consulting on the younger man's first three major arch-bridge designs. The strong visual content of Ritter's teaching related applied mechanics to structural form and thus included aesthetics in the education of civil engineers. This approach encouraged his students to visualize the flow of forces within a structure as well as the way in which different forms could change that flow. Maillart eventually developed into a master of design in reinforced concrete; he took that new material and made it into an art form parallel to, but independent of, sculpture and architecture. Each of these three-dimensional arts expresses form through construction in space, but each has a different function: sculpture usually works as symbolic expression, with little connection to spatial use and control of forces; architecture, by contrast, has as its primary role the use of space, with some connection to symbolic expression and control of forces; and structure's central task is the control of forces that arise from the influences of society and nature. The use of space is an important secondary consideration in structure, but it applies in a more restricted sense there than to architecture since the structural artist does not design directly for intensive human use (houses, churches, or offices, for example). Rather, the structural artist designs bridges, towers, and vaults where the use is less intensive for individual people and more so for cars, industry, and large coverings over stadiums, auditoriums, or hangars. The structural artist also creates symbols that reflect the ideals of a democratic society—elegance disciplined sternly by public accountability and safe performance. Maillart was the first engineer to sense that the full expression in concrete structures could be efficient (safe performance with minimum materials), economical (accountable to the public welfare or to private industry with competitive costs), and elegant all in the same construction. How did this come about?

George Kubler, a distinguished art historian, has written about the two inequalities of birth, "one governing the allotment of its temperament, and the other ruling its entrance into a sequence."[1] An artist thus is born with a special talent and appears at the right time. Maillart came to maturity just as reinforced concrete became an acceptable substitute in construction for all other previous structural materials—stone, wood, and steel. He entered the field after the pioneers had developed useful methods for design and construction but before anyone had dared to invent new forms that departed radically from the aesthetic of those earlier materials. He might not have succeeded, however, had he not had an unusual education. As Kubler noted, "[U]ndiscovered talents abound as well among people whose schooling failed to gear with their abilities, as among people whose abilities were unrequited in spite of their talent."[2] In both of these requisites, Maillart was fortunate to come under the powerful influence of Wilhelm Ritter. Ritter's teaching meshed with Maillart's talent, and Ritter's consulting advice to Maillart's clients prevented the rejection that might have discouraged a young designer.

Maillart had a strong natural ability in mathematics, not as a mathematician interested in proofs and theorems, but as an engineer engaged in predicting full-scale performance. He also had a deep aversion to the overuse of mathematics, the use of complex formulations as abstract decoration applied to the description of performance.[3] His talent for visualization—the ability to see how structural elements carried loads and how they would appear as a whole—expressed itself in planar sketches, not in sculptural modeling. As a young boy, he was attracted to chess, a flat board game that requires planar visualization, but not to sports like soccer or tennis, which are three-dimensional and physical. He did, however, have a strong physical constitution as well as the ability to focus intently and concentrate alone on one thing. This ability was balanced by an urge to play. Not only did he master chess, but he also played card games: he became skillful at bridge and even invented a card game while trapped in Russia during World War I. Although his play was serious, as in competitive games, a joyful side to his personality emerged when he was in the company of his few close friends and when he wrote letters to his wife and children.

In his book *Homo Ludens*, the Dutch historian Johan Huizinga (1872–1944) elevates "playing man" to a level of cultural significance equal to that of "knowing man" (*Homo Sapiens*) and of "making man" (*Homo Faber*). Huizinga's description of play coincides with Maillart's attitude: "It creates order, is order. The least deviation from it 'spoils the game,' robs it of its character and makes it worthless. The profound affinity between play and order is perhaps the reason why play… seems to be to such a large extent in the field of aesthetics. Play has a tendency to be beautiful. It may be that this aesthetic factor is identical with the impulse to create orderly form, which animates play in all its aspects."[4] Maillart's career would center on structures that had to obey the rules of nature but within those rules allowed for the creation of new forms that grew from an aesthetic sensitivity.

Robert Maillart was born in Bern on February 6, 1872. His mother, Bertha Kupfer Maillart (1842–1932), was German-Swiss, and his father, Edmond Maillart (1834–1874), was Belgian. Robert was the fifth of six children, the first

Figure 2
Ritter's beam problem as recorded in Maillart's class notes.

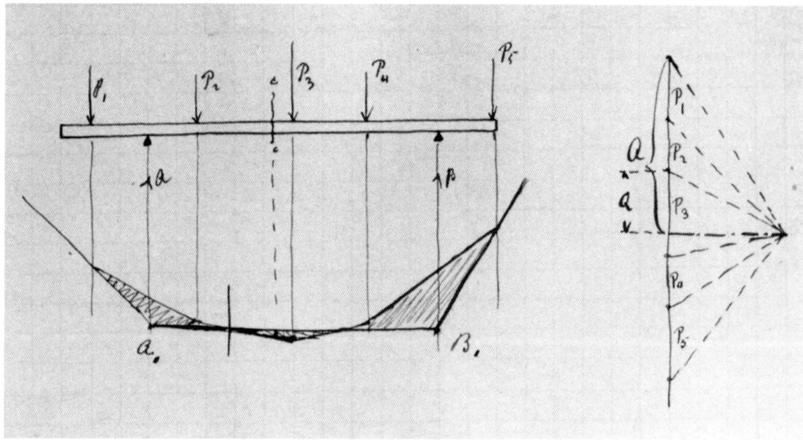

three of whom were born in French-speaking Geneva and the last three in German-speaking Bern. Thus, the Maillarts were culturally mixed: part of a native Swiss family (the Kupfers), and part of a displaced Belgian family (the Maillarts). Maillart's father spoke French as his native tongue and his mother German; together his parents conversed in French. But Robert's native language was German, which he always spoke with his mother. She was his confidante, and from early childhood he developed a respect for her that sharply curtailed any kind of boyish misbehavior. Robert became a Swiss citizen in 1886. His great-grandfather, Philippe Joseph Maillart (1764–1856), was a well-known engraver and painter.

The Kupfer and Maillart families were Calvinists, and this strict Protestant background played a part in Robert Maillart's extraordinary self-discipline. Brought up in French Switzerland, Robert's father had studied theology, presumably with the intention of becoming a minister, but a speech impediment deterred him, and he entered the banking business instead. On May 21, 1864, he married Bertha Kupfer of Bern, and the couple settled in Geneva. Shortly after the birth of their third child, the family moved to Bern, where the other three children, including Robert, were born. A mere seven months after the birth of the last child, Edmond died suddenly (April 24, 1874), leaving his widow with five small children (one girl having died in 1873) and little money. Robert was only two years old when his father died. With neither social position nor family wealth, Maillart grew up feeling that he was on the fringe of Swiss society. He and his brothers made their careers on their own merit, and they all married non-Swiss women.

Between 1885 and 1889, Maillart attended the Bern gymnasium, where he showed talent in mathematics and drawing. He excelled in both freehand sketching and technical drawing. In 1889, he passed the State examinations with a score of 4.8 out of 6. His grades qualified him for admission to the Federal Polytechnical Institute in Zurich, later renamed the Federal Institute of Technology—Eidgenössische Technische Hochschule (ETH). Being too young to enter, he went to Geneva for a year to study mechanics in watchmaking school. This experience foreshadowed a critical aspect of his talent: he earned a 4 (out of 6) in practical work (watchmaking) and a 6 in design and in mechanics. His aptitude was obviously not for making things with his own hands, but rather for design and analysis of how things work.

In the fall of 1890, Maillart moved to Zurich to enter the ETH. He carried with him a strong bond to his family and an independence of mind despite his restrictive upbringing. He had reflected on and developed many ideas on career, virtue, and religion, ideas that because of his natural reserve had not been tested in discussion. In the future, his engineering ideas would also not be tested in discussion but rather in a full-scale performance.

Maillart entered the ETH in the fall of 1890 and graduated with an engineering diploma in March 1894. He received grades in thirty-seven courses and followed twenty-eight other courses that were ungraded; four of the grades he received were from Ritter. Maillart's grades were unremarkable. He earned a 6 (the highest grade) only in the theory of ordinary differential equations and suffered a 2 (probably failure) in the theory of machines. Ritter's grades for Maillart were two 5s in graphic statics, a $4\frac{3}{4}$ in stone, wood, and iron bridges, and a $4\frac{1}{2}$ in bridge construction. Compared to Ritter, who received almost straight 6s between 1865 and 1868, Maillart's average was 81%, probably a B minus on an American grading scale. His grade average from Ritter was slightly less.[5] In spite of middling grades, Maillart received an excellent education, thanks to the visual nature of Ritter's teaching and his approach to structures.

During the spring semester of 1892, Maillart followed Ritter's first course on graphic statics; from this comes figure 2, found in Maillart's class notes.[6] The drawing represents a beam supported at two points, *a* and *b*, and loaded with five concentrated loads P_1–P_5. The two outermost loads (P_1 and P_5) are applied to the cantilever parts of the beam (the parts which each have one unsupported end). The sketch below gives a bending moment diagram which plots the size of the bending at each point over the beam. For example, the vertical line within the shaded area at B_0 gives, to some scale, the magnitude of the bending at

Figure 3
Illustration of a typical mathematically derived moment diagram for Ritter's beam problem.

Figure 4
Diagram of the effect of downward forces on a string.

Figure 5
Diagram of the effect of upward forces on a string.

Figure 6
Superimposing figures 4 and 5 creates an alternative moment diagram for Ritter's beam problem. The moment diagram is then inverted.

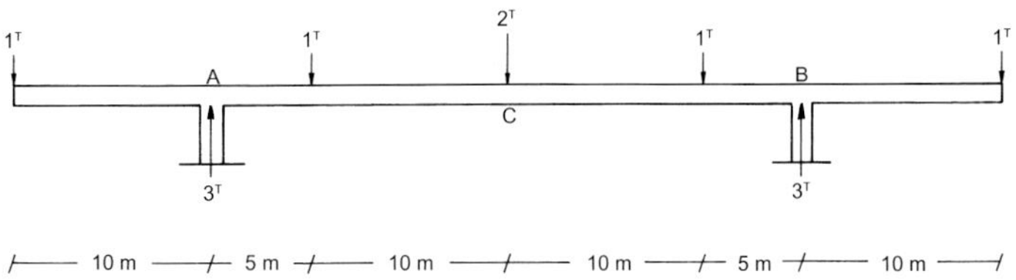

3A

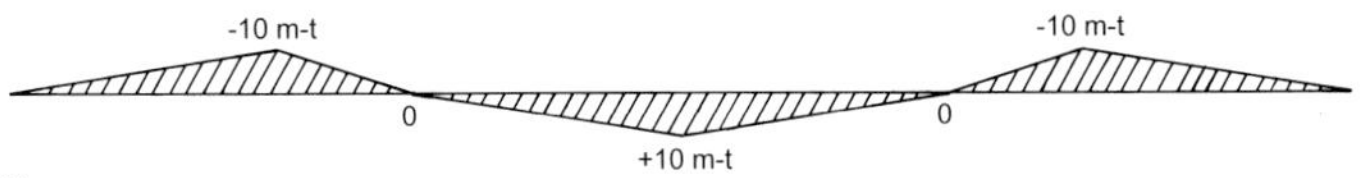

3B

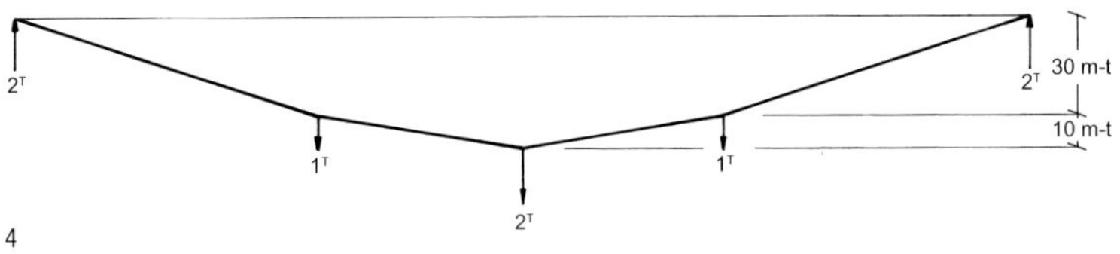

4

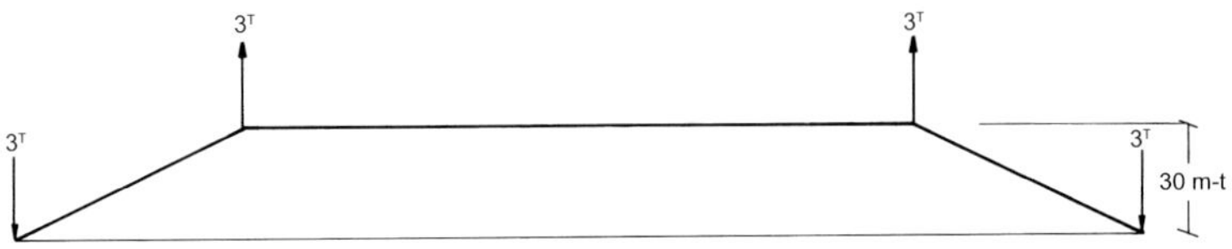

5

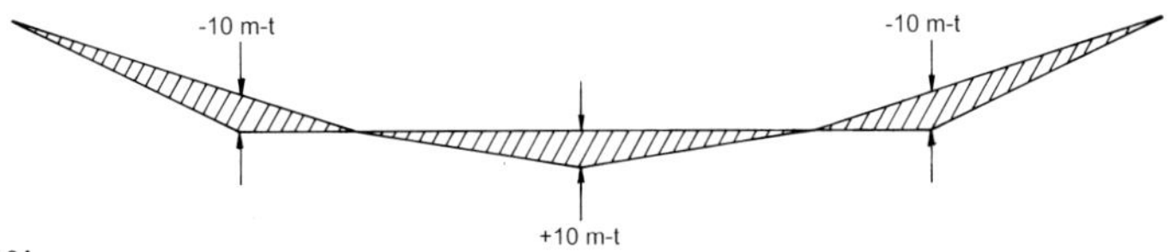

6A

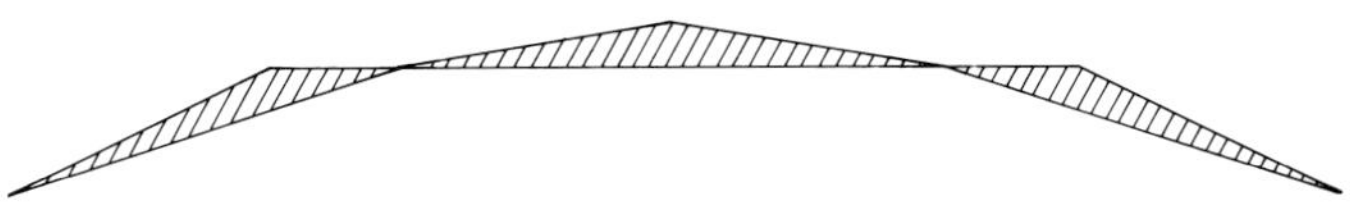

6B

Figure 7
Diagram of the Tavanasa Bridge, Switzerland.

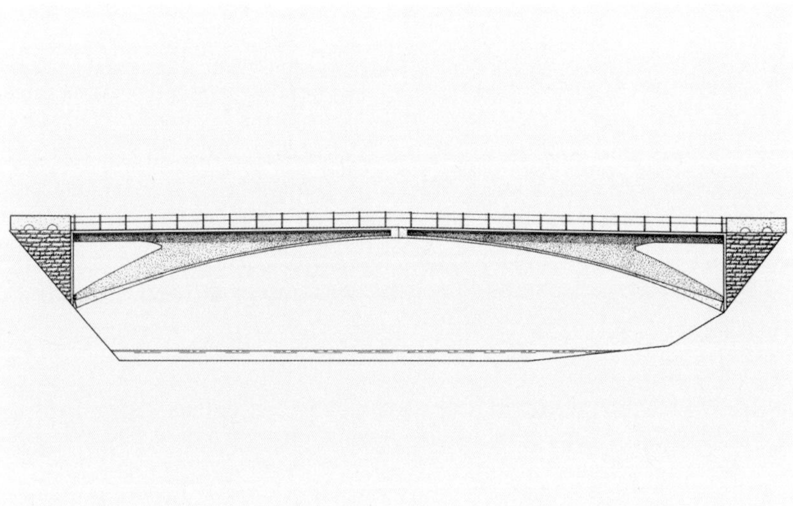

point *b*, that bending being the product of the load P_5 and its distance from the support *b*. The diagram to the right provides graphically the means for drawing the sketch of bending moments. What is important to recognize here is not the exact method Ritter used, but rather to see that there is no algebraic formula involved. Such formulas could easily be used to draw that diagram, and the loaded beam in fig. 3A illustrates roughly the same problem solved algebraically and plotted in fig. 3B, although the formulas are omitted (for example, the bending moment at B is one ton x 10 meters or 10 meter-tons, the formula being $M_B = P_5 \times 10$).

The great virtue of Ritter's approach was its visual relationship to structural form, which we can see by means of the examples shown in figures 4–7. We shall use a graphic solution (not the same as Ritter did in figure 2, but with similar results), which starts with imagining a string held at its two ends and loaded with three weights (fig. 4). The string takes a shape based on the location and size of the loads. Since the total load is four tons and the weights are symmetrically placed, the force needed to hold up the string at each end is two tons. Next we imagine another string pulled up by two loads as shown in figure 5. The string must be held down at each end in this case. When we superimpose these two strings by inverting the second and drawing it over the first, the result is a bending moment diagram (fig. 6A) that numerically has the same values as that in figure 3, but its shape resembles Ritter's diagram shown in figure 2. Inverting that new diagram (fig. 6B) creates an arch form that, with a small change in the loading, would give the profile of Maillart's 1905 three-hinged arch bridge at Tavanasa (fig. 7).

Figure 8

Maillart's class notes on Ritter's four categories of arch bridges: unstiffened, stiffened, trussed, and truss-stiffened arches.

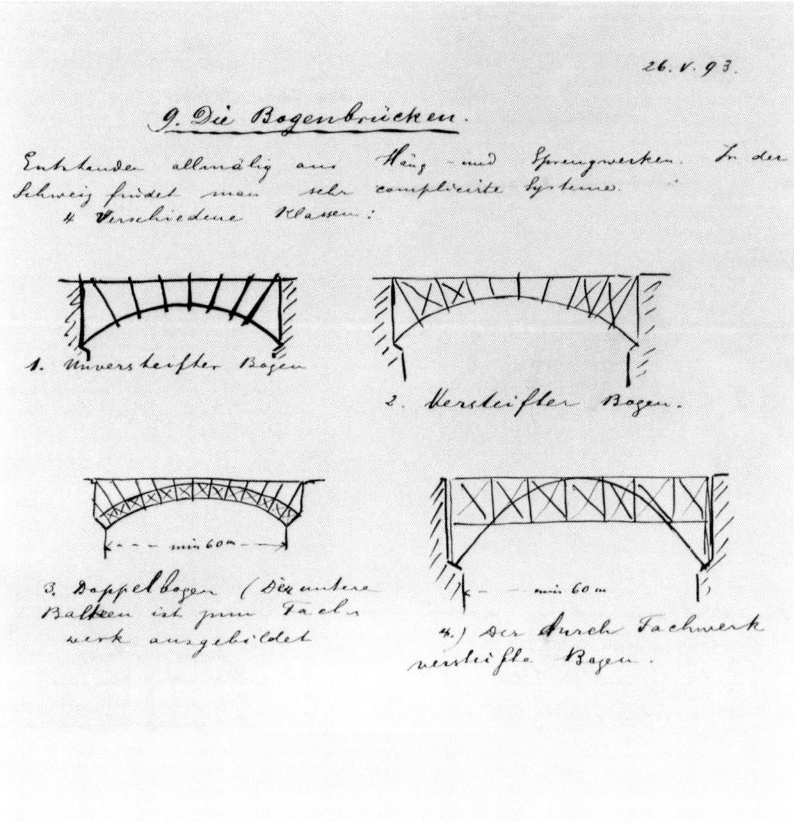

Figure 9

Maillart's class notes on an arch with a truss-stiffened deck, based on American covered bridges.

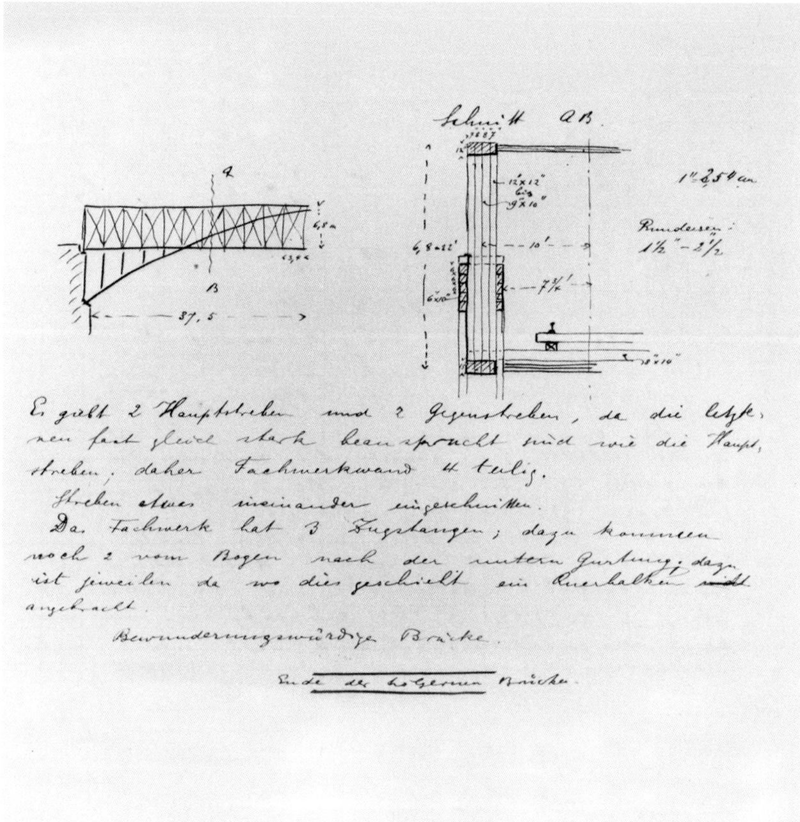

Ritter was teaching students to determine the forces in a structure (structural mechanics) by a method that at the same time encouraged visualization of form; hence, the method allowed him to connect mechanics to aesthetics. Ritter also sought to link both to actual practice by describing built designs. For example, Maillart's notes from the lectures on arch bridges (fig. 8) show that Ritter identified four classes: (1) unstiffened arches, (2) stiffened arches, (3) trussed arches, and (4) truss-stiffened arches. This section appears under "wooden bridges." As described in chapter 1, Ritter had earlier written about stiffened arches, describing how the straight deck of a bridge could serve to stiffen the arch supports so that the arch could be elegantly thin.[7] Another page of Maillart's lecture notes contains Ritter's details of an arch stiffened by a trussed deck (fig. 9), the form of which comes directly from the American practice with wood covered bridges, observed by Ritter during his trip in 1893 to the United States. Maillart carefully copied down the form from the lecture and at the bottom of the page exclaimed "wonderful bridge" [*Bewunderungswürdige Brücke*].[8]

From his graduation in 1894 until the founding of his own company in 1902, Maillart gradually developed from an engineer of standard work to a designer of startling originality. In early 1897 Maillart took a position with the city building department in Zurich, where he began to design the Stauffacher Bridge, a major structure in that city. The city initially proposed a series of metal or stone designs, most of which envisioned several spans with a support in the middle of the Sihl River, but Maillart submitted a comparatively long-span concrete design. The prevailing ideal for urban design was to build in stone even though it had become obvious by the 1890s that steel bridges were far less costly. The city called in Professor Ritter to consult on the projects. An eleven-page report that focused heavily on aesthetics resulted.[9] His choice was Maillart's single-span concrete arch. Ritter's early support gave the twenty-six-year-old engineer the rare opportunity to make a design new to Switzerland's largest city and to show his superiors that it could be less costly and more aesthetically acceptable. It was less expensive than the

Figure 10
Stauffacher Bridge over the Sihl River, Zurich, Switzerland.

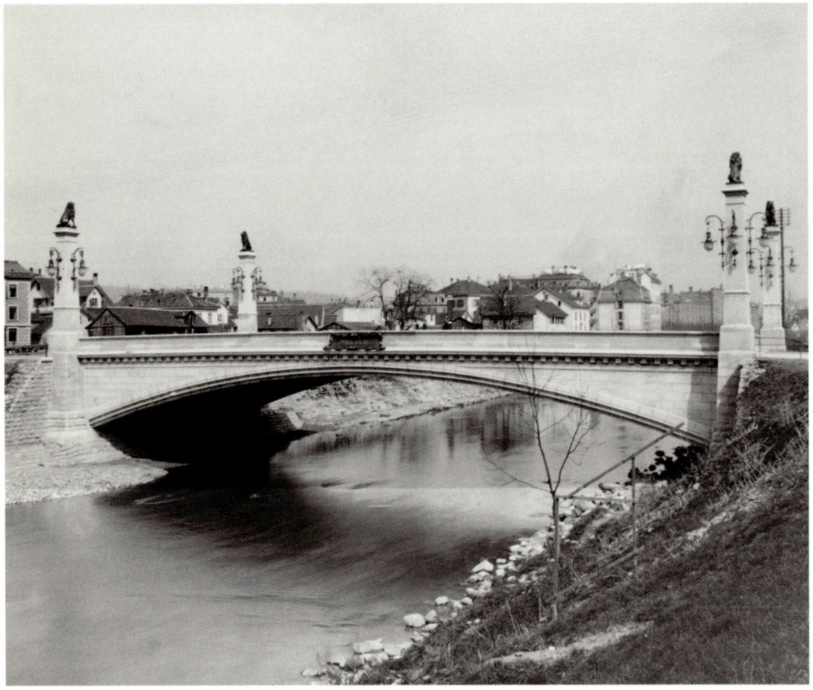

Figure 11
Stauffacher Bridge, showing the three-hinged arch hidden by a purely decorative stone wall.

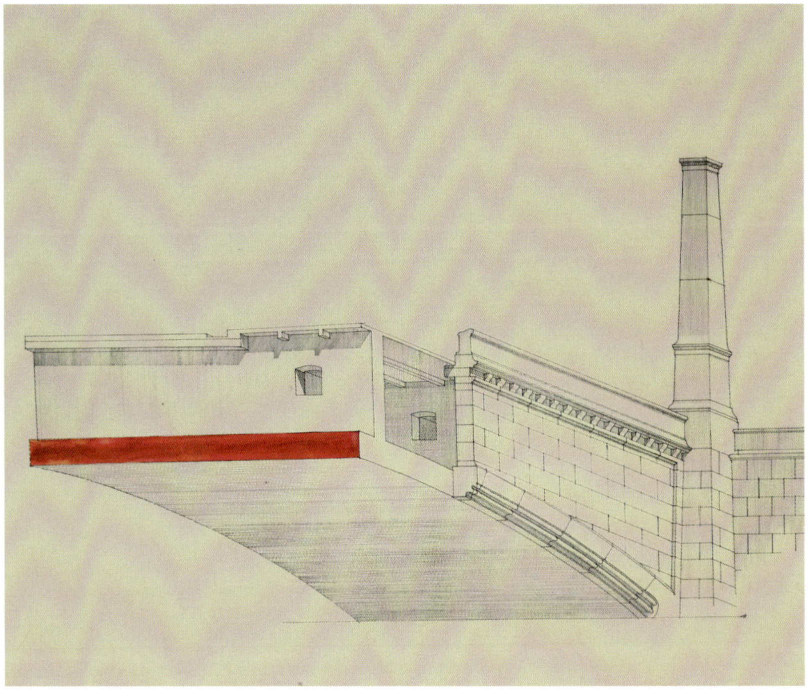

other designs, but its appearance was better only in the sense—as Ritter had argued—that it crossed the Sihl River in one sweeping curve and that it avoided the skeletal look of steel (fig. 10). The city architect, however, covered over Maillart's three-hinged arch with fake stone walls to satisfy urban taste (fig. 11). Ritter had made no recommendation for exposed concrete, and once Maillart knew that the design would be hidden, he was not stimulated to think about it as art. Indeed, the concrete design itself was more like stone than reinforced concrete since it was built of concrete blocks without any metal reinforcement.

As the Stauffacher Bridge took shape, Ritter and Maillart were moving deliberately toward the same goal, a clear understanding of the new material. Early in 1899, Ritter published a series of three articles in *Schweizerische Bauzeitung* [the Swiss Building Journal] on the Hennebique system for reinforced concrete.[10] François Hennebique (1842–1921), from northern France, had by the end of the nineteenth century developed a complete and patented system of designing structures in reinforced concrete that was in use throughout Europe. What Ritter did in 1899 was to bring that system out of a proprietary realm and into Swiss professional practice.

Describing how Hennebique designed slabs, beams, columns, and arches, Ritter sought to provide a clear understanding of performance. But primarily he was educating the Swiss profession on this new material so that engineers would not have to rely on a patented system. Still, the Hennebique concessionaires had so much field experience that Ritter wanted to connect the theoretical part of his treatise to his discussion of construction and performance. It was no accident that the deck of Maillart's Stauffacher Bridge was designed by a Hennebique concessionaire, demonstrating its practicality for beams and slabs. The combination of Ritter's support, his emphasis on the Hennebique system, and its use at Stauffacher led Maillart to focus seriously on the study of reinforced concrete. In fact, he accepted employment with the very firm that built the deck of his bridge.

Maillart spent the next two years as chief engineer for the young firm of Froté and Westermann, which was part of the Hennebique network. Along with many small projects, Maillart designed the first all-reinforced concrete bridge in Zurich, over Hadlaub Street, where his calculations followed closely the methods presented by Ritter in his articles that modified the Hennebique

Figure 12
Cross section of Zuoz Bridge, the first concrete hollow-box structure ever built (1901), Zuoz, Switzerland.

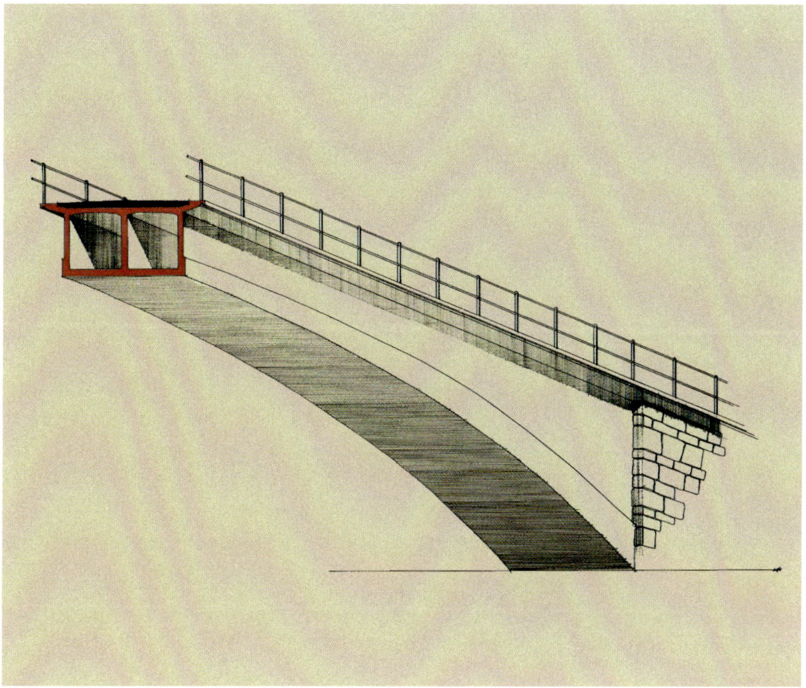

approach.[11] He also worked on the large sanatorium in Davos, called "Schatzalp," and his first publication was an article in 1901 on Hennebique works.[12] These years were a time of study, imitative designing, and field experience, but they also included something radically new, the Zuoz Bridge.

Being active in the Graubünden, Froté and Westermann knew that the canton was planning a new steel-truss bridge at Zuoz. The firm sent Maillart there on August 16, 1900, to inspect the proposed site and make a presentation to the community council.[13] Maillart argued that his concrete bridge was competitive with steel and had the added advantage that costs for concrete would be paid to local suppliers and not, as for steel, to manufacturers outside the canton. He also emphasized that "because concrete outwardly resembles stone, it will have a simple but elegant appearance and will unquestionably bring honor and embellishment to the community." He stressed that his design would have a fine finish achieved not by plastering over the concrete, but rather by a smooth forming and a special concrete mix. Maillart explained that the structure would have the virtues of stone but without its great weight because the concrete sections are reduced to a minimum. In spite of this minimization, the strength of Maillart's hollow-box arch is far greater than his solid (95-centimeter-thick) Stauffacher arch. At the critical section midway between the crown and the support, the strength of the Zuoz arch is similar to that of a solid 130-centimenter-thick stone arch but with the approximate weight of a 40-centimeter-thick stone arch. In addition, the stresses in the thinner Zuoz arch are less than those in the Stauffacher arch. Thus, Maillart's bridge was stronger, less expensive, and, in his view, more pleasing aesthetically than comparable bridges being built at this time.[14]

Maillart made a strong case for the aesthetics of his first major design innovation that expressed his central idea: form is more important than mass. The curved arch, the longitudinal walls, and the horizontal roadway slab "will together make the arch," he explained, where the bottom curved slab or arch is connected to the three vertical walls, which themselves connect to the top roadway slab. This construction forms two hollow boxes, the first such structure ever built in concrete (fig. 12). Its three elements (arch, walls, and deck) together carry the load, as contrasted with the Stauffacher Bridge, where the entire load is carried by the arch alone (see fig. 11).

If Maillart's new design was a radical departure, so was his construction plan, for which he designed scaffolding to carry only the thin-arch weight. Once the arch concrete hardened, it would carry the walls and deck. This idea originated with the ring method of building stone arches used in the Graubünden for the Solis Bridge (1900), the first long-span arch in Switzerland to be constructed in three concentric rings of stone. In its construction, only the first layer was carried by the wooden scaffold, the second being carried by the layer already in place.[15] In addition to reducing costs on materials, therefore, Maillart saved on scaffolding expenses at Zuoz by following the Solis example.

Negotiations over price took more than a month. On September 18, Froté and Westermann received word that they had been awarded the contract with the requirement that the structure must satisfy their consultant–Wilhelm Ritter.[16] The district engineer wrote to Ritter in early February 1901 asking when his report would be ready.[17] Ritter's response was unusual. "For about two weeks," he explained, "I have been busy checking the design for the Inn Bridge at Zuoz which you sent me. The problem you have posed is not an easy one; otherwise, I should have finished with it much sooner."[18] The difficulty arose because there was no precedent for Maillart's idea of building the arch slab, the longitudinal walls, and the roadway slab together to form the arch. Ritter, one of the world's leaders in arch-bridge analysis, was confronted by a problem to which there was no available solution. He did not send in his report until May 13, but when he did, he recommended that the design be approved with no change.

At that time Swiss public authorities did not accept bridges until they had been load tested. Besides assuring structural soundness, these tests have two major benefits: they yield data for the design professionals, and they provide a festive occasion for the local community. Undoubtedly because of the bridge's novelty, the district engineer asked Ritter to design and direct the load test personally. Ritter, the district engineer, the building superintendent, and Maillart (as the representative of the builder) together carried out the test program. This consisted of measuring the vertical movements of the bridge: first, under its own dead load after the scaffold was lowered; second, under traffic loads simulated by layers of gravel on the roadway; and third, under heavy truck weights. Ritter reported all movements to be satisfactorily small, and the only difficulty arose from several small cracks that appeared near the middle of the bridge next to the crown hinge. Maillart agreed to minor alterations, and Ritter's report closed with this summary: "The fundamental ideas upon which the bridge has been designed and built can in certain respects be characterized as new. The bridge, being an innovation, will doubtless find many imitators. Under the circumstances, that not everything was perfect should not surprise us; that the resulting defects are insignificant brings honor to the builder. After all is said and done, the community of Zuoz has gained through the new bridge a commodious, reliable, and relatively maintenance-free facility."[19] This seemingly matter-of-fact statement from Ritter, a man of few words, represented substantial praise from Switzerland's leading structural consultant. The greatest teacher of bridge design thus handed over the future to his most promising student. Time has proven the result of this last close collaboration to be one of the most original concrete structures ever built: the first reinforced-concrete bridge in the Graubünden, the first concrete hollow-box bridge anywhere, and the first concrete arch built by the ring method (fig. 13).

Zuoz was a major event in the history of concrete structures, but it is not a masterpiece of structural art. Maillart's vision had shifted from stone to concrete but had not yet broken with the visual traditions of stone. That would require many more years. Two thousand years of stone building does not fully leave the imagination without a great struggle, one that engaged Maillart during the first five years of the new century and that would reappear a quarter-century later.

At Zuoz, during completion of the bridge, Maillart met a young Italian woman, Maria Ronconi, and proposed marriage almost immediately, on August 1, 1901. His subsequent marriage in November convinced him of the need to make a good life for his wife and future family by starting his own business, which he did on February 1, 1902. The following year he entered a contest to design and build a two-span bridge at Billwil, which he conceived in the same form as Zuoz. The owner asked Wilhelm Ritter to report on the project with reference to the lowest bidders. Ritter recommended Maillart, and once again the teacher played a major role in the student's early career.[20] With that design nearly completed, Maillart returned to Zuoz in the fall of 1903 at the request of his former client, who was worried about cracks that had appeared in the bridge walls near abutments (fig. 14).[21]

Maillart was now on his own as Ritter was unable to work after his last report on Billwil in August 1903. His student faced a new problem, which would bring into question the aesthetic tradition of bridge construction dating at least as far back as Roman works of the first century A.D. Following the Stauffacher Bridge, Maillart had designed Zuoz with vertical walls, extending right to the abutments, which he imagined to be part of his hollow box. Such full-depth walls had been a crucial feature of Roman bridges because of their circular profile, even though they are structurally inefficient and sometimes unsafe. The Roman designers knew this danger and corrected it by placing the extra loading of rubble masonry on the arches near the abutments. Full-height fascia walls were then essential to contain this loose material and thus prevented cracks from harming the arch. Of course, they could have shaped the arch differently, but the circular profile was easier to build because all the voussoirs (stone pieces) could be cut to the same wedge shape along the radii of the circle. The ancients traded construction economy for material efficiency with the consequence of full-height walls. During the nineteenth century, such walls gradually became useless, but the image of "appropriate" form persisted even for Maillart until he saw the cracks at Zuoz. He realized immediately that those cracks were not dangerous for his reinforced concrete bridge, and he further recognized that the material there was structurally useless. Maillart had designed his arch as three-hinged, like Stauffacher, so that all of the load was carried into the abutment hinges (like a door hinge that allows the arch to rotate slightly under load to avoid local stresses) located at the junction of the curved lower slab and the foundation (fig. 15). He now understood that the arch's internal forces, while distributed over the entire box section at the crown, gradually moved down the walls to focus entirely at the abutment hinges. This represented a new

Figure 13
Zuoz Bridge, completed in 1901.

Figure 14
Cracks near the abutments in the Zuoz Bridge observed by Maillart in 1903.

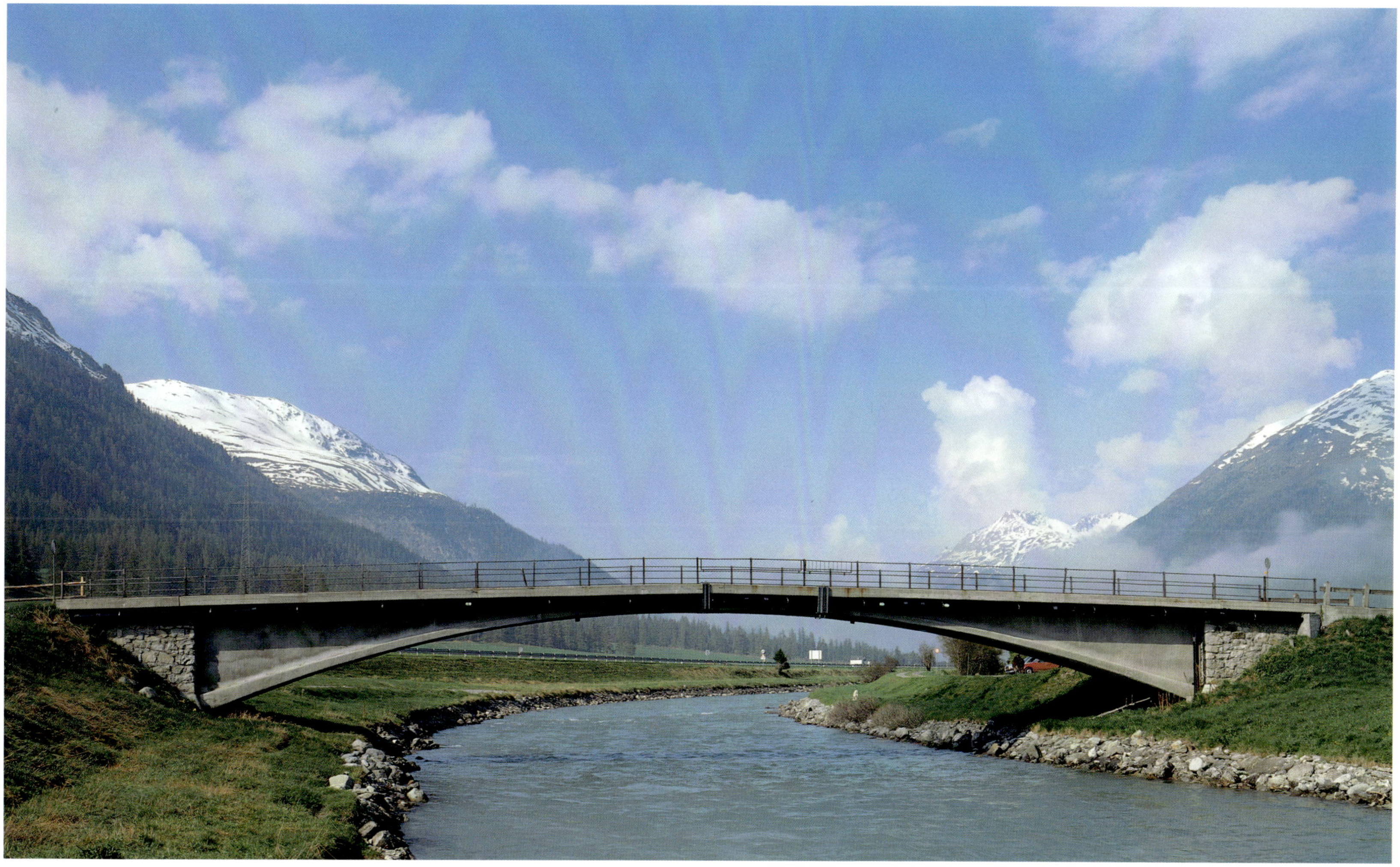

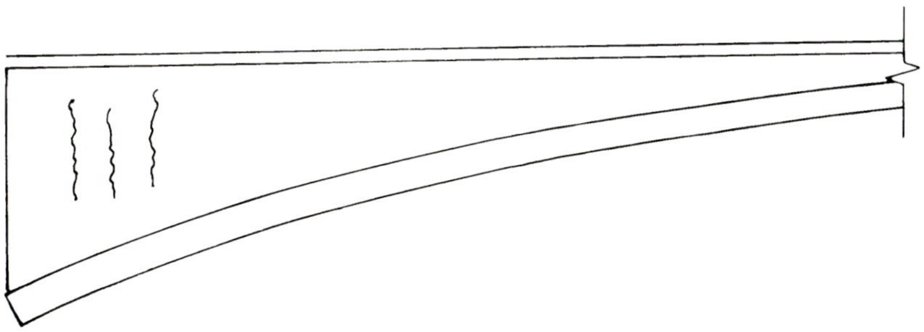

Figure 15
Zuoz Bridge, where the load is carried into the abutment hinges at the junction of the curved lower slab and the foundation.

Figure 16
Maillart's drawing for the Tavanasa Bridge competition, 1904, showing his understanding that forces were concentrated at the abutment hinges.

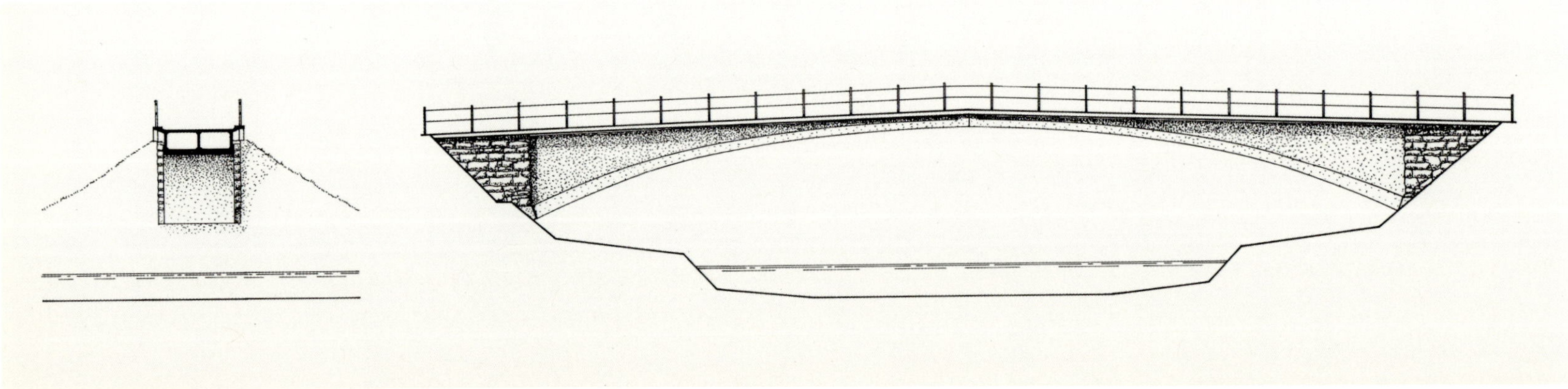

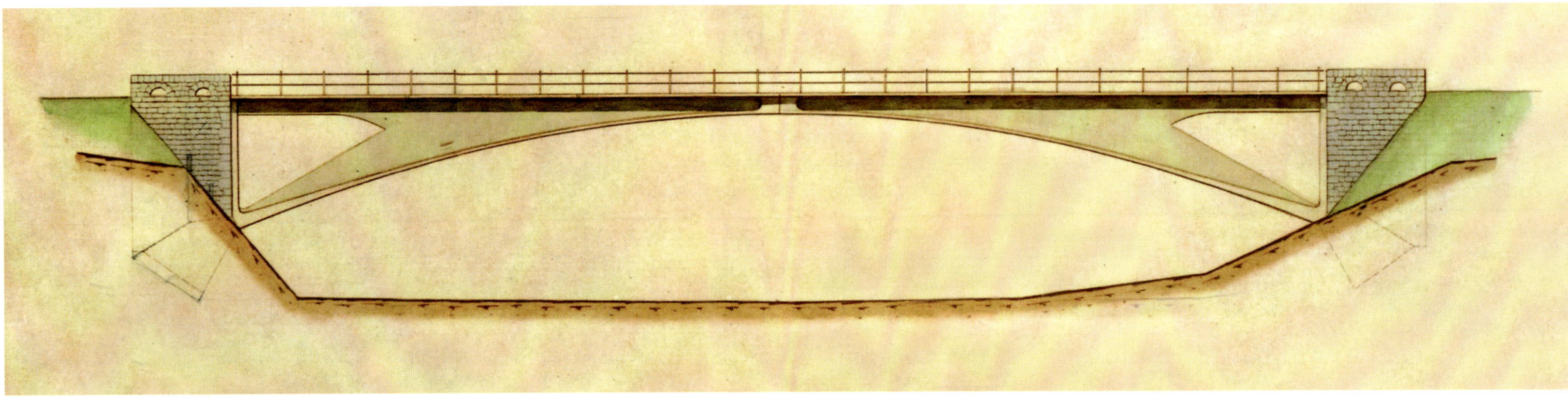

image and one that suggested a new form as he thought about bridge design during 1904. He did not conceive this new idea until after a visit in August to the site for a new bridge at Tavanasa over the Rhine, once again in the Graubünden. His competition drawings of October 1, 1904, show this new form for the first time (fig. 16). He won the competition, built the bridge in 1905, and thereby produced the first masterpiece of structural art in reinforced concrete (fig. 17).[22]

Since this is our first example of structural art at the highest level, it is advisable to explain the three-part basis for this judgment. First, the structure was even lighter than the Zuoz Bridge but no less strong; it was highly efficient because it used a relatively small amount of material for a given strength. Second, the bridge was economical; it resulted from a design-construction competition in which cost was a major factor in awarding the contract. Third, the designer presented a new form that has been judged elegant by aesthetically sensitive people in such examples as Max Bill's book on Maillart (1949) and the exhibition on Maillart at the Museum of Modern Art in New York (1947).[23] Maillart was far ahead of both the engineering profession and the art world that revered imitative urban design. There was no publication in Switzerland on his bridge until 1914 and no chance for him to continue developing his new idea until after the Tavanasa was destroyed in an avalanche on September 25, 1927.[24] Maillart therefore had to turn to other types of structures in order to keep his construction business growing and to keep his creativity active. For the most important example of this partial retreat, Maillart, quite literally, went underground to design and build an almost surreal work in 1912.

Beginning in 1908, Maillart turned to buildings and especially to standard industrial works that promised a wide market as well as a demonstration of

Figure 17
Tavanasa Bridge, Tavanasa, Switzerland.

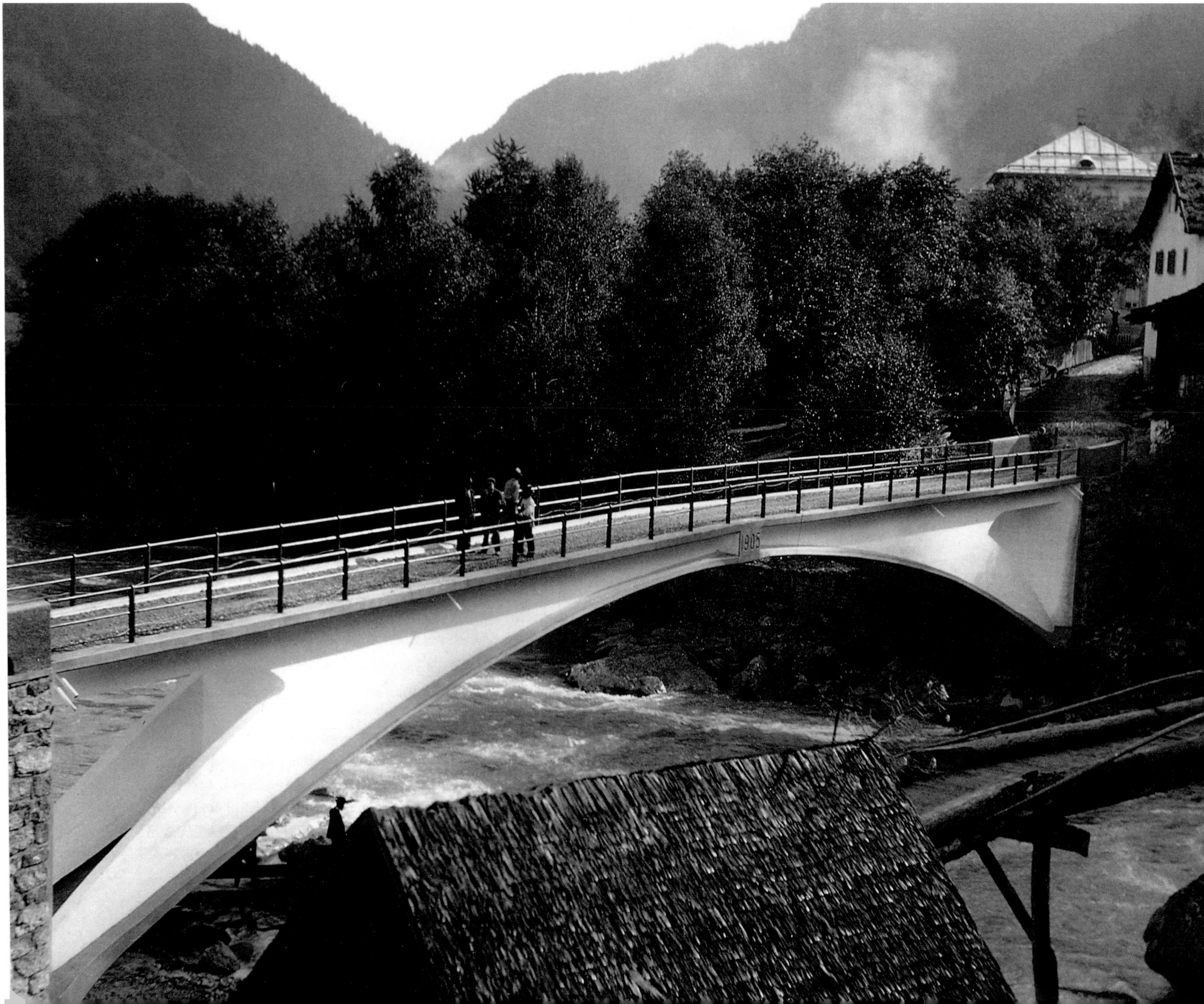

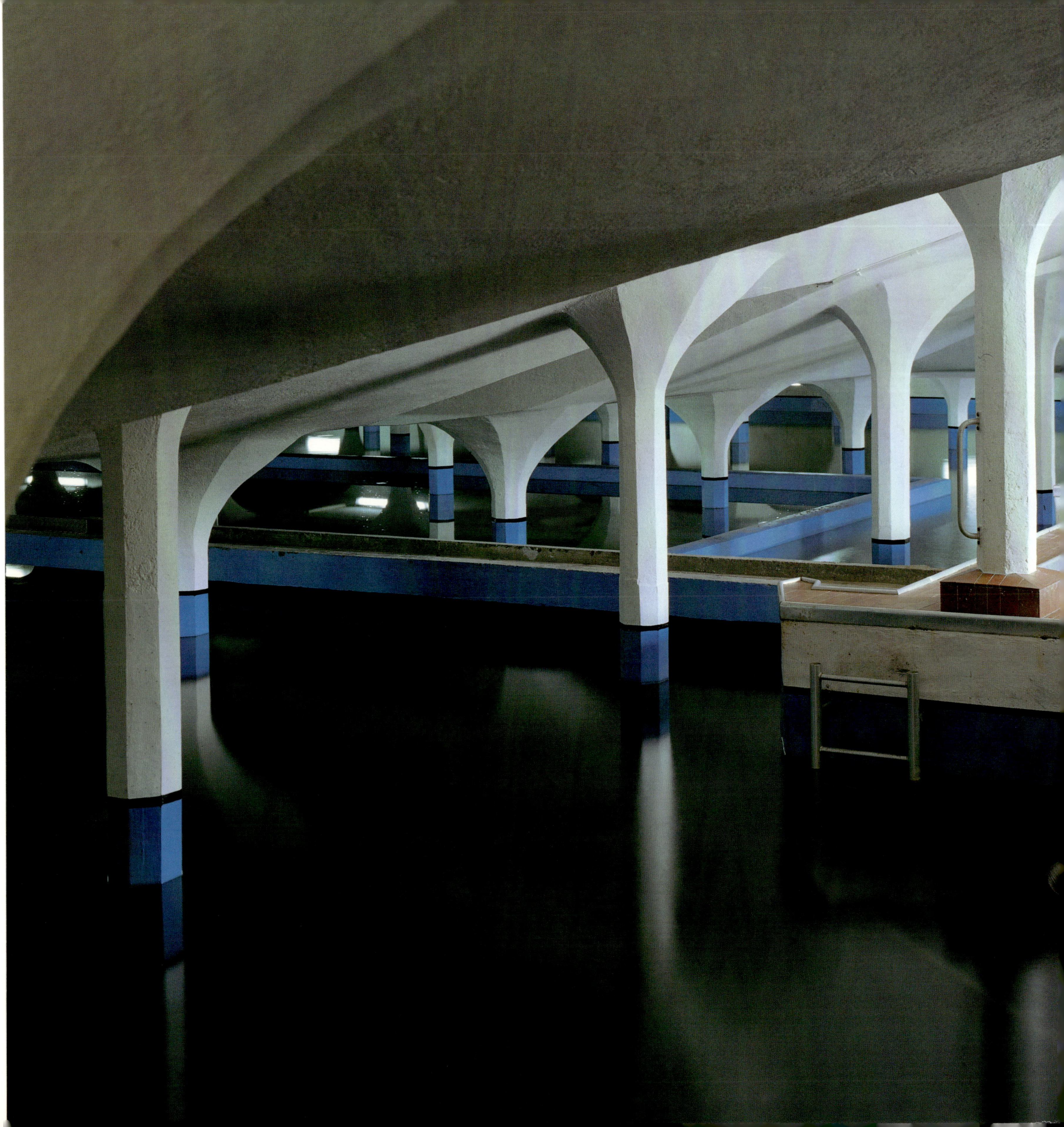

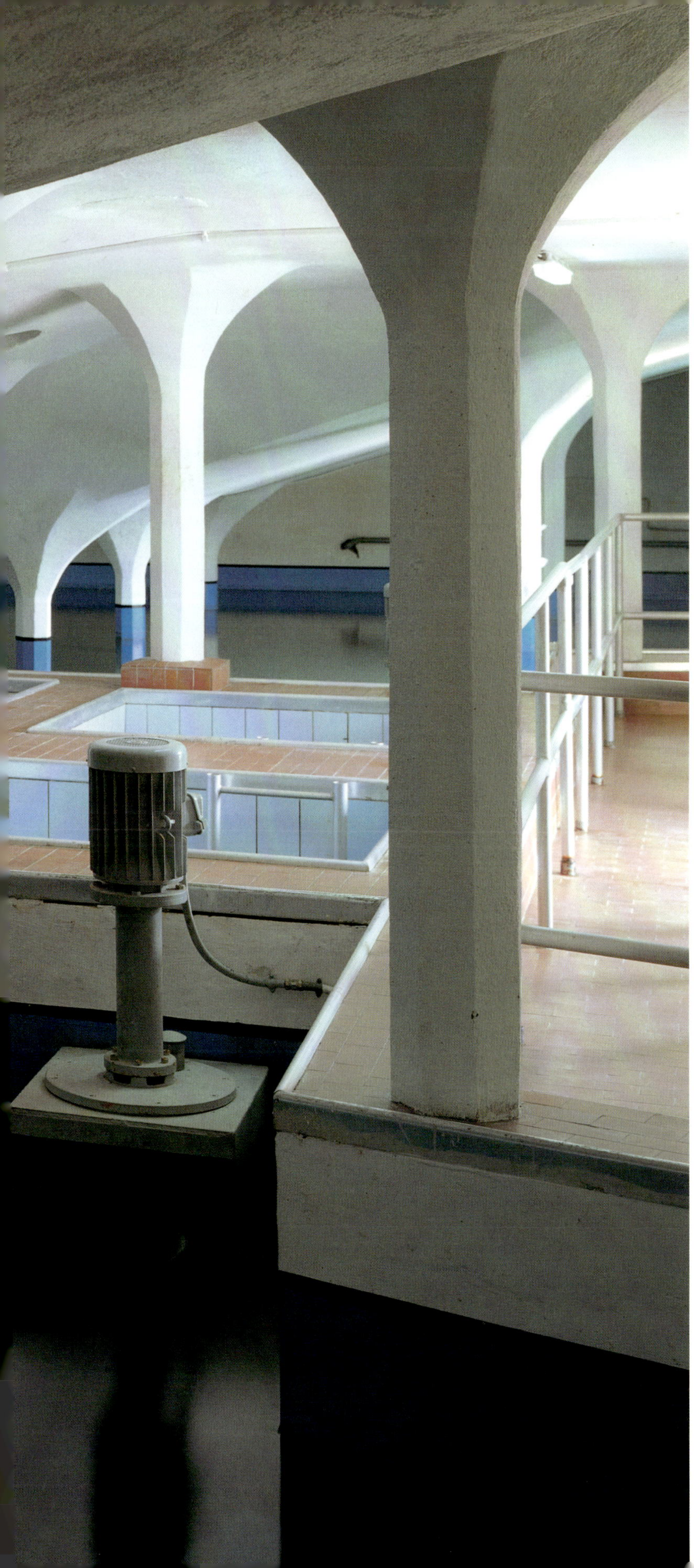

Figure 23
Maillart's columns supporting the sloping roof of the water filtration plant in Rorschach, Switzerland.

Not only was Maillart designing strange new forms; he also succeeded in amassing a fair fortune in so doing. His business was profitable, and at places like Rorschach he also enjoyed the design challenge. But after 1912, business would become dominant and even entice him out of Switzerland into Spain and then Russia.

By 1912 Maillart had a wife, three young children, a large house in Zurich, and a rapidly expanding construction business (fig. 24). More and more he traveled to the rapidly industrializing Romanov empire, and in early 1914 he convinced his wife, Maria, to bring the family and the servants along with him to Riga for the summer while he worked on large projects there and in St. Petersburg. After some hesitation, Maria consented and the Maillarts settled on the Riga coast in July 1914. The timing was bad. The war caught them, and they were effectively trapped in Russia for more than four years during the great upheaval. Even worse, Maillart's wife died in 1916 after they had moved from Riga to St. Petersburg and then to Kharkov. His Swiss business collapsed, the Bolshevik Revolution destroyed his fortune in Russia, and he barely escaped the revolutionaries in late 1918. In early 1919, Maillart and two of his children arrived back in Switzerland penniless and homeless. He would have to begin again, this time living with relatives in Geneva. Just as marriage had encouraged him to start his own business and to develop major innovations, the loss of wife and capital forced him into an almost monastic living style in 1919 and into the profession of design rather than the business of building. It was a sad time in his personal life, but a liberating era for his professional career.[27]

Although Maillart was a brilliant builder, he was not as talented in managing a company. He had done it but at the cost of pursuing design. The structural artist especially needs to think hard and usually alone about the complexities of designing and constructing elegant works. Maillart had no real option: he was deep in debt, which ruled out the capital-intensive business of a building contractor. By 1920, he had begun to solicit design work, based on his strong prewar reputation, and he opened an office in Geneva. During the early 1920s and without much work, he turned his mind to a series of engineering problems, which he presented for publication. Considering his preoccupation with construction over the preceding twenty years, the extent and depth of these writings is remarkable. In one case, he developed a new understanding of a major unsolved problem in engineering mechanics, known today as the theory of the shear center—the location for a load on a steel beam that allows it to bend downward without twisting. The discovery, credited to Maillart, was a problem that had baffled the academic research community, and his clear and elegant solution is still admired by researchers.[28] He had a deeper motive than merely becoming respected academically; he was concerned that academic research was having a negative influence on design freedom. By 1923 he had begun to find opportunities to exercise that freedom even though unsympathetic professors at his alma mater would oppose him vigorously over the next two decades.

Maillart's "rediscovery" of his design imagination began in 1923 when, through a builder friend, he obtained the commission to design a series of small mountain bridges in the Wägital Valley in the canton of Schwyz. For the Flienglibach Bridge there, he recognized the possibility of a new form owing to the need for a deep deck beam or parapet along with an arch design. He recalled Wilhelm Ritter's lectures that introduced the idea of arch and deck sharing the otherwise dangerous live loadings. Ritter had explained a simple yet accurate method of determining the relative bending in deck and in arch and had especially emphasized the principal design conclusion that "the stronger the beam [deck] and the weaker the arch, then the greater will be the bending moment in the beam [deck] and the smaller it will be in the arch, and vice versa."[29]

Here was an invitation to make the arch as thin as it could be built because with a strong parapet reinforced properly as a beam, there would be barely any live-load bending in the arch. Indeed, Maillart's calculations of 1923 showed that the live-load bending was negligible in the thin arch he designed for Flienglibach.[30] Here again the function followed the form; the form of this deck-stiffened arch directed the bending into the deck rather than the arch. Maillart's aesthetic motivation for the thinnest possible arch led him back to Ritter's original idea and resulted in a new form for bridges that he would develop fully over the next ten years.

Following Flienglibach, Maillart produced two other designs in 1924, which signaled the full recovery of his design imagination and prepared him for the last and most creative period of his life. During the summer he began to study a roof for the single-story shed beside a flat-slab warehouse he had designed in Chiasso. This roof could easily have been framed in a standard manner and built inexpensively, but Maillart was not in the mood for such trivial forms. Here was a structure that would be fully seen and deserved a structural expression. Many writers have speculated on Maillart's thinking and on what influenced his unique solution to this common problem.

Figure 24
Maillart family, Zurich, Switzerland, 1913.

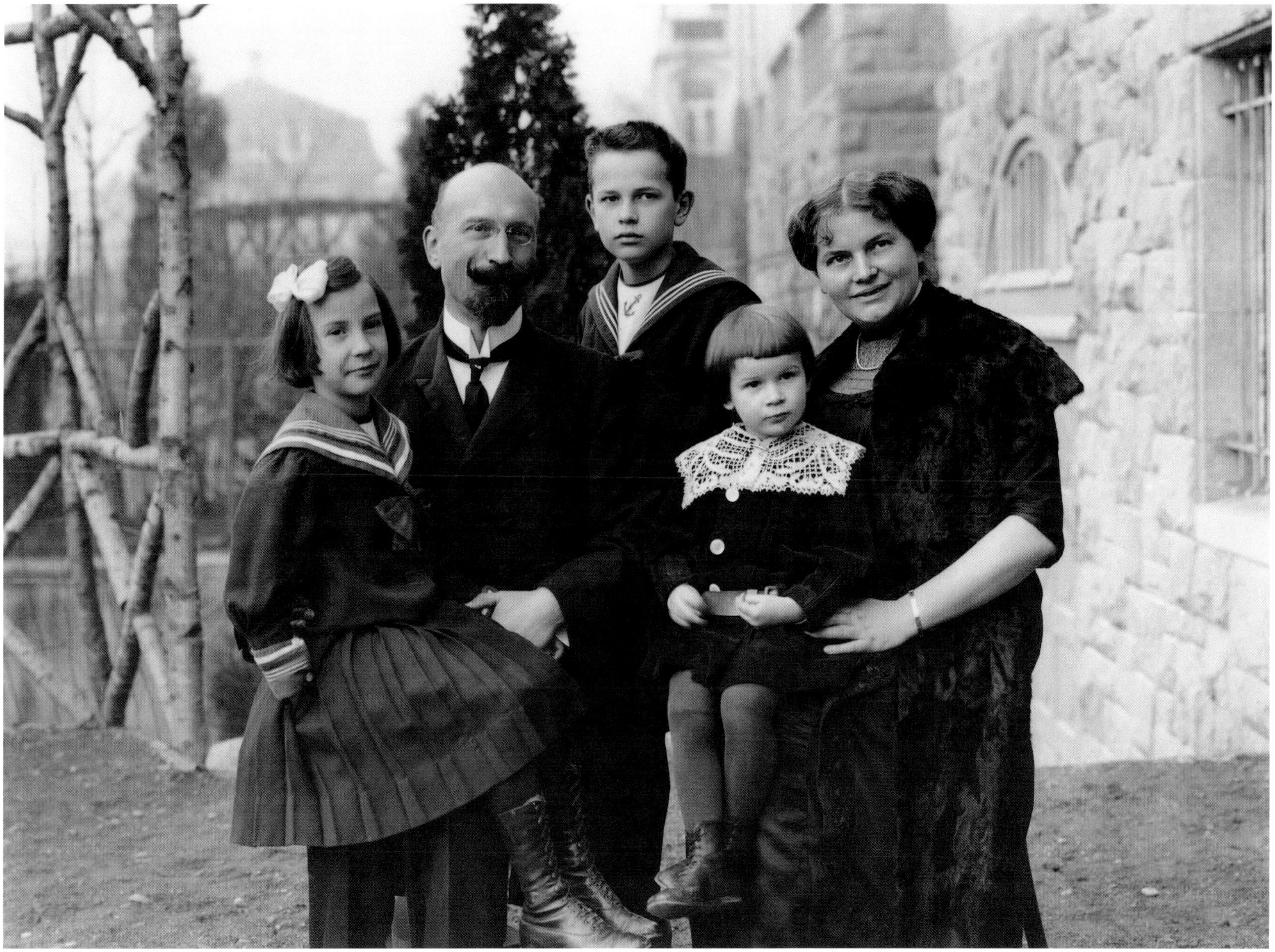

Figure 25
Diagram of Chiasso shed cross section.

Figure 26
Diagram of the Chiasso shed and string polygon.

Figure 27
Shed with two inverted arches under a pitched roof, completed in 1924, Chiasso, Switzerland.

Figure 28
Chiasso shed column showing the stiffening flange.

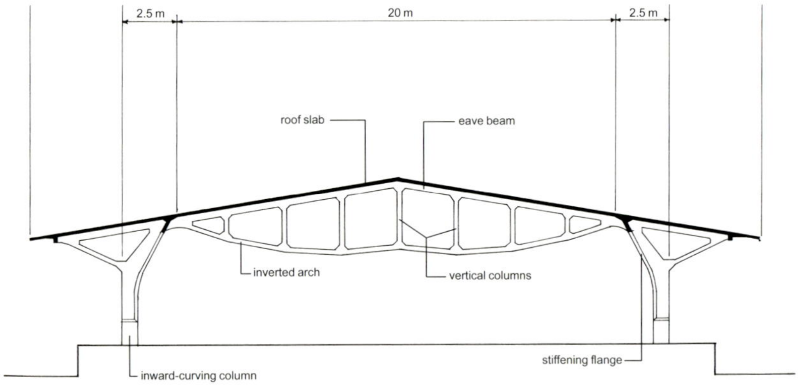

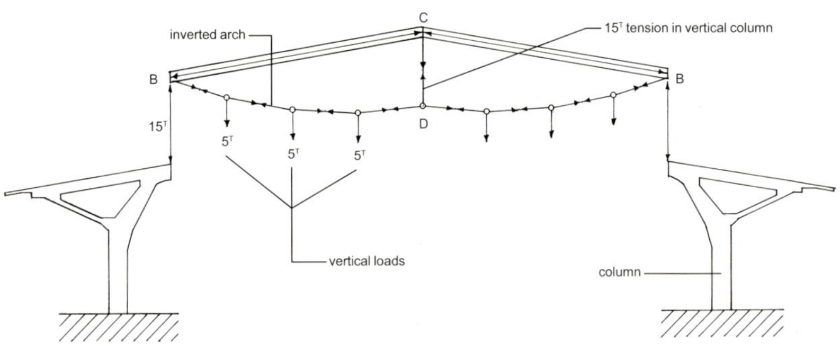

Maillart's design calculations, however, settle the issue; he saw the structure as two inverted arches stiffened by the pitched roof above (fig. 25).[31] It was an inverted deck-stiffened arch or a roof-stiffened cable (encased in concrete). To explain the profile, a string is attached between points B and D from which hang three weights of five tons each (fig. 26). The string forms a polygon of four straight lines to carry the fifteen tons, half to point B and half to point D. By having a kink at midspan, Maillart directed the forces at D to travel up to the gable peak at point C and then down the eave beams to point B. Thus, all the external loads are carried by internal forces acting along the axes of the elements. This is the principle of the truss, but Maillart had achieved it without the usual diagonal elements. The hanging arch, however, could be subjected to the dangerous effects of loading only half the span. Maillart, realizing this danger, also recognized that in connecting the inverted arch to the gable by vertical columns, he could arrange to have the stiff eave beam and roof slab together carry the bending in the same way as the parapets did at Flienglibach. He therefore achieved his primary goal, an expression of thinness, by inventing a form that could work like a standard truss. But he avoided the cluttered effect of the diagonal elements in trusses by having the eave beam and roof-slab stiffener perform their function (fig. 27).

Maillart played with this roof form beyond anything done in the Wägital. The inverted arches do not reach the columns directly but end 2.5 meters from them, being supported on a slanted-arm extension of the vertical column. He thereby shortened the roof span from 25 meters (the spacing between columns) to 20 meters (the distance between inverted arch ends). By this choice, Maillart reduced the internal forces in the arches and eaves by thirty-six percent. But this reduction required a large column in order to prevent it from bending inward because the entire weight of the roof span must come onto the arm extension 2.5 meters from the center line of the column. Maillart's solution was partly to balance that inward bending by the outward bending of the cantilevered roof, which extends 4.2 meters outside the interior span. He designed the cantilever to be made of two light elements, an extension of the eave beam and a thin slanted element of the same size and shape as the elements in the inverted arch. But this cantilever was much too light to balance the weight of the entire span of 20 meters, so Maillart provided a stiffening flange on the inside of the column, hardly visible in the profile view, which permits the inward-curving column arm to carry the bending easily (fig. 28). Moreover, he used that flange as part of a thin curving beam in the transverse direction to provide stiffness against motion perpendicular to the profile or span direction.[32]

Because of this two-way integration of form, the structure is visually intriguing as one walks beneath it and sees it transformed from a strong profile span to an interconnected set of elements curving in two directions (fig. 29). Sadly, today an ugly shed extension completely obliterates the famous profile, but from inside one can still see Maillart's structure in fine shape and in startling contrast to the thoughtless structure in front. From within, the inverted arch and its vertical elements almost seem to hang from the solid roof slab, and we must keep the profile in mind as we try to interpret the closer and more detailed views still possible. Along with the striking simplicity of the elements, closer inspection reveals small haunches or filletlike connections between verticals and horizontals, as well as the raw edges of the lateral arched flanges. An industrial work with no special polish or finish, the structure demonstrates that beautiful form does not depend upon ornate, facaded, or veneered coverings.

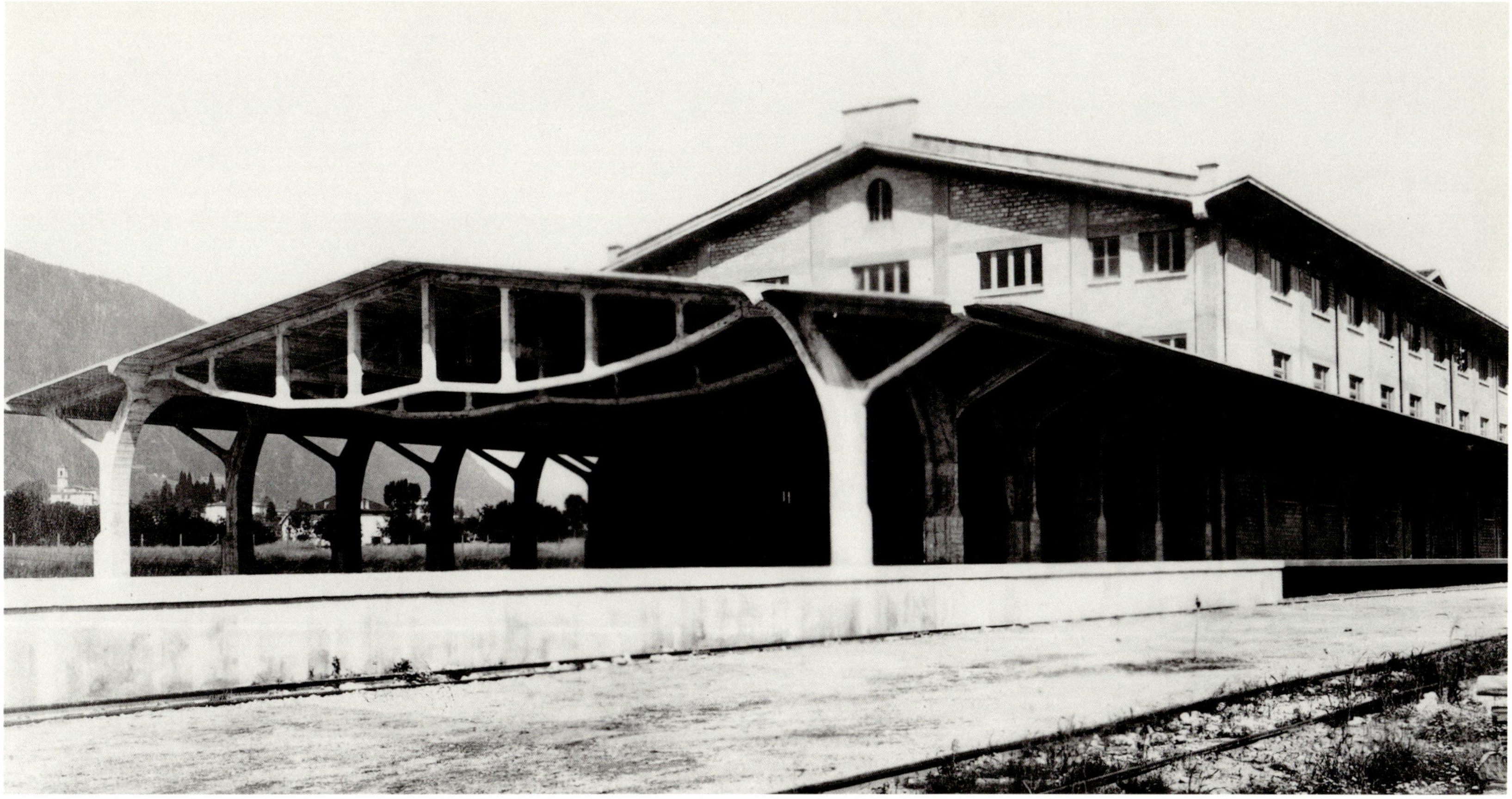

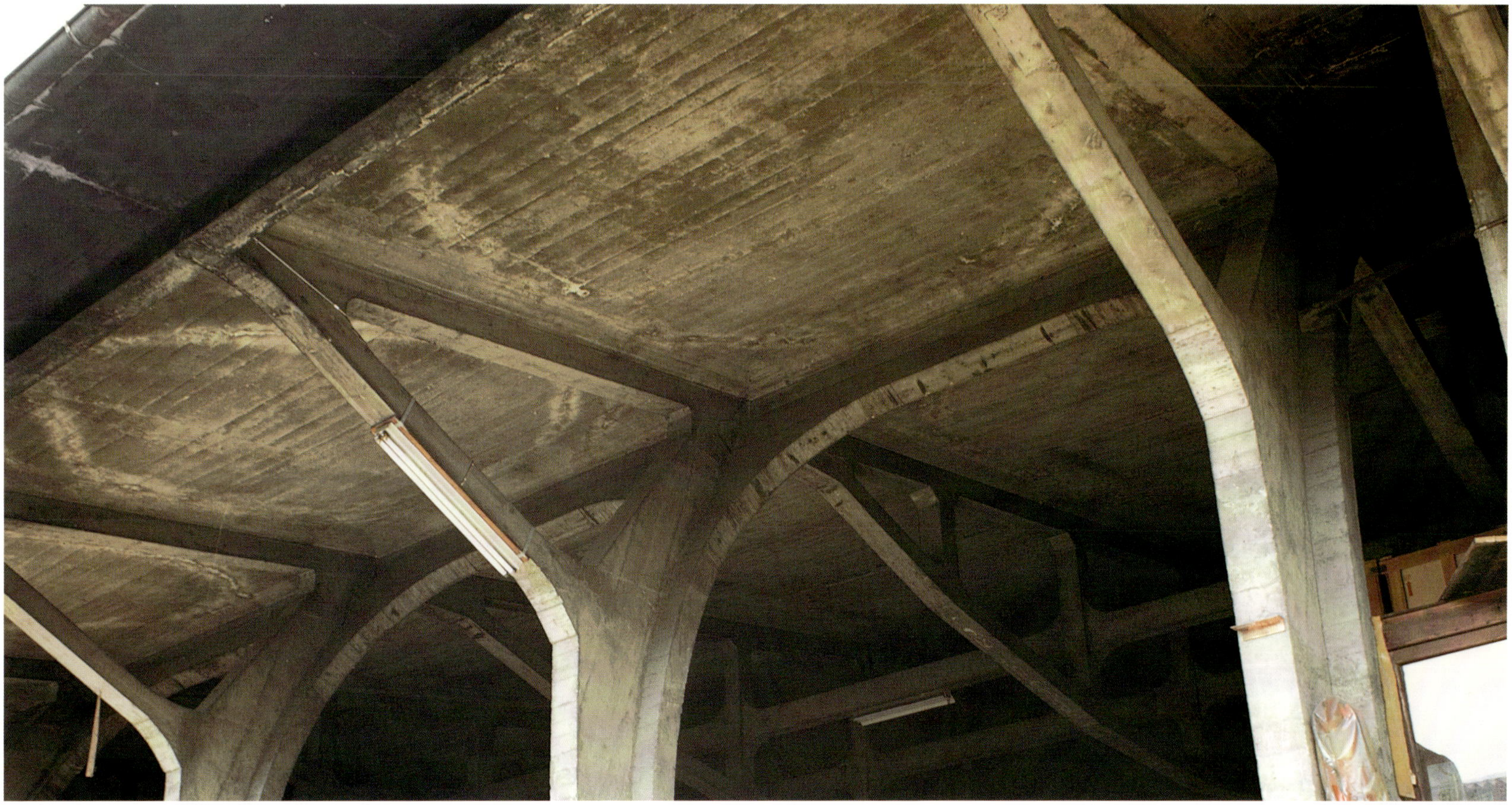

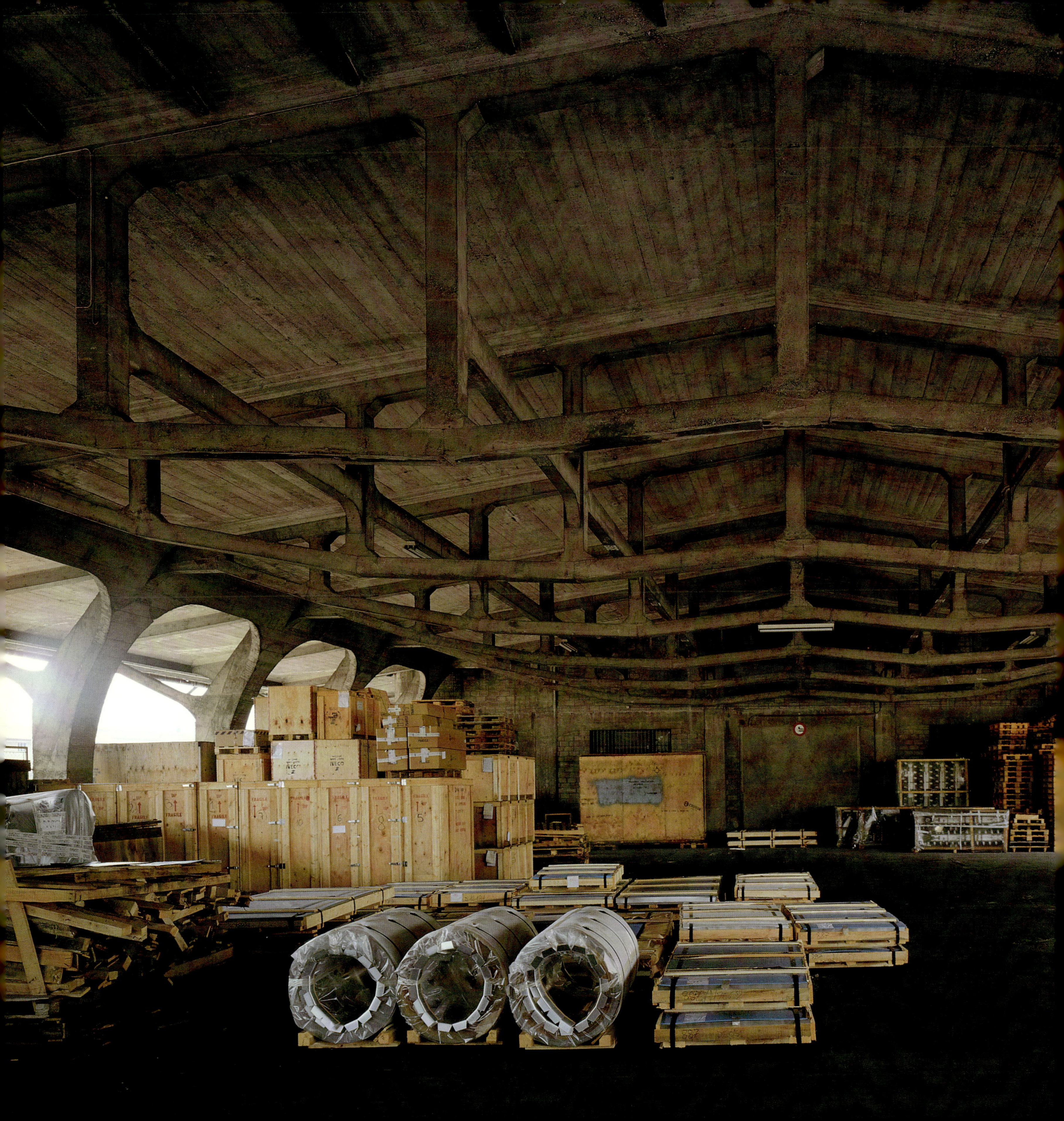

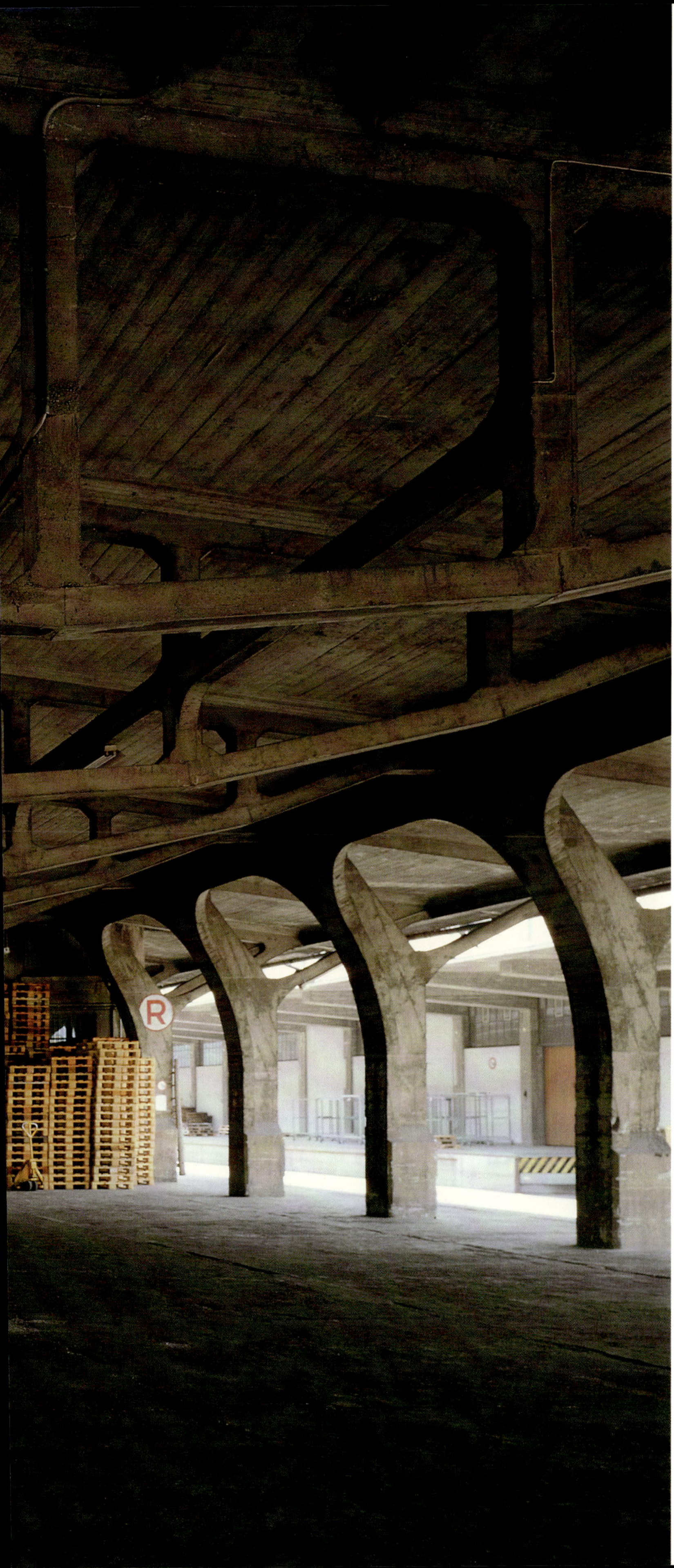

Figure 29
Interior of the Chiasso shed.

Figure 30
Châtelard aqueduct, built in 1925, Châtelard, Switzerland.

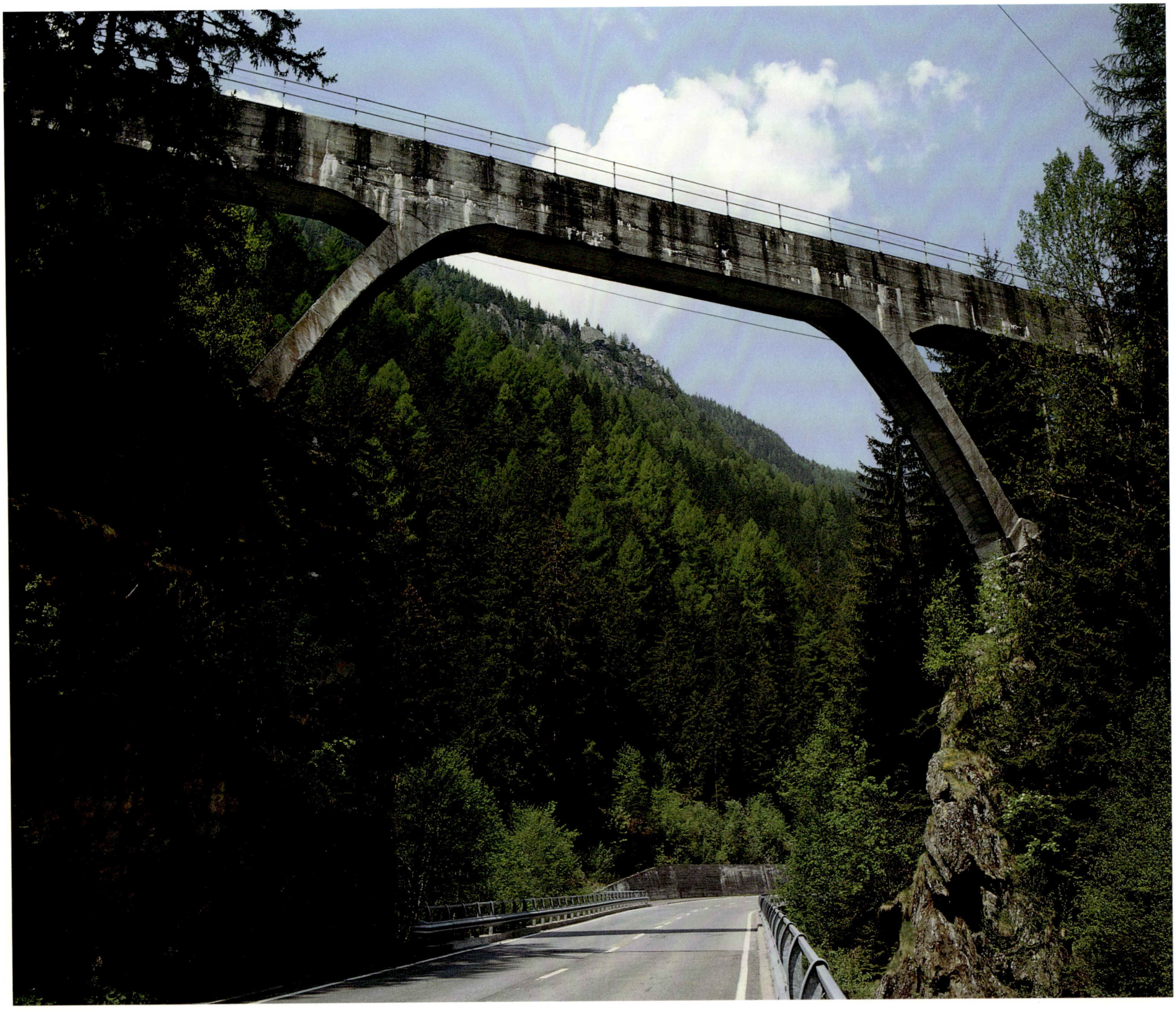

Figure 31
Railway engineers' rejected proposal for the Châtelard aqueduct.

Figure 32
Maillart's accepted proposal for the Châtelard aqueduct.

Figure 33
Cross section of the Châtelard aqueduct, revealing the thinness of the columns.

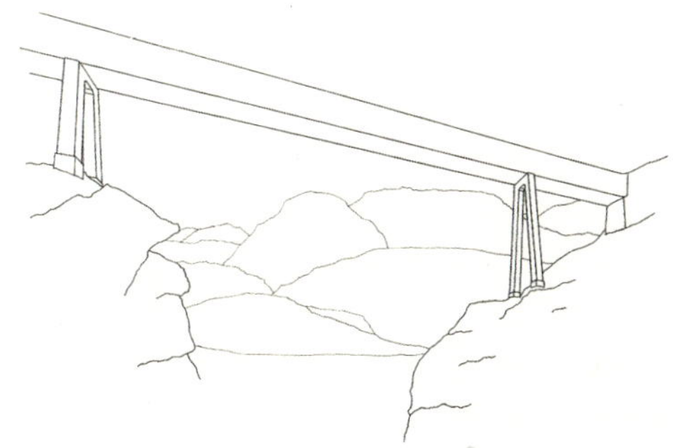

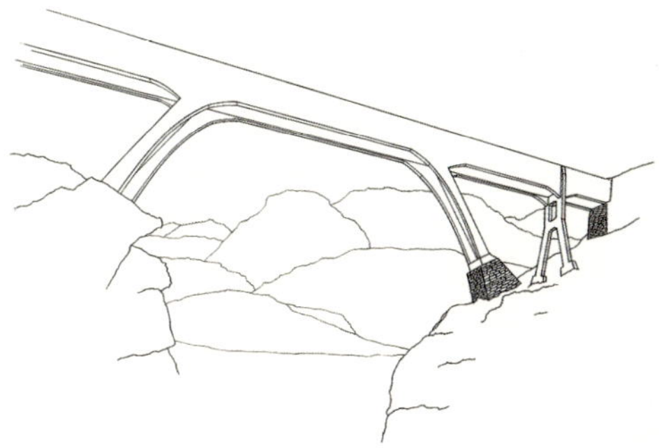

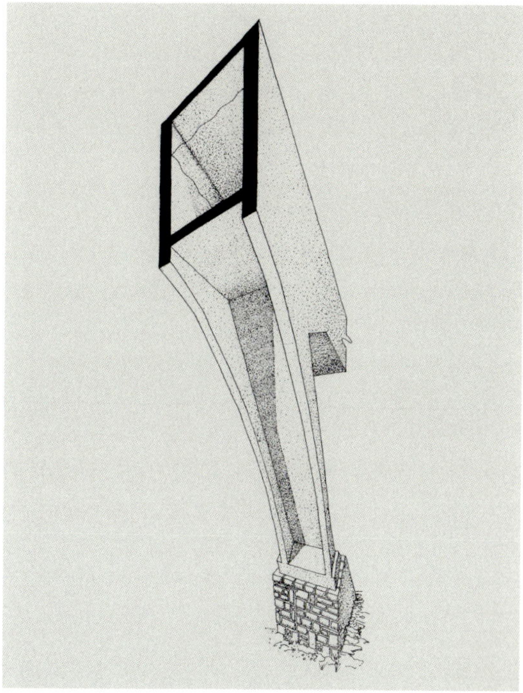

By the end of 1924, with the Wägital bridges completed and Chiasso in design, Maillart began to study a design for the Swiss Railways, which planned an aqueduct for its power plant high up at Châtelard near the Forclaz Pass leading from Switzerland to Chamonix, France (fig. 30). Here the problem was to carry water in a rectangular box section across the roadway about 25 meters above it. The railway engineers had made a design (fig. 31); Maillart, however, proposed his own design (fig. 32), which was accepted, and built in 1925. The channel section, 2.2 meters high and 2.65 meters wide, required a box of great depth just to contain the water. This meant that Maillart could not design a light, open, horizontal structure as he had done at Chiasso.[33] He had to start with a heavy-looking beam profile. A deck-stiffened arch would make no sense here, in spite of having a natural stiffening beam, because that beam could span the valley without the need of an arch at all. Maillart, therefore, decided to support the beam on light, slightly curved, slanted columns, whose thin, finely shaped profile contrasted with the deep horizontal beam. The span over the roadway exemplifies an unusual and dramatic profile that Maillart designed with a flat, smooth concrete face, but it is much thinner than it looks. The profile does not reveal that the columns are merely thin flanges on each face connected by a narrow wall (fig. 33). As we move beneath the aqueduct, this slender column form becomes visible, as do the approach-span supports that in profile are light cross walls, really cross frames.

This relatively long aqueduct is much narrower than any of Maillart's bridges. (Châtelard has a length to width ratio of twenty-five whereas Tavanasa's ratio is fourteen.) To provide lateral stability, Maillart decided to spread the legs of the open cross walls (see fig. 32). These cross frames go straight down for about two meters below the box, then widen at the base to about double the width at the top. Designing the frames to be made of slender elements and to meet the box with a slight haunch in the box wall, he carried the vertical line of these frames up the side of the box to its top, thus stiffening the box where it is heavily stressed at the support and also providing a visual emphasis to that sensitive place. He even deepened the horizontal beam slightly by extending walls below the base of the water conduit. The extensions provide the haunches at the supports and flow together with the column flanges of the same width (see fig. 33). Since he could not get any form-play with the base of the aqueduct because it had to be smooth and straight to carry the flowing water, he created walls that perform a visual integration and serve to transfer forces more smoothly as well.

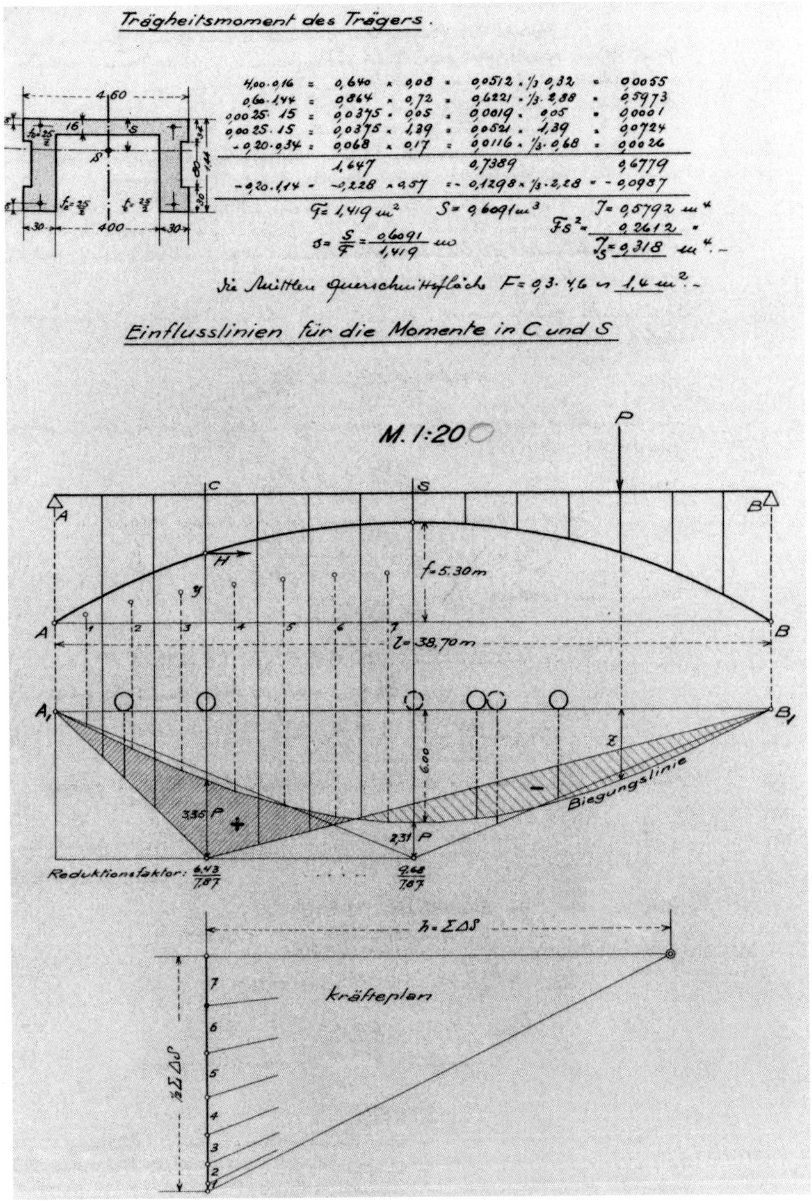

Figure 34
Maillart's calculations for the Flienglibach Bridge, demonstrating Ritter's teaching of graphic statics, the first of four pages calculating the influence of live load, 1923.

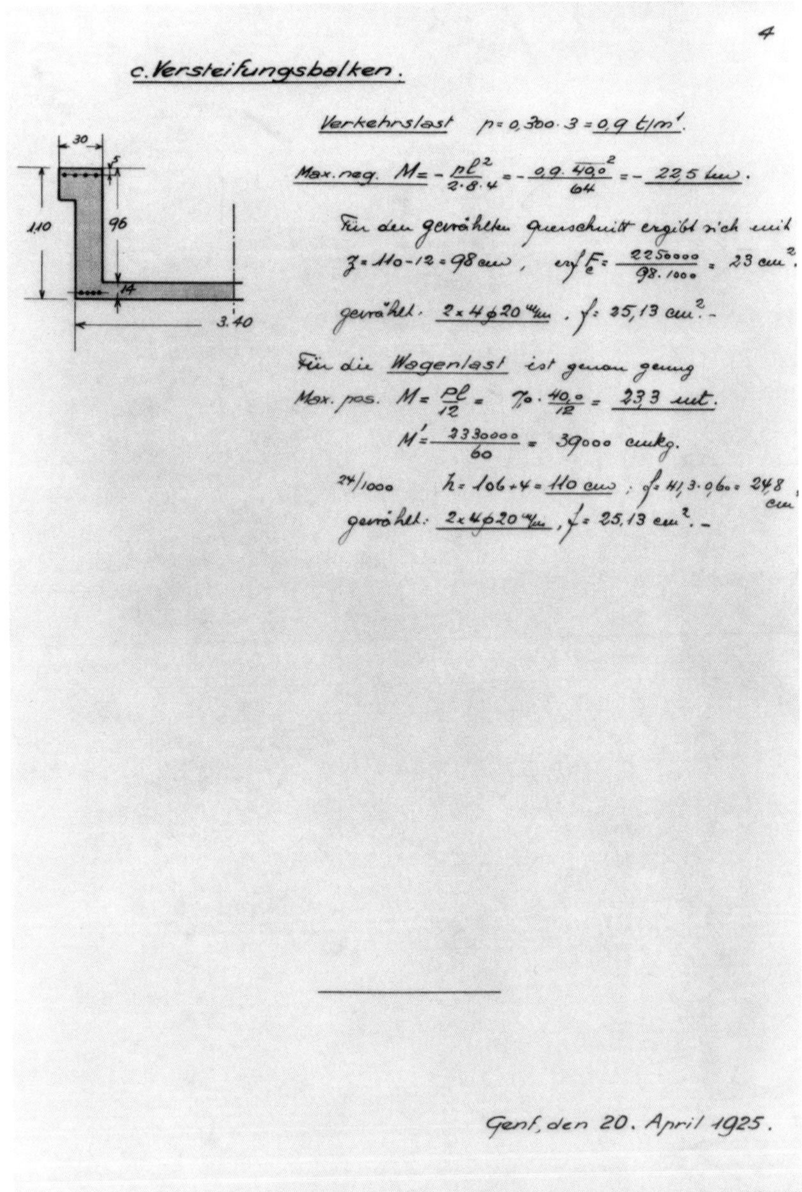

Figure 35
Maillart's concise calculations for the influence of live loading on the deck-stiffening arch of the Valtschielbach Bridge, 1925.

Ritter had approved of Maillart's hollow-box form and had stimulated his deck-stiffened arch form so that by 1925 the former student was now poised to carry these two seminal ideas onto the highest levels of structural art. From 1925 to 1933 he perfected his Flienglibach form, and from 1930 to 1936 he transformed the Tavanasa design in surprising new directions. Both original bridges have disappeared, but their progeny survive to show how the thinking of a structural artist continually evolves. Until the public and its political leaders recognize the new art of structural engineering, however, all of Maillart's works will be threatened with destruction or defacement. Flienglibach was removed after Maillart's death to make way for a wider roadway, while a powerful avalanche destroyed the Tavanasa Bridge in 1927 and served to revive his imagination of the three-hinged, hollow-box arch design. These two original bridge designs illustrate Maillart's ideals of strong structural expression without wasting materials or money. His revival in the 1920s began with the deck-stiffened arch that reached a mature expression in 1925 with the Valtschielbach Bridge. Here, he improved upon Flienglibach by creating a thinner arch and a more slender parapet, but his thinking (as seen through calculations) evolved in a more fundamental way. In Flienglibach he solved the problem of the live loading by a relatively complex analysis built upon Ritter's writings, using graphic statics along with a mathematical analysis including four pages of numerical work (fig. 34). Much of this analysis is not clearly described: Maillart's references to Ritter and to methods commonly used in Switzerland enabled him to leave out many steps.

At Valtschielbach the analysis is totally different. Instead of four pages, Maillart used less than half a page, and the method is technically clear (fig. 35). He abandoned the so-called more rigorous analysis in favor of one that is just as good and far simpler. For example, he used a complex combination of graphic statics and algebraic formulas to calculate for Flienglibach the bending in the deck and arch under the concentrated loading of a heavy truck located one quarter of the span length from the support. For Valtschielbach he states: "For the truck load it is accurate enough [to use] Pl/12 = 7.0 x 40/12 = 23.3 meter-tons" for the bending moment. This simple formula gives almost identical results for the Flienglibach as did the much more complex approach, which Maillart had used for that bridge.[34] For Maillart, however, it was an essential part of his thinking that the more complex preceded the simpler; his earlier experience led him to search for a clearer and more concise calculation. In every case, these bridges were carefully load tested and demonstrated the validity of Maillart's methods.

Once again, he was not satisfied with the design of the Valtschielbach (fig. 36). Between 1923 and 1934, Maillart completed twelve deck-stiffened arch bridges, and finally in the last two he arrived at his greatest masterpieces of structural art in this form: the Schwandbach Bridge of 1933 is the finest such example he or anyone else ever created (fig. 37). The bridge has almost the same overall dimensions as both Flienglibach and Valtschielbach, but the differences in detail are essential to Maillart's evolution. This becomes clear in a comparison of the Valtschielbach Bridge (1925) and the Schwandbach Bridge (1933). Both form a U-shaped crossing in plan, both are technically equal as deck-stiffened designs, and both achieve remarkably thin arches (figs. 38, 39).[35]

The earlier bridge in plan crosses the valley in a straight line and kinks at both ends; the later design makes a smooth continuous curve (fig. 40). Whereas the Valtschielbach has stone, Romanesque-arch approach spans, the Schwandbach eliminates all reference to the aesthetic of stone and fully integrates the approach spans with the stiffening beam on the deck over the arch. The earlier arches are smoothly curved in elevation, while the later arches are polygonal with slight breaks where the vertical walls rest. And finally the arch at Schwandbach follows the roadway curve on the concave side and splays outward as it reaches the abutments on the convex side. Maillart joined the arch and deck by trapezoidal walls that cover the full arch width and help create the dramatic side view of this bridge (fig. 41). Having settled the issue of satisfactory analysis with the Valtschielbach Bridge, Maillart was free to think carefully about the conceptual design form and gradually come to the exemplary proportions of the Schwandbach Bridge.

His last deck-stiffened arch was for a footbridge over the Töss River near Winterthur, where a local engineer had made a design, crediting it to Maillart's system and sending it to him for review. Maillart completely redesigned it with the result that an ordinary design became an exemplary work of structural art (fig. 42). The Töss Bridge expresses lightness through its thin arch and slender cross walls which connect the stiffening-deck beam to the arch. This efficiency, a consequence of Maillart's deck-stiffened arch innovation, goes together with a construction economy. Another concrete footbridge, built several years later also over the Töss River, cost about fifty percent more than Maillart's design. This second work had a good structural form without decoration, but it lacked the elegance of Maillart's design. It was not designed by an artist. With the Töss Bridge built, Maillart's work in that form ended, but his three-hinged hollow-box arch form was reaching its peak.

Figure 36
Valtschielbach Bridge, built in 1925, Donath, Switzerland.

Figure 38
Elevation and plan of the Valtschielbach Bridge.

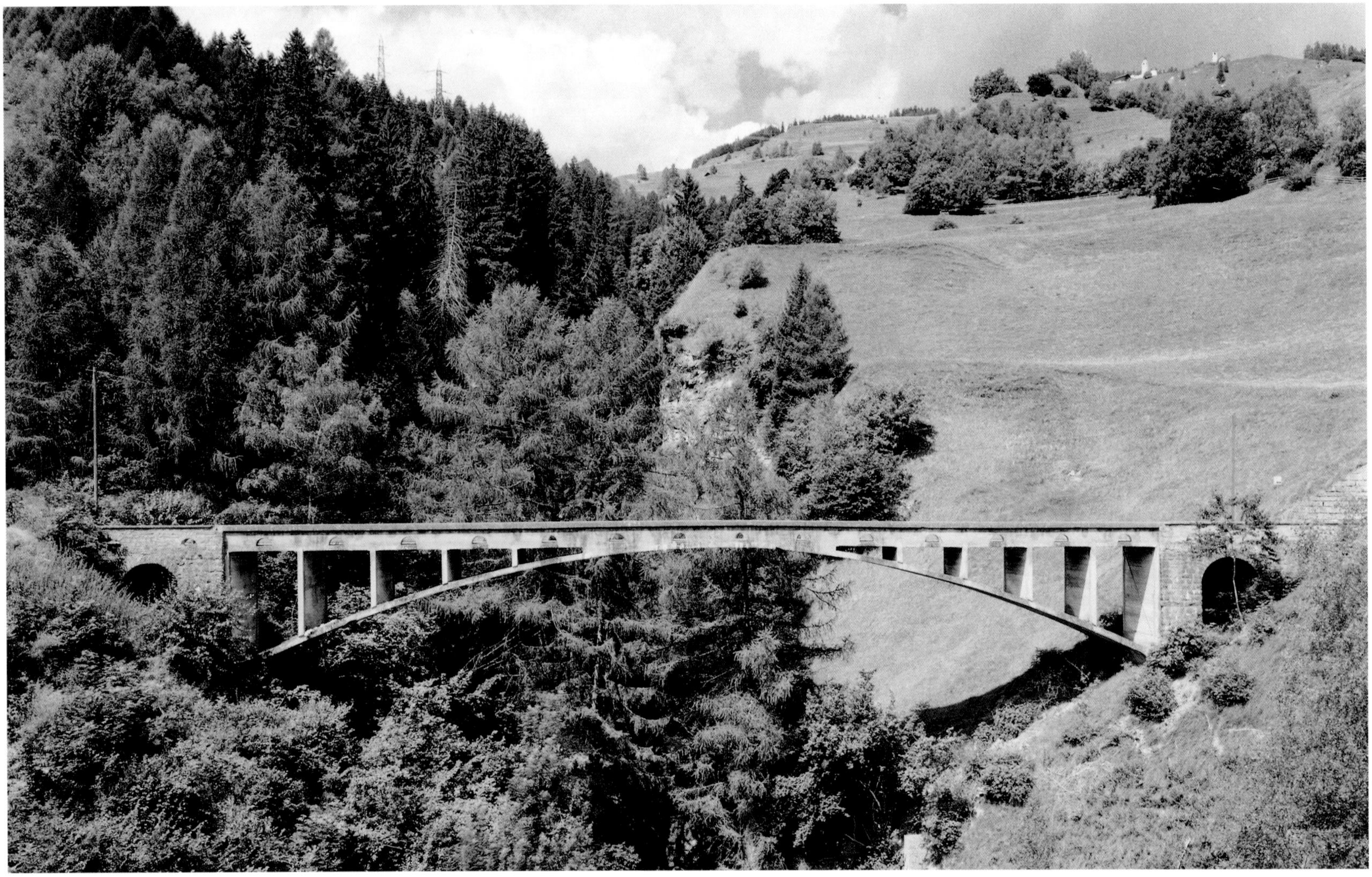

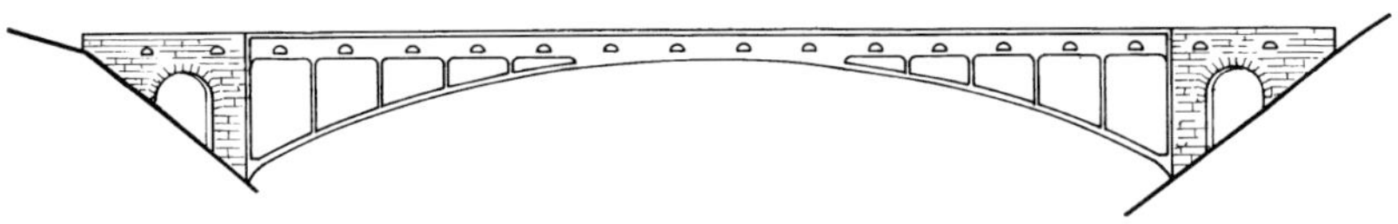

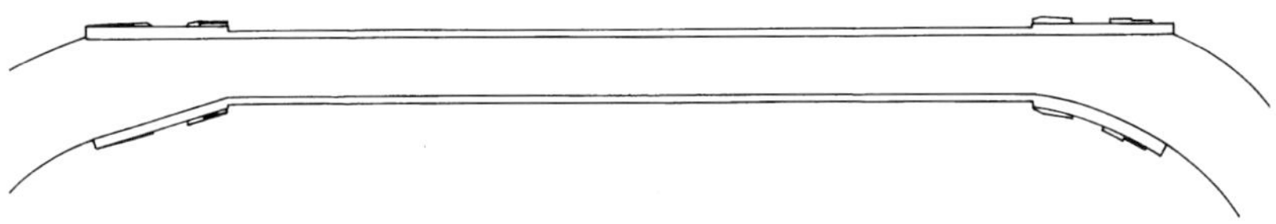

Figure 37

Schwandbach Bridge, built in 1933, Hinterfultigen, Switzerland.

Figure 39

Elevation and plan of the Schwandbach Bridge.

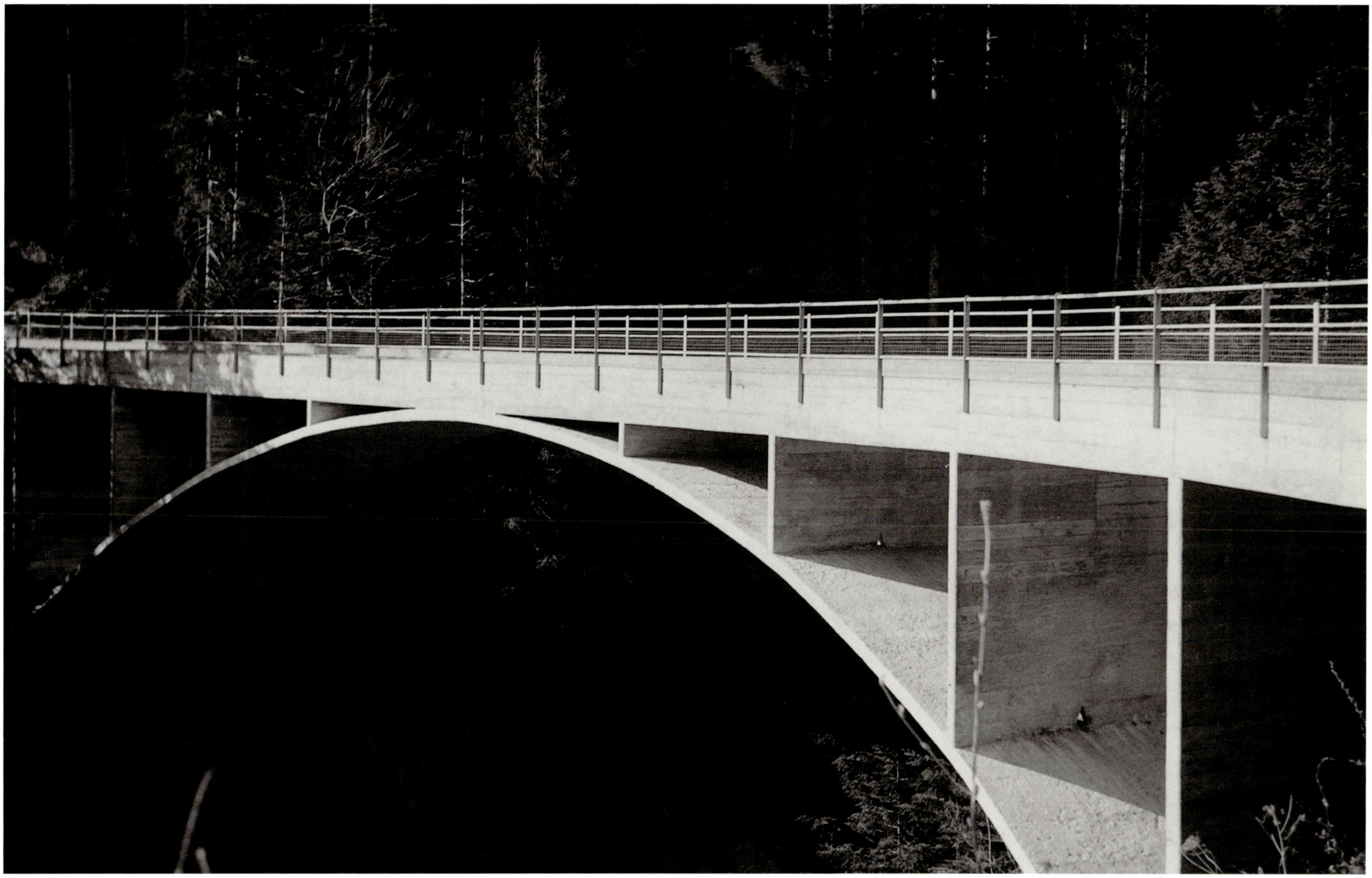

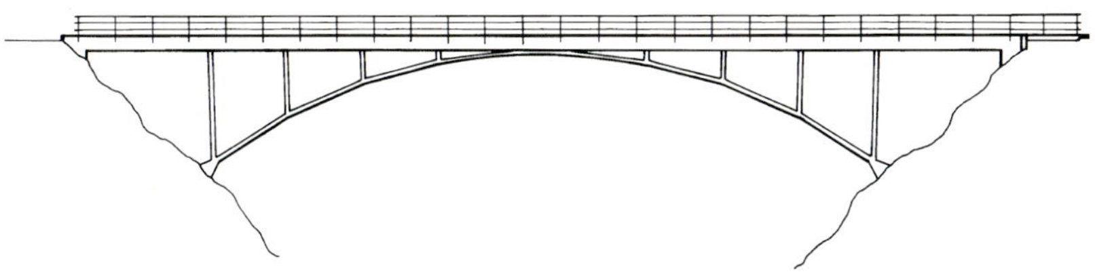

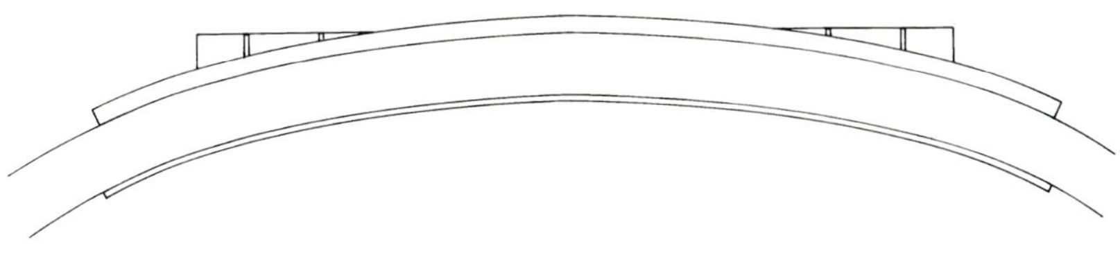

Figure 40
Schwandbach Bridge, showing the trapazoidal cross walls.

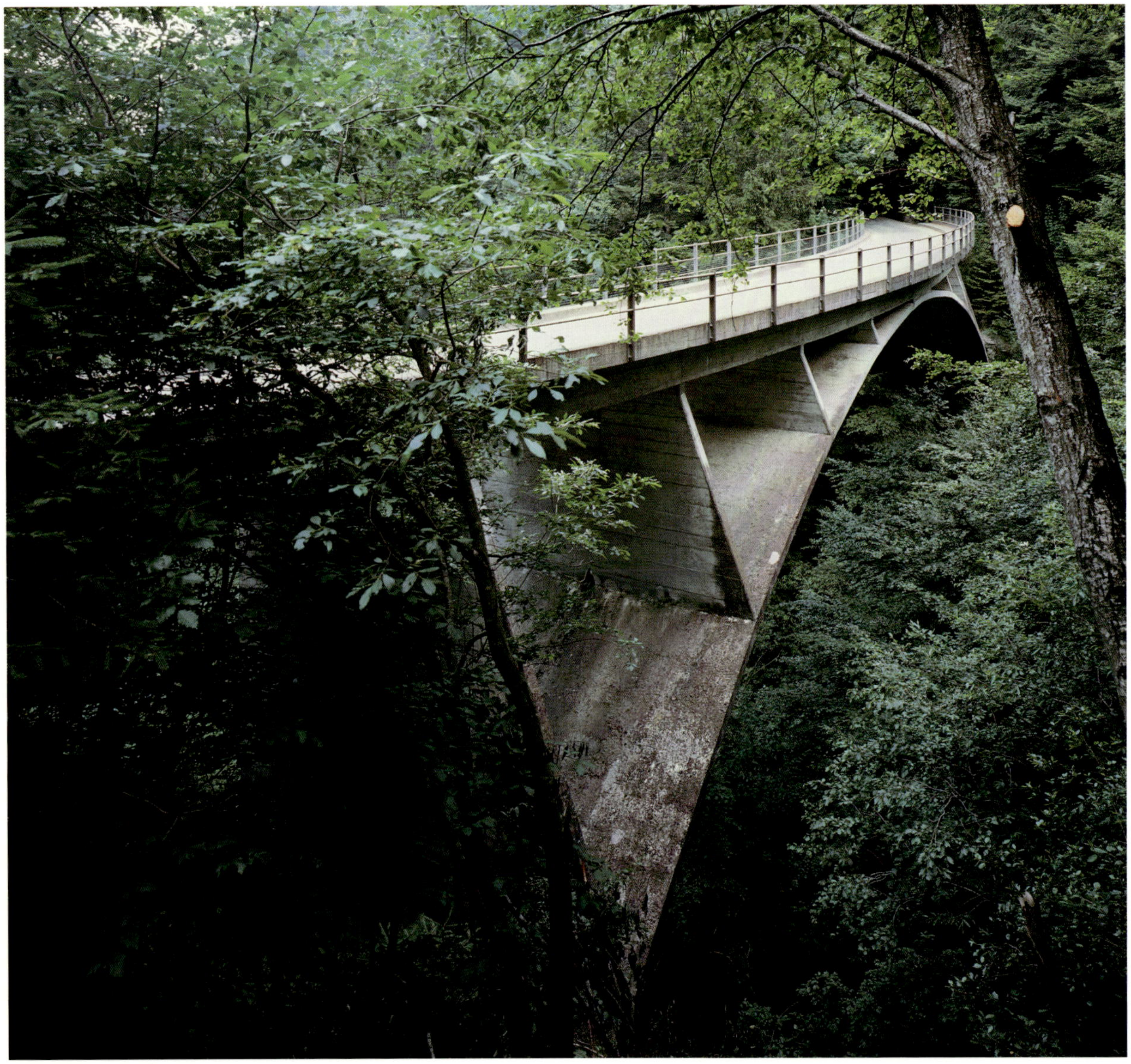

Figure 41
Model of the Schwandbach Bridge, revealing Maillart's evolution toward smoother, more continuous curves. (Model by Leon Barth, 1972)

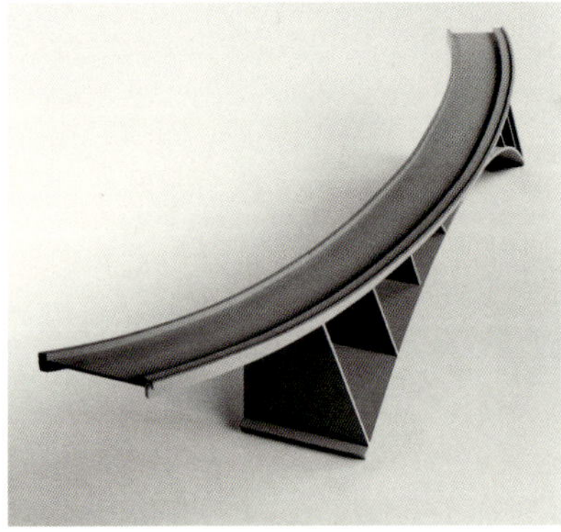

Figure 42
Töss Bridge, Maillart's last deck-stiffened arch, Wülflingen-Winterthur, Switzerland, completed in 1934.

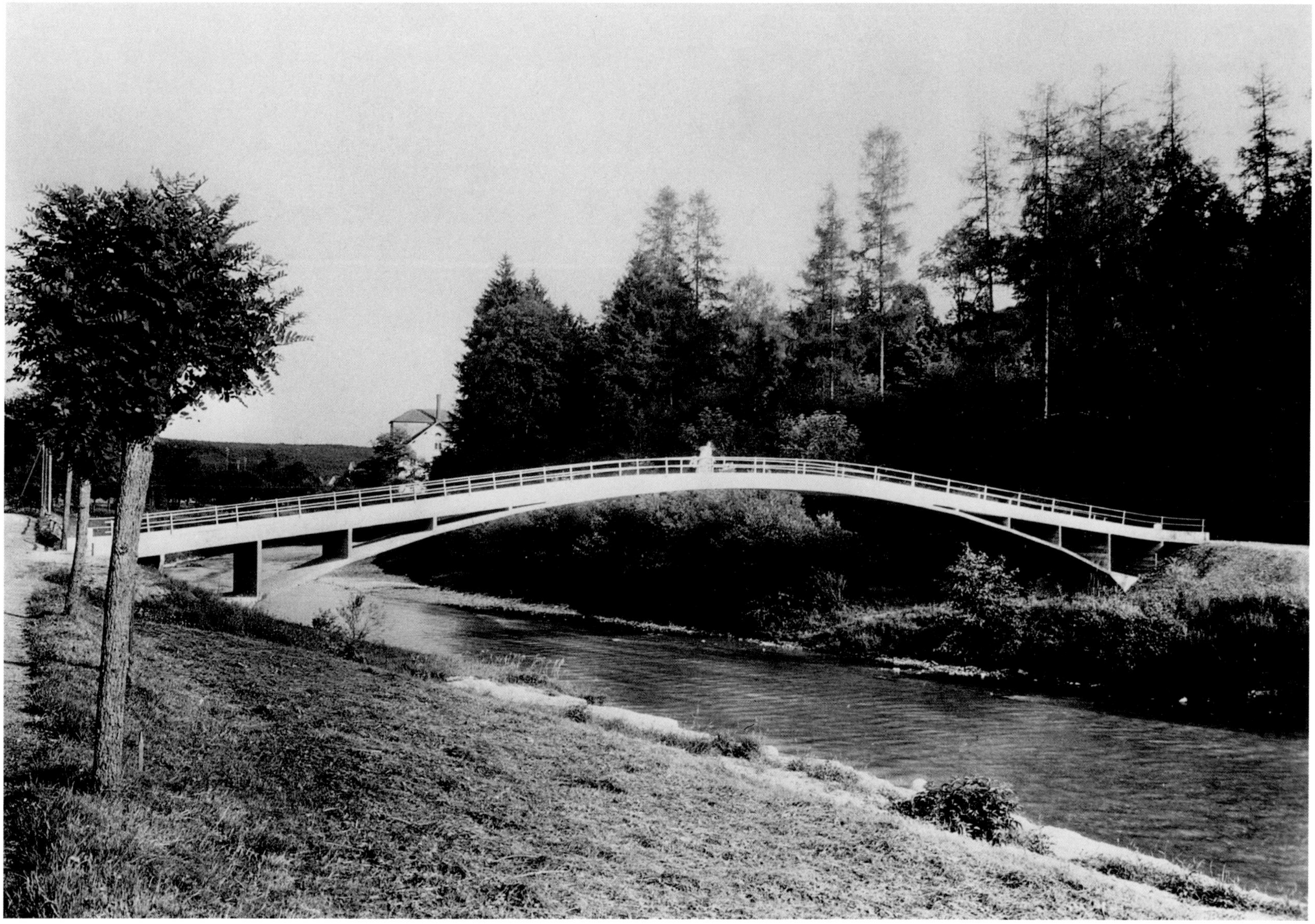

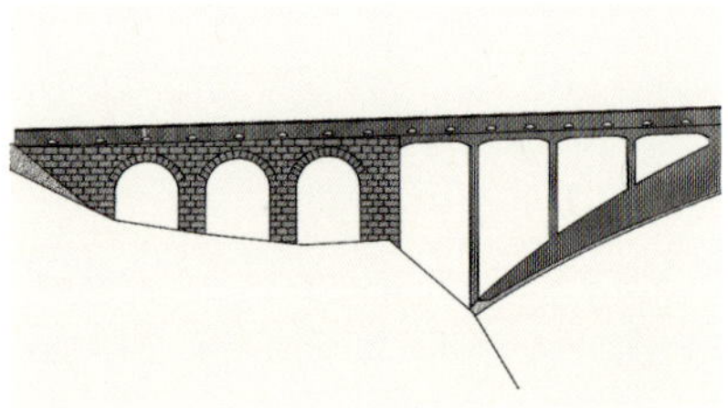

Figure 43
Maillart's and Florian Prader's proposal of Romanesque approach spans for the Salginatobel Bridge.

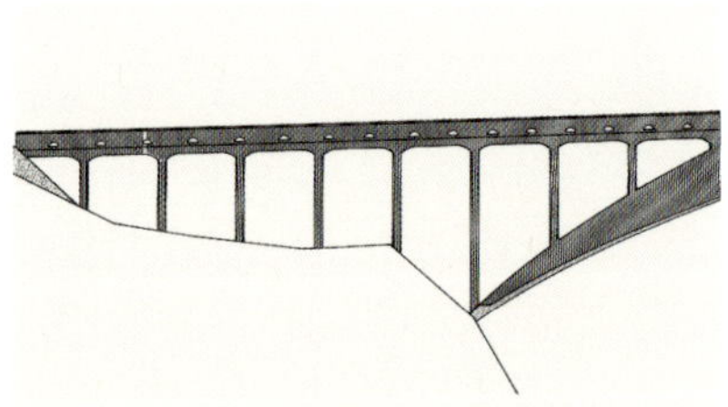

Figure 44
Maillart's and Florian Prader's simpler proposal of "pure form" approach spans for the Salginatobel Bridge.

Figure 45
Salginatobel Bridge, built in 1929–30, Schiers, Switzerland.

Between 1930 and 1940, Maillart would complete six three-hinged arch bridges. Two are the finest ever built in that form: the Salginatobel Bridge (1930) above Schiers in the Graubünden and the Vessy Bridge (1936) on the outskirts of Geneva.

Following the destruction in 1927 of the Tavanasa Bridge, Maillart made a concerted but failed attempt to get the contract for its replacement. His three proposals show how he was beginning to rethink one of the major visual defects in his Tavanasa design, which persisted even in his Valtschielbach Bridge (1925), notably the stonelike forms at the arch abutments. The opportunity to overcome this error arose when a new competition, also in the Graubünden, turned his attention in late summer of 1928 to the high alpine crossing over the Salgina Brook. With the builder Florian Prader, his close friend and a Graubünden native, Maillart submitted a Tavanasa-like design for this much longer span (90 meters) with two alternatives: one with Romanesque approaches like Valtschielbach (fig. 43) and the other with a pure structural form visually integrated with the main span deck (fig. 44). This pure form was the least expensive bid of nineteen competing designs and thus went into construction early in 1929.[36] It would become Maillart's most famous structure and in 1991 would be named an International Historic Civil Engineering Landmark—until then only the thirteenth such project and the first concrete bridge to be so recognized (fig. 45).[37]

This evolution from Zuoz to Tavanasa to Salginatobel shows how Maillart slowly changed his ideas, always guided by his strong technical background, his graphic approach learned from Ritter, and his visual observations of previous structures. By diagramming the live load for two trucks at the quarter span (fig. 46A), Maillart's form emerges from the bending moment diagram, not unlike the graphic statics example from Ritter's lecture (see fig. 6). The algebraic solution to arch bending commonly led to diagrams that look entirely different and not at all suggestive for design even though numerically the diagram is entirely correct. Maillart did not arrive at the Tavanasa form by looking at the suggestive diagram; he came to it through observing the cracks at Zuoz, but his graphic studies provided a discipline that allowed him to see the correctness of his new form as well as to rectify an earlier error.

Happy as he was with the Salgina form, it also had an error that Maillart observed several years later upon returning to the site (fig. 47). Again, the aesthetic of stone had left a residue at Salgina, where the Roman tradition of smoothly curving the arch's underside was evident. Maillart later wrote: "Even the latter [Salginatobel Bridge] cannot lay claim to complete sincerity of form. Indeed, if both constant [dead] and shifting [live] weights are taken into consideration, the extreme curves of exerted pressures form two lenticular surfaces whose lower contours meet at an acute angle. It was only in the bridge at Felsegg built as recently as 1933 that I had the chance of realizing a truly logical form."[38] Figure 46 helps to explain Maillart's meaning. The bridge is loaded by only one truck so that the bending moment (fig. 46B) is greater at the loaded point than it is when two loads are applied (fig. 46A). When two cases are plotted separately (fig. 46C), the diagram has an acute angle at the center hinge (fig. 46D). This logic led Maillart to realize that he should express this angle through a slightly broken arch that in addition makes clear that the structure has a discontinuity at the crown—it can rotate there even though its physical rotation is too small to observe visually.

This change in vision reached its most mature form at the Vessy Bridge (1936) with its extreme flat form, characterized by the high ratio of span length to vertical rise of the crown, reaching 11.9 (fig. 48) compared to 6.9 at the Salginatobel. Tavanasa, Rossgraben, and Felsegg all have the ratio of 8.6. Maillart also introduced a narrow vertical space at the crown to emphasize further the hinge and the broken arch. He made two more innovations with this out-of-the-way bridge. The first is in the cross walls, which at the Salginatobel were thin solid walls. At Vessy they suddenly became lively X-shaped forms, which are both aesthetically motivated and technically logical, like the broken arches (fig. 49). There is a deep affinity between Maillart's form and a painting by Paul Klee, a fellow Bernese artist (fig. 50). Klee taught in the 1920s at the Bauhaus in Germany and was attracted at times to building forms. Maillart had never seen Klee's *Double Tent* (1923) but, like Klee, he often sought playful forms. Maillart's forms were also structurally reasonable, as shown by the idealized structural form taken from his calculations, which imply a bending moment diagram of the same shape. The cost of construction for the bridge was even lower than the estimate. The technical, constructional, and artistic synthesis gives a proper summary of Maillart's achievement.

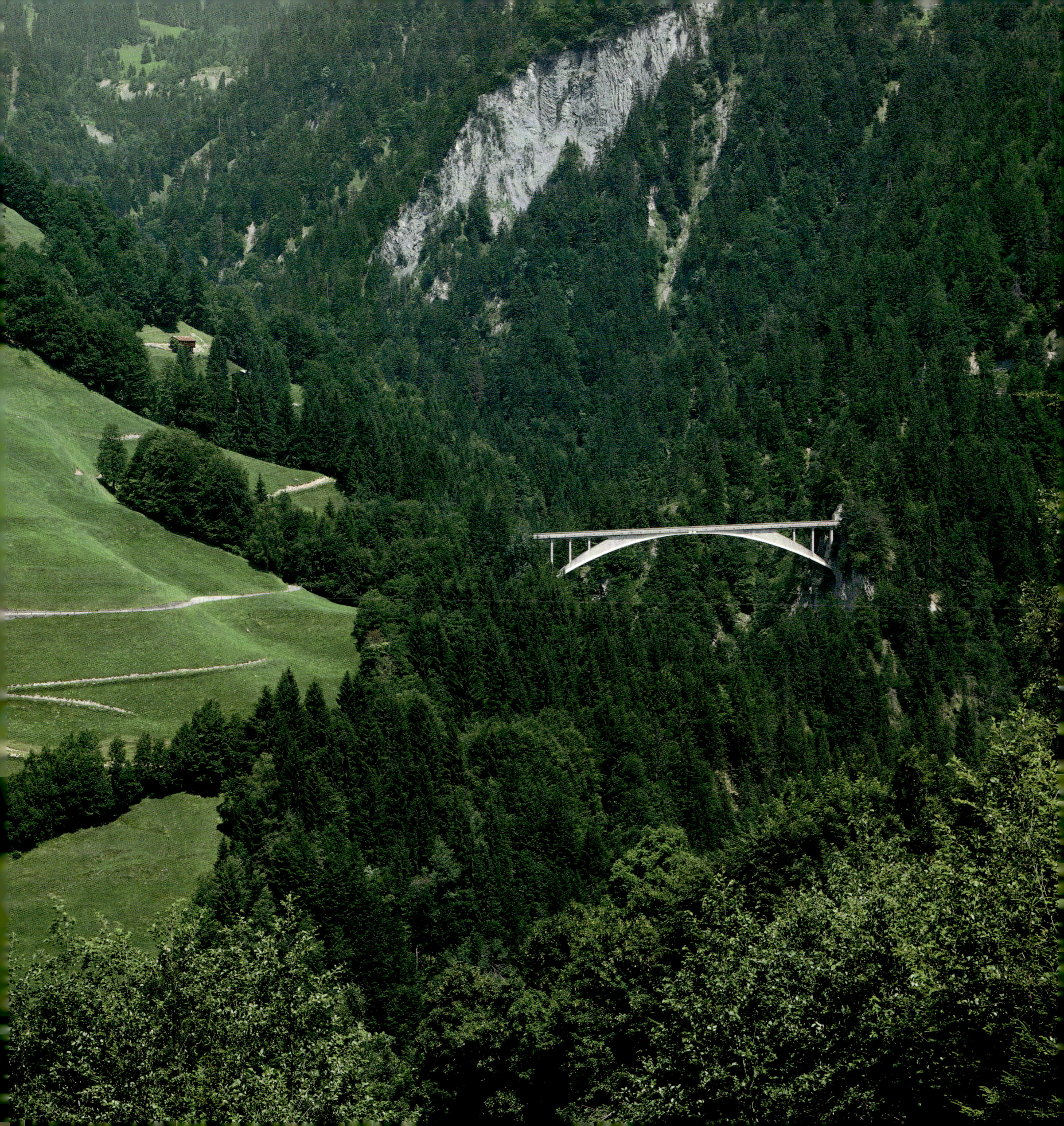

Figure 47

Maillart returning to contemplate the Salginatobel Bridge, early 1930s.

Figure 46 (A)

Live-load diagrams and effects on the bending moment: (A) two trucks.

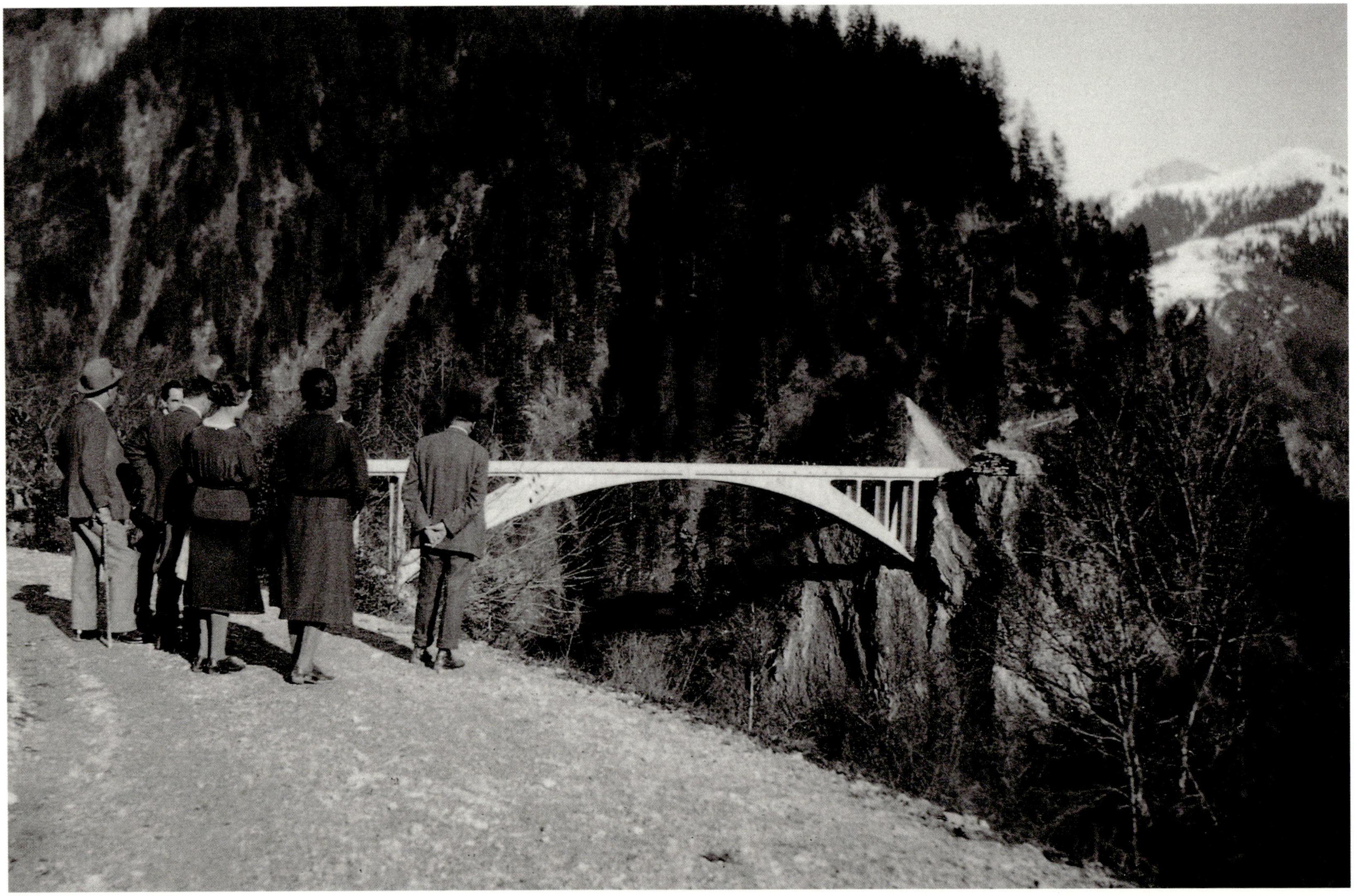

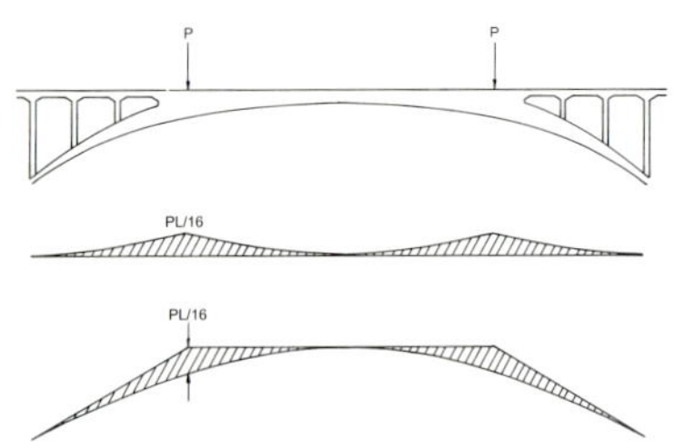

A

Figure 48

Vessy Bridge, with a high ratio of span length to height, built in 1936, Vessy, Switzerland.

Figure 46 (B, C, D)

Live-load diagrams and effects on the bending moment: (B) one truck on left side; (C) one truck on right side; (D) plotting separately the two cases of one truck on either side and combining them to show an acute angle at the center hinge.

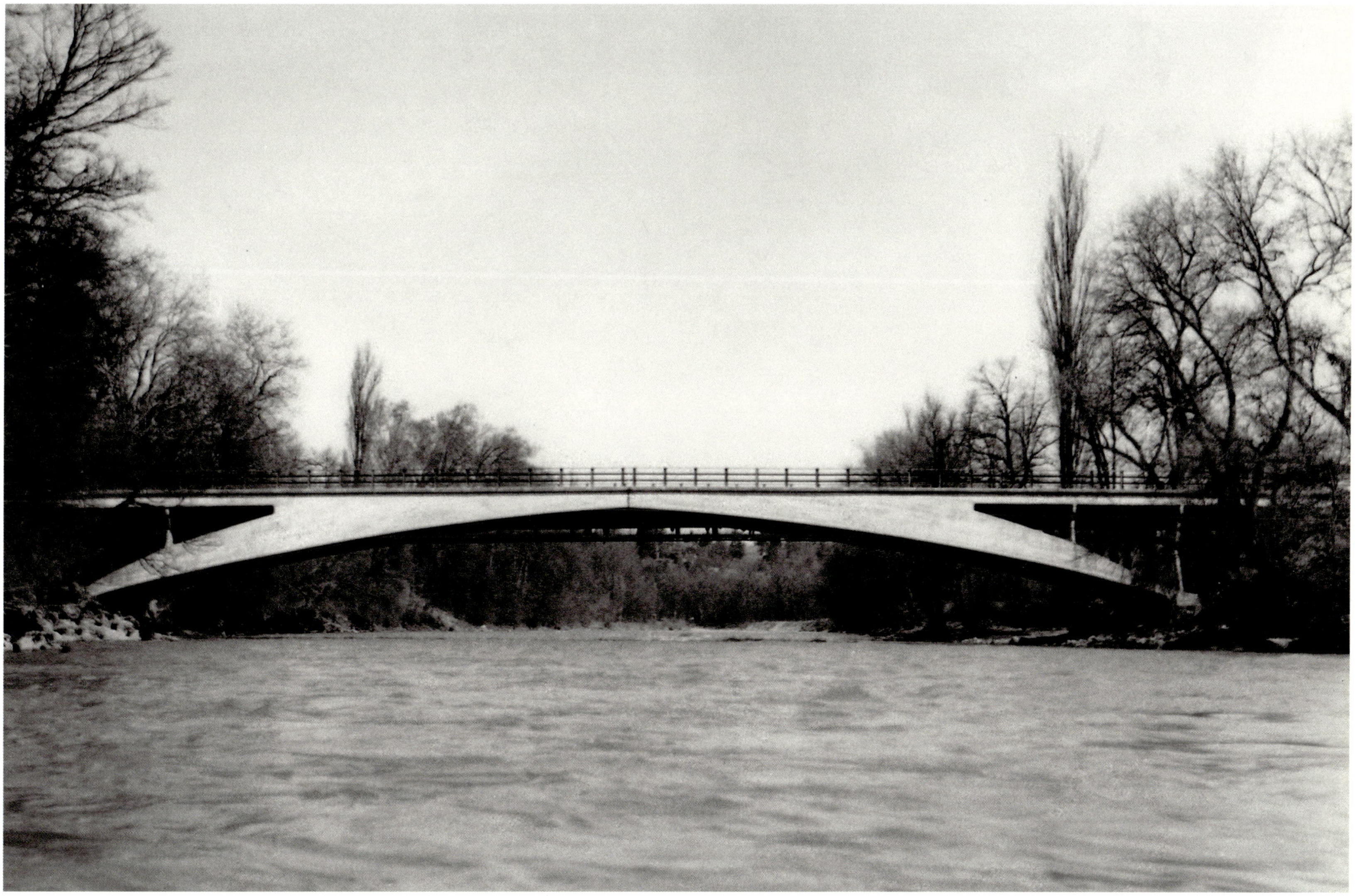

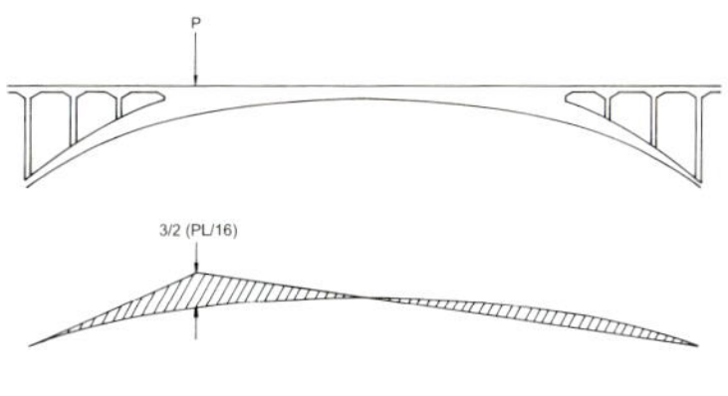

B

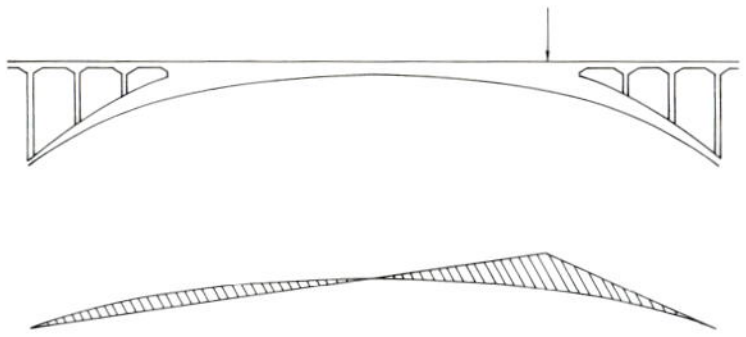

C

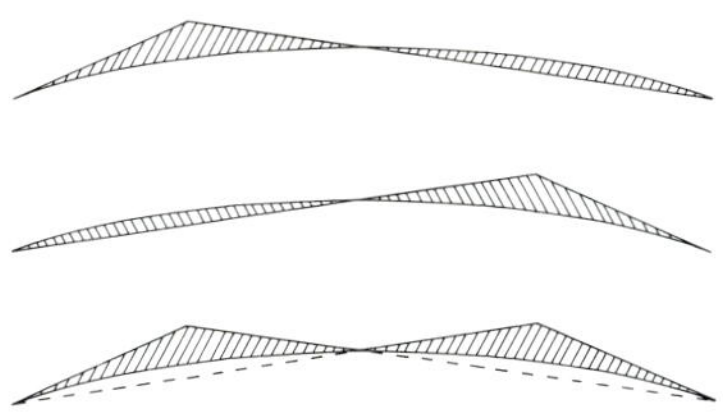

D

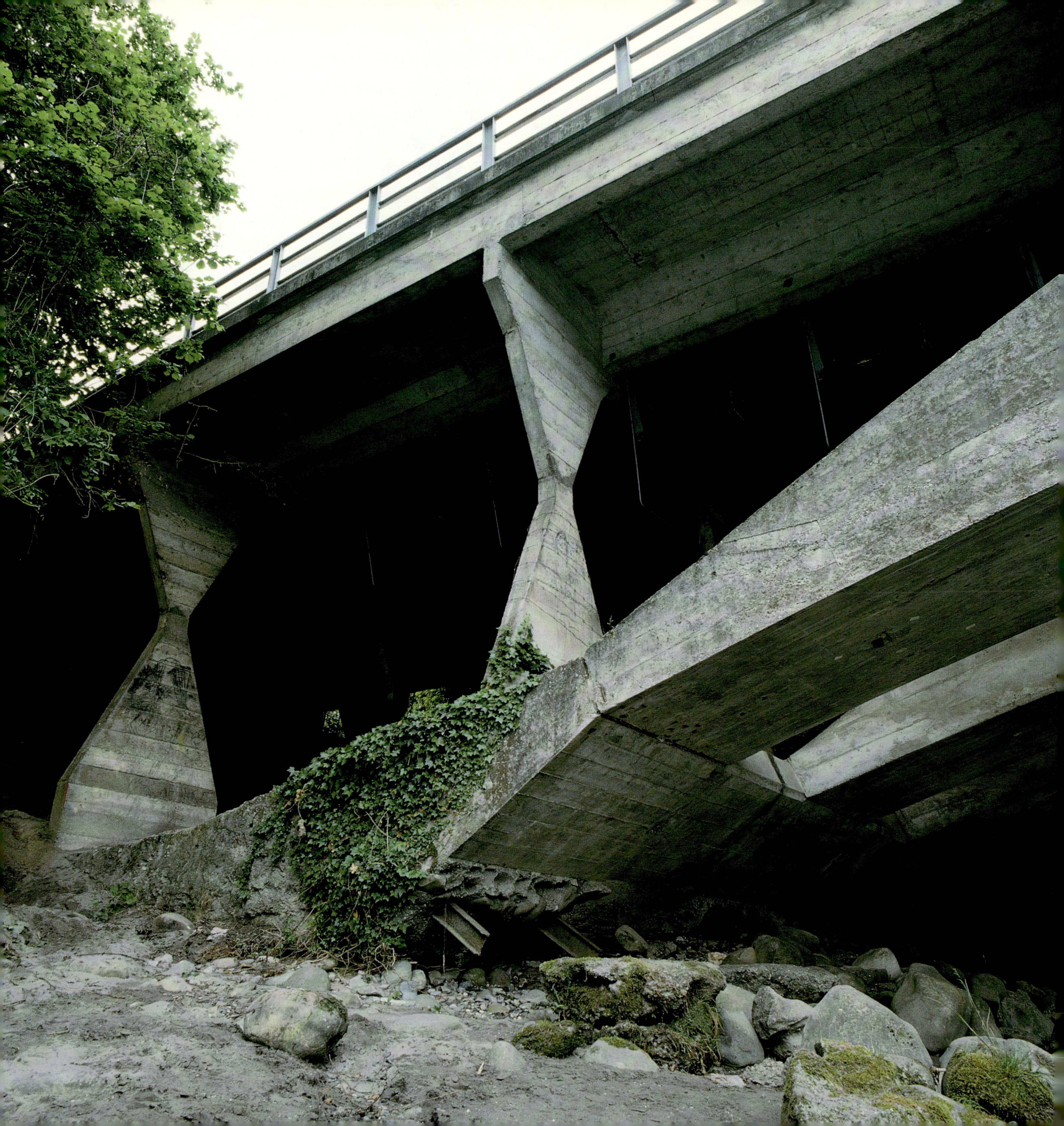

Figure 49

X-shaped cross walls of the Vessy Bridge.

Figure 50

Paul Klee, Swiss, 1879-1940. *Double Tent*, 1923. Watercolor, 19 1/3 x 12 1/2 inches. The Rosengart Collection, Lucerne.

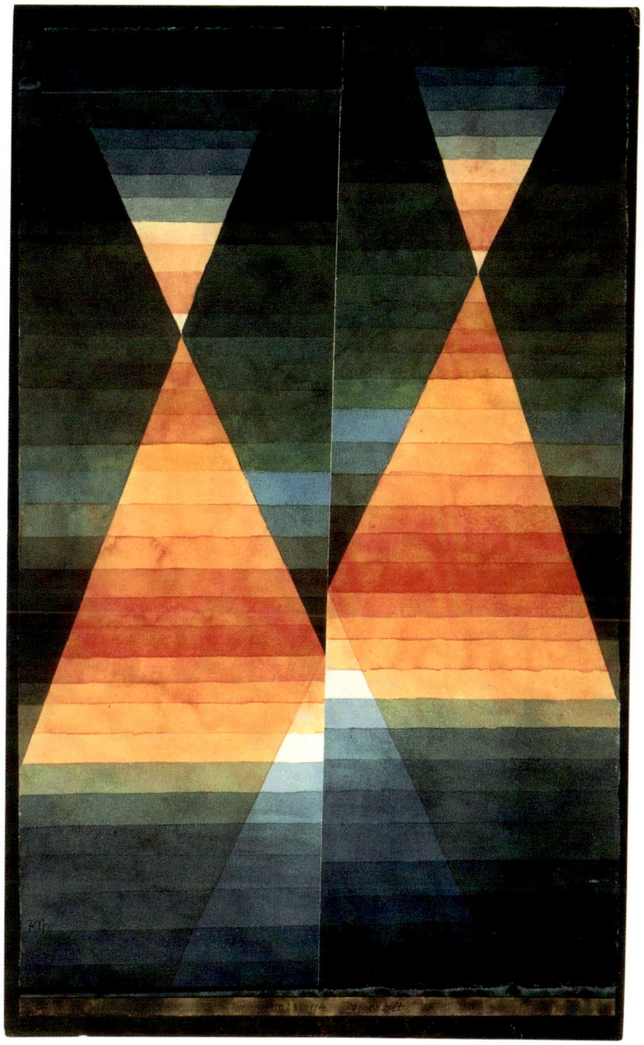

Figure 51
Maillart's last bridge, the Lachen Bridge, built in 1940.

Figure 52
Detail of Lachen Bridge showing aesthetic play with form boards above the hinge.

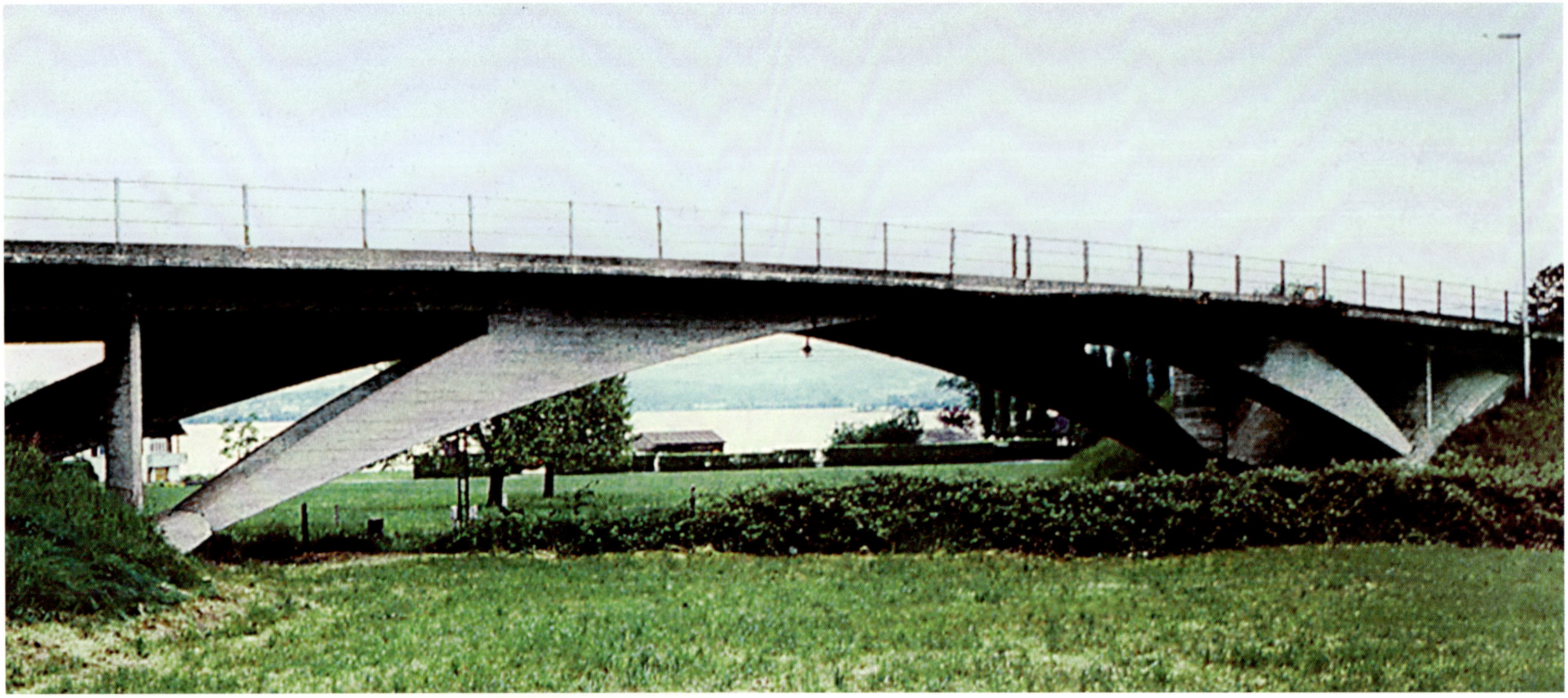

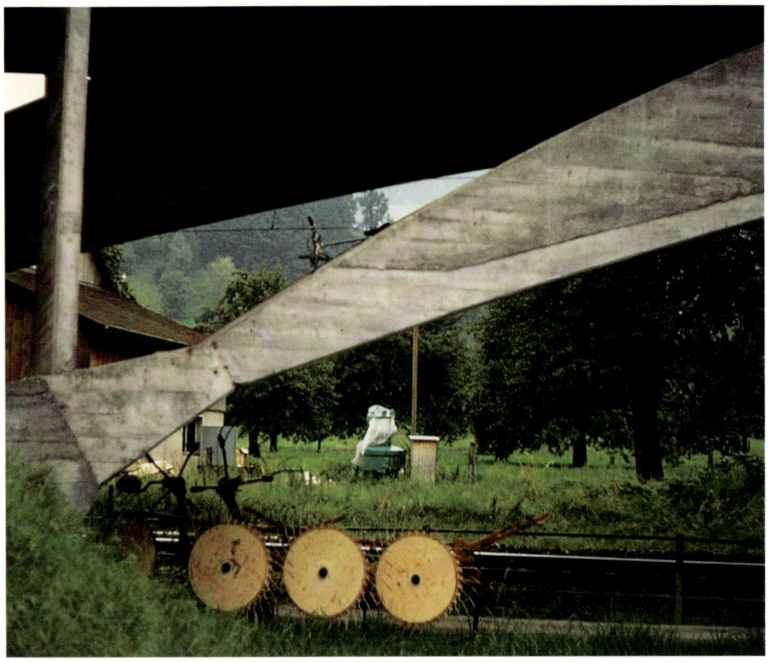

Figure 53
Cement Hall, Zurich, Switzerland.

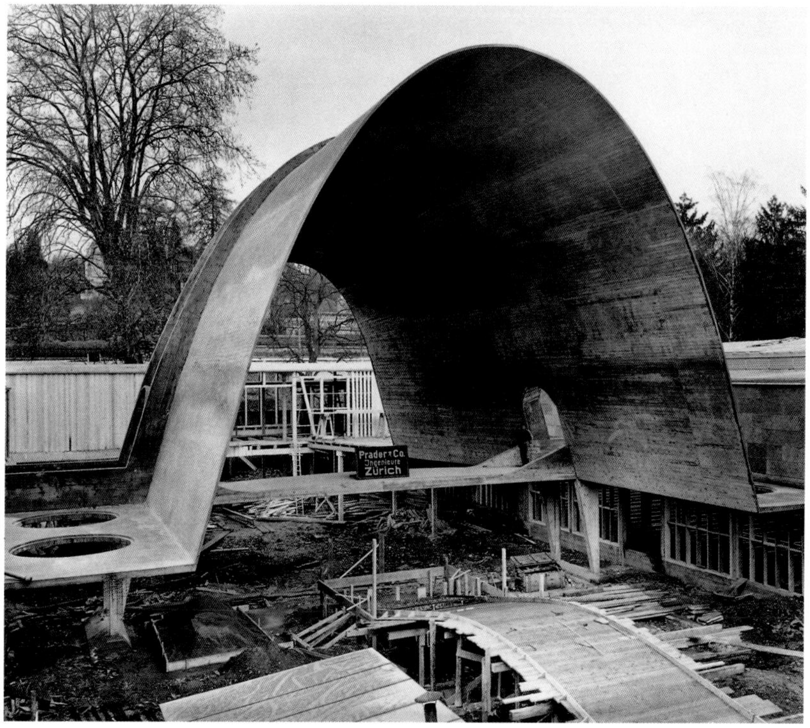

Figure 54
Cement Hall cutaway diagram showing diagonal reinforcing steel to enforce cantilever action in the lower part of the thin shell.

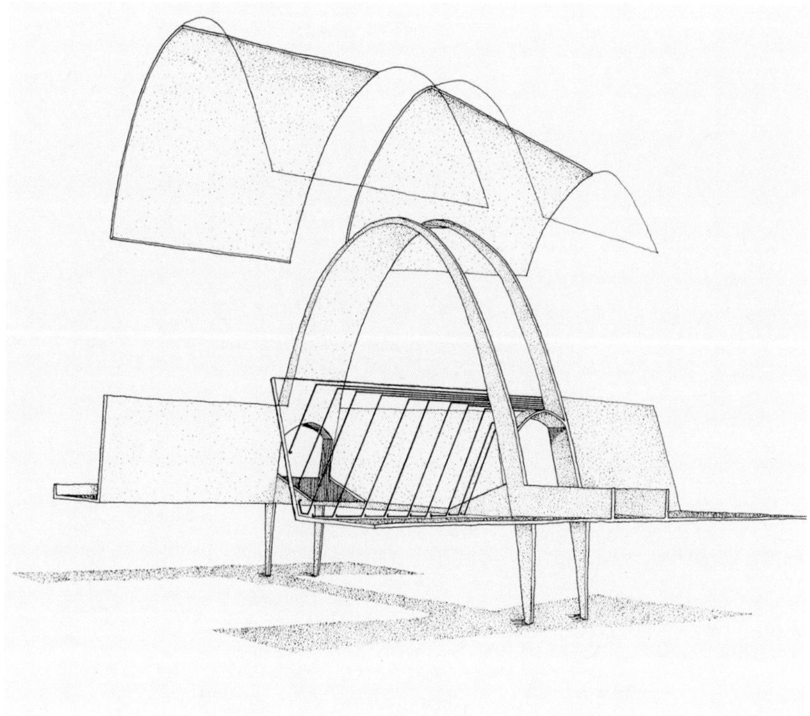

The second innovation, shown with the X-forms, is most clearly illustrated by the facade of the arches where the lower part is smoothly formed, whereas above it is patterned (see fig. 49). These two surfaces are in the same plane, but they depict a difference in form. The lower surface is part of the arched slab that runs the full width, and the upper surface forms the thin wall that creates a hollow box over the central part of the span and a U-shaped section over the outer parts. Maillart is playing with form but remains within the discipline of structural art. Vessy was built at low cost and has performed well for two-thirds of a century.[39] Maillart's last bridge, built at Lachen in 1940, is the best example of this play of form (figs. 51, 52).

It is perhaps symbolic that Maillart's great works are in wilderness locations away from places that might offend "culture leaders." Salginatobel goes to a town with fewer than one hundred inhabitants, the closest town to Schwandbach is Hinterfultigen (a Swiss term for nowhere), and Vessy is at a place called *au bout du monde* (at the end of the earth). To create great structures in prominent places, the engineer and the general public would benefit from an education in this grand tradition of structural art. They would then call for bridges that carry forward Maillart's ideals rather than those that produce the costly and illogically formed structures that often have replaced the equally expensive designs that mimicked ancient forms.

At the end of his life, Maillart had the chance to move from the wilderness to the heart of Swiss culture and wealth with the design of a unique roof. The national exhibition of 1939 took place in Zurich, where Maillart created a thin-shell concrete roof (fig. 53) for an exhibition building sponsored by the Swiss Society for Cement, Chalk, and Gypsum Manufacturers. Nine years before, he had written an article entitled "Mass or Quality in Concrete Structures" in which he placed thinness and lightness of form on the side of quality, and mass on the side of inefficiency and defective performance. He connected that contrast to aesthetics by proclaiming that "there is no doubt that light, slender structures will one day be praised by the layman for being as beautiful or even more beautiful than massive ones."[40] The Salginatobel, Schwandbach, and Vessy bridges all have borne out that prediction. But more than any other structure, Maillart's Cement Hall seems to have best symbolized quality over mass.

Figure 55
Construction of the Cement Hall, showing the diagonal reinforcing bars designed to enforce cantilever action.

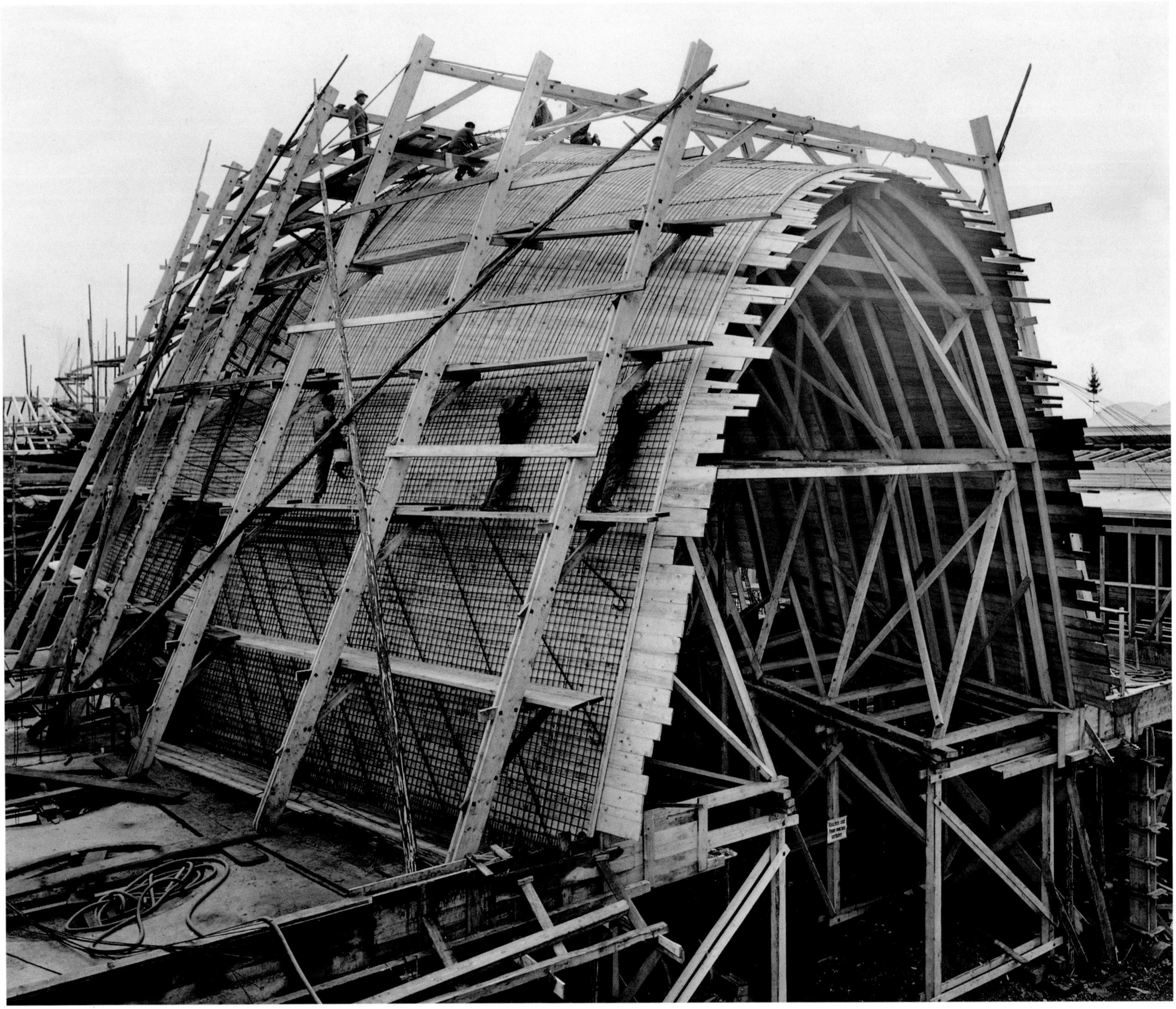

Figure 56
Destruction of the Cement Hall, with Maillart in the foreground, center, 1940.

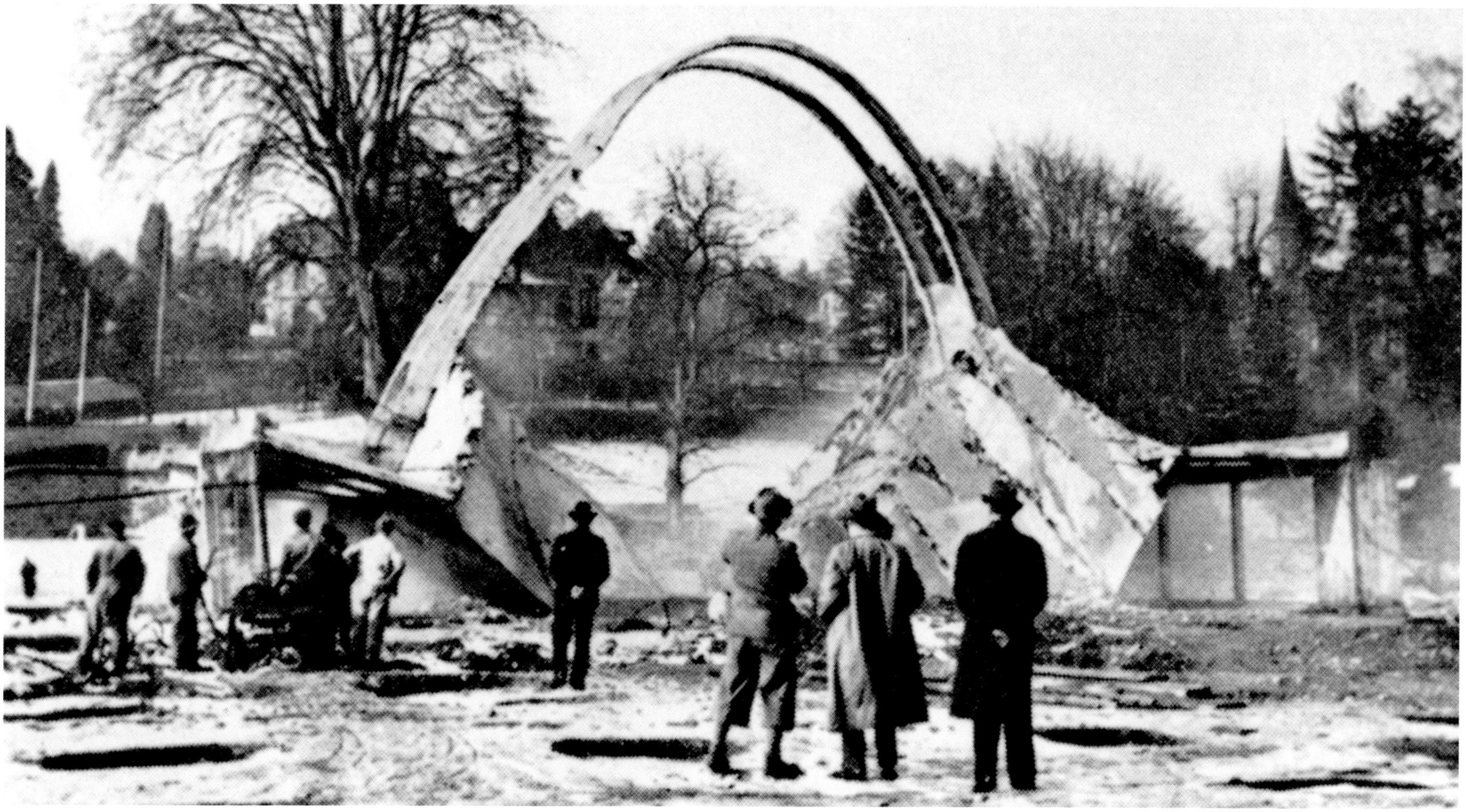

This Cement Hall represents a variation on Maillart's major theme of thin expressive curves stiffened to prevent poor performance and to add visual interest. The thin-arched shell could never stand alone without the two stiffening arches rising above the central walkway. But the thin-arched shells needed vertical support, which the horizontal edge slabs could not provide. Maillart gave the daringly thin form a magical appearance of floating freely. He achieved this startling effect by reinforcing the lower, almost straight sides of the thin-arched shells so that they acted as cantilevers to carry the shell weight back to the stiffening arches (figs. 54, 55). He designed the structure so that its load-carrying function followed its reinforced-concrete form. The form defined the function. Its planned destruction after the fair (fig. 56) proved its extraordinary strength by the carefully instrumented load test carried out in early 1940 just weeks before Maillart's death.[41]

Making such new forms was a legacy of Maillart's that some artists, architects, and writers have recognized. The Swiss artist Max Bill wrote a fine book in 1949 on Maillart; Sigfried Giedion, a Swiss architectural historian, publicized the works, and partly through Giedion, the Museum of Modern Art mounted a one-man exhibition on Maillart in 1947. Also, from 1924 to 1940, Maillart's good friend Mirko Ros, a professor at the ETH, wrote up a series of load tests on his structures.[42] Following Maillart's death in 1940, however, there was very little interest in Maillart within the engineering profession outside Switzerland throughout the twentieth century. Three educational developments following World War II have perpetuated this unfortunate neglect in the United States: the ideology that defines engineering as "applied science" rather than as design; the research emphasis on methods of analysis as opposed to the study of designs and designers; and the almost total disregard for the aesthetics of structure.

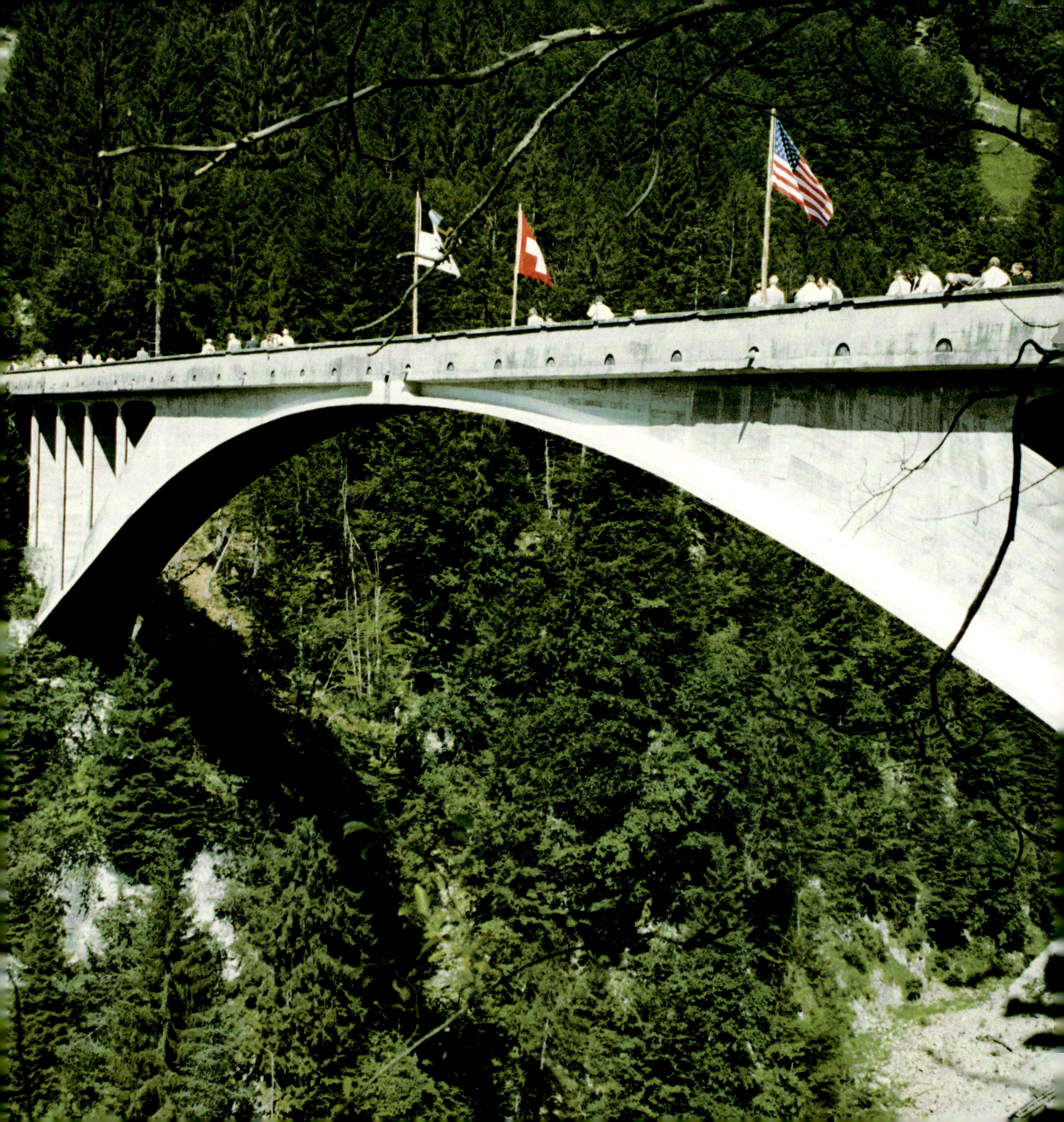

Figure 57
Salginatobel Bridge on the occasion of its dedication as an International Historic Civil Engineering Landmark, August 21, 1991.

Twentieth-century engineering education rested on the assumption that scientists made discoveries and engineers applied these discoveries to the needs of society. From this viewpoint, the pure sciences and mathematics constituted the so-called fundamentals and all engineering design was based upon a thorough grounding in those formal subjects. In engineering schools, such subjects were followed by courses in engineering science, a term that implied applications of pure science but couched in analytic methods similar to those in the fundamentals.[43] As the century proceeded, fewer and fewer teachers had design experience, and more and more went directly from undergraduate to graduate studies leading to a doctoral degree. Teaching focused heavily on increasingly complex mathematical analysis, which came to dominate academic research; this, in turn, became the primary goal of educators in the most prestigious engineering schools. Academic researchers wrote papers intended largely for other researchers, and the focus on practice continued to diminish. Because of this narrowing education, there arose a desire to broaden a student's perspective by requiring courses in the humanities and social sciences, but these almost never were connected in any significant way to engineering. The overall result was that during the last half of the twentieth century, engineering education became literally dehumanized and scientized. In structural engineering, this meant that the study of actual designs and of exemplary designers disappeared from the curriculum. Engineering students were left with no way to evaluate structures apart from mathematical calculations. Issues of correct form, competitive construction costs, or elegant design were rarely discussed, and teachers interested in those subjects found little financial support for research.

Maillart was once told that he was well known in American educational institutions, but it turned out that with minor exceptions, only in schools of architecture was he known at all. Only after education, eight years of practice, and four years of full-time teaching did I first hear the name Maillart, in 1964, and then only from students in the school of architecture at Princeton. Finally, in 1991 the American Society of Civil Engineers agreed to designate Maillart's Salginatobel Bridge as the thirteenth International Historic Civil Engineering Landmark (fig. 57).[44] Ten years later, in a short book on the history of all engineering, the editor decided to put the Salginatobel Bridge on the cover as an illustration of the most exemplary work.[45] Wilhelm Ritter would have approved, just as he did of Maillart's great innovation at Zuoz—without which the Salginatobel structure would probably not have been conceived.

Notes

1. G. Kubler, *The Shape of Time* (New Haven, 1962), 7.

2. Ibid.

3. The discussion of Maillart's temperament and background comes from D. P. Billington, *Robert Maillart: Builder, Designer, and Artist* (New York, 1997), 1–6. The Maillart family name is mentioned as far back as 988 in records of the region of Liège. In 1017, Jean Coley, *dit* Maillart, commanded a victorious army for the bishop of Liège but was blinded in battle. This event is presumed to have been the origin of the French version of Blind Man's Bluff, known in French as "Colin-Maillard." For his aversion to complex formulations, see R. Maillart, "Review of *Elastizität und Festigkeit* by C. Bach, Berlin, 1920," *Schweizerische Bauzeitung* [hereafter *SBZ*] 77, no. 18 (April 30, 1921): 203–4.

4. J. Huizinga, *Homo Ludens: A Study of the Play Element in Culture* (Boston, 1955; completed in 1938; published in Switzerland, 1944), 10.

5. Copies of the transcripts of Ritter and Maillart are in the Princeton Maillart Archive (hereafter PMA).

6. Lecture notes by Robert Maillart, copies in the PMA. See also W. Ritter, *Anwendungen der Graphischen Statik*: Part 3, *Der Kontinuierliche Balken* (Zurich, 1900), 7. The discussion involving figures 3–7 presents the idea of seeing forces (bending moments, in this case) as related to forms (arches here), but in the example of figure 2, that was not Ritter's goal. Rather, he was demonstrating a method which he would later connect more explicitly to the idea of appropriate form and aesthetics.

7. W. Ritter, "Statische Berechnung der Versteifungsfachwerke der Hängebrücken," *SBZ* 1, no. 1 (Jan. 6, 1883): 6–7; no. 2 (Jan. 13, 1883): 14; no. 3 (Jan. 20, 1883): 19–21; no. 4 (Jan. 27, 1883): 23–25; no. 5 (Feb. 3, 1883): 31–33; no. 6 (Feb. 10, 1883): 36–38.

8. W. Ritter, *Der Brückenbau in den Vereinigten Staaten Amerikas* (Zurich, 1895). See also lecture notes by Robert Maillart, copies in the PMA.

9. W. Ritter, "Consultation over the Design of a Stauffacher Bridge in Zurich," Zurich Tiefbauamt, August 9, 1898, p. 1, PMA. Ritter based part of his report on the precedent of an earlier bridge. See W. Ritter, "On the New Construction of the Coulouvrenière Bridge in Geneva," *SBZ* 27, no. 14 (April 11, 1896): 100–101.

10. W. Ritter, "Die Bauweise Hennebique," *SBZ* 33, no. 5 (Feb. 4, 1899): 41–43; no. 6 (Feb. 11, 1899): 49–52; no. 7 (Feb. 18, 1899): 59–61.

11. The bridge design is described by H. Schleich, "Die Drahtseilbahn des Rigiviertels in Zürich," *SBZ* 38, no. 17 (Oct. 26, 1901): 179–81. The Hadlaub Bridge stood without any significant difficulties for more than seventy-five years before being removed in 1978. Maillart never wrote about this bridge himself, but he did describe it near the end of his life in a retrospective lecture given in Basel in 1937, PMA.

12. R. Maillart, "Das Hennebique-System und seine Anwendungen," SBZ *37*, no. 21 (May 25, 1901): 225–26.

13. Froté and Westermann [Maillart], "Report for a Bridge over the Inn near Zuoz Designed as an Arch Bridge of Reinforced Concrete," Zurich, August 13, 1900, PMA.

14. See discussion in D. P. Billington, *Robert Maillart's Bridges* (Princeton, 1979), 23–28.

15. Froté and Westermann, "Solisbrücke: Notiz zur Statischen Berechnung," Zurich, May 4, 1901, PMA. The design was by Hans Studer. See also H. Studer, "Steinerne Brücken der Rhätischen Bahnen," in *Schweizerische Ingenieurbauten in Theorie und Praxis*, Internationaler Kongress für Brückenbau und Hochbau (Zurich, 1926).

16. Froté and Westermann in Zurich to Maillart in Paris, Sept. 25, 1900, PMA.

17. E. Ganzoni in Samedan to Ritter in Zurich, Feb. 3, 1901, PMA. Possibly Ganzoni sent the design with this letter; more likely, it had been sent earlier, and the letter was a reminder.

18. Ritter in Zurich to Ganzoni in Samedan, March 6, 1901, PMA. Ritter's report is not with these documents, but it is referred to in the building contract "Bauvertrag, Zuoz" of July 10, 1901, PMA.

19. W. Ritter, "Report on the Load Test of the New Inn Bridge near Zuoz," Zurich, winter 1901–2, pp. 11–12, PMA.

20. Ritter in Zurich to Bersinger in St. Gallen, June 17, 1903, PMA.

21. Maillart in Zuoz to his wife Maria in Oleggio, Sept. 8, 1903, and letter report from Maillart in Zurich to Schucan in Chur, Sept. 26, 1903, PMA. Maillart wrote about these ideas much later; see his "The Construction and Aesthetic of Bridges," *The Concrete Way* (May–June 1935): 305. The statement reads: "The realization that as a link between the arch and the platform, solid spandrels serve no useful purpose, except in the middle of the bridge, and that close to the abutments, they exert a useless dead weight which is positively a potential danger, has led to the practice of slotting triangular cavities out of them."

22. Drawing nos. 332 and 333, "Brücke über den Rhein bei Tavanasa," Sept. 30 and Oct. 1, 1904, PMA.

23. M. Bill, *Robert Maillart* (Zurich, 1949) and E. Mock, *The Architecture of Bridges* (New York, 1949). Mock's book, published by the Museum of Modern Art, was prompted by the exhibition on Maillart.

24. J. Solca, "Die Rheinbrücken bei Tavanasa und Waltensburg," *SBZ* 63, no. 24 (June 13, 1914): 343–46. For the destruction of the Tavanasa Bridge, see C. Jegher, "Zur Zerstörung der Rheinbrücke bei Tavanasa," *SBZ* 90, no. 15 (Oct. 8, 1927): 194–95.

25. R. Maillart, "Eine Schweizerische Ausführungsform der unterzuglosen Decke—Pilzdecke" [A Swiss-Developed Form of Beamless Slab—the Mushroom Slab], in *Schweizerische Ingenieurbauten*. See also Billington, *Robert Maillart*, 49–52.

26. Maillart in Zurich to Maria in Karlsbad, June 22, 1911. The city design is shown in a drawing labeled R27/17 and dated St. Gallen, Jan. 20, 1912. Maillart's design appears in a drawing entitled *Filter-Neubau im Riet*, labeled R27/18 and dated St. Gallen, March 1912. Maillart's name does not appear on this last drawing, but it shows his design as clearly seen in Maillart photos numbered 491, 492, 493, 495, 497, 500, 501, 502, and one unnumbered photo, all in the Zurich Maillart Archive.

27. Billington, *Robert Maillart*, 84–96.

28. R. Maillart, "Zur Frage der Biegung," *SBZ* 77, no. 18 (April 30, 1921): 195–97. For a review of Maillart's role in the center of shear development, see E. Reissner, "A History of the Center-of-Shear Concept: Maillart's Work and Ramifications," *Maillart Papers* (Princeton, 1973), 77–96.

29. Ritter, "Hängebrücken," 6–7.

30. Maillart & Co., "Statische Berechnung, Flienglibach Bridge," 4, PMA.

31. See Maillart's calculations of 1929 in which he bases the form for the Sihlhölzli gym roof on the hanging string and refers to Chiasso as a model: R. Maillart, "Submissions Berechnung: Turnhallen Sihlhölzli," Zurich Maillart Archive, Oct. 15–Nov. 1, 1929. The sketch appears on page 1, which is undated; page 5 bears the date of October 15, 1929. This roof structure is not a Vierendeel truss because the verticals are clearly too slender to make up for the lack of diagonals.

32. For a summary of these ideas, see R. Mark, J. K. Chiu, and J. F. Abel, "Stress Analysis of Historic Structures: Maillart's Warehouse at Chiasso," *Technology and Culture* 5, no. 1 (Jan. 1974): 49–53.

33. R. Maillart, "Note explicative concernant une variante pour l'aqueduc sur l'eau noire" [submitted to the Federal Railroad Office in Bern], Zurich and Geneva, Oct. 3, 1924.

34. "Brücke über den Valtschielbach: Statische Berechnung," no. 4094/4, Geneva, Maillart and Company, April 20, 1925, 4, PMA. The complete calculations appear in a mere three-and-a-half pages, of which only the last half page is for deck stiffening.

35. D. P. Billington, "Bridge Aesthetics: 1925–1933," *Final Report, Eleventh Congress of the International Association for Bridge and Structural Engineering, Vienna, 1980* (Zurich, 1981): 47–52.

36. Billington, *Maillart's Bridges*, 77–90.

37. D. P. Billington, "Maillart and the Salginatobel Bridge, Switzerland," *Structural Engineering International* 1, no. 4 (Nov. 1991): 46–47.

38. R. Maillart, "Construction and Aesthetic of Bridges," *The Concrete Way* (May–June 1935): 303–4.

39. Billington, *Robert Maillart*, 202–5.

40. R. Maillart, "Masse oder Qualität im Betonbau," *Beitrag zur Denkschrift anlässlich des 50-Jährigen Bestehens der Eidgenössische Materialprüfungsanstalt an der Eidgenössische Technische Hochschule, Zürich*), reprint, *SBZ* 98, no. 12 (Sept. 19, 1931): 149–50.

41. "Die Zementhalle der LA," *SBZ* 115, no. 4 (Jan. 27, 1940): 50; no. 5 (Feb. 3, 1940): 62. For the detailed report, see M. Ros, "II. Ergebnisse der Belastungsversuche an der Zementhalle der Schweizerischen Landesaustellung Zürich, 1939," *Bericht* 99, *Eidgenössische Materialprüfungsanstalt Zürich*, addendum 2 (August 1, 1940): 21–60.

42. Giedion's first writing on Maillart appeared in *Cahiers d'Art* 5 (1930): 146–52 and was later incorporated into his *Space, Time, and Architecture: The Growth of a New Tradition*, 5th ed. (Cambridge, Mass., 1967), 450–76; M. Bill, *Robert Maillart*, 3d ed. (Zurich, 1969); and M. Ro, "Versuche und Erfahrungen an Ausgeführten Eisenbeton-Bauwerken in der Schweiz, 1924–1937," in *Bericht* 99, *Beilage zum XXVI Jahresbericht des Vereins schweizerischer Zement- Kalk- und Gipsfabrikanten* (Zurich, 1937), 346–77.

43. The argument against engineering as applied science was convincingly presented in a major 1973 symposium subsequently published in *Technology and Culture* 17 (1976): 621–742.

44. Billington, "Maillart and the Salginatobel Bridge," 46–50.

45. J. Rae and R. Volti, *The Engineer in History*, rev. ed. (New York, 2001), cover.

CHAPTER 3

OTHMAR H. AMMANN: MASTER OF STEEL BRIDGES

David P. Billington and Jameson W. Doig

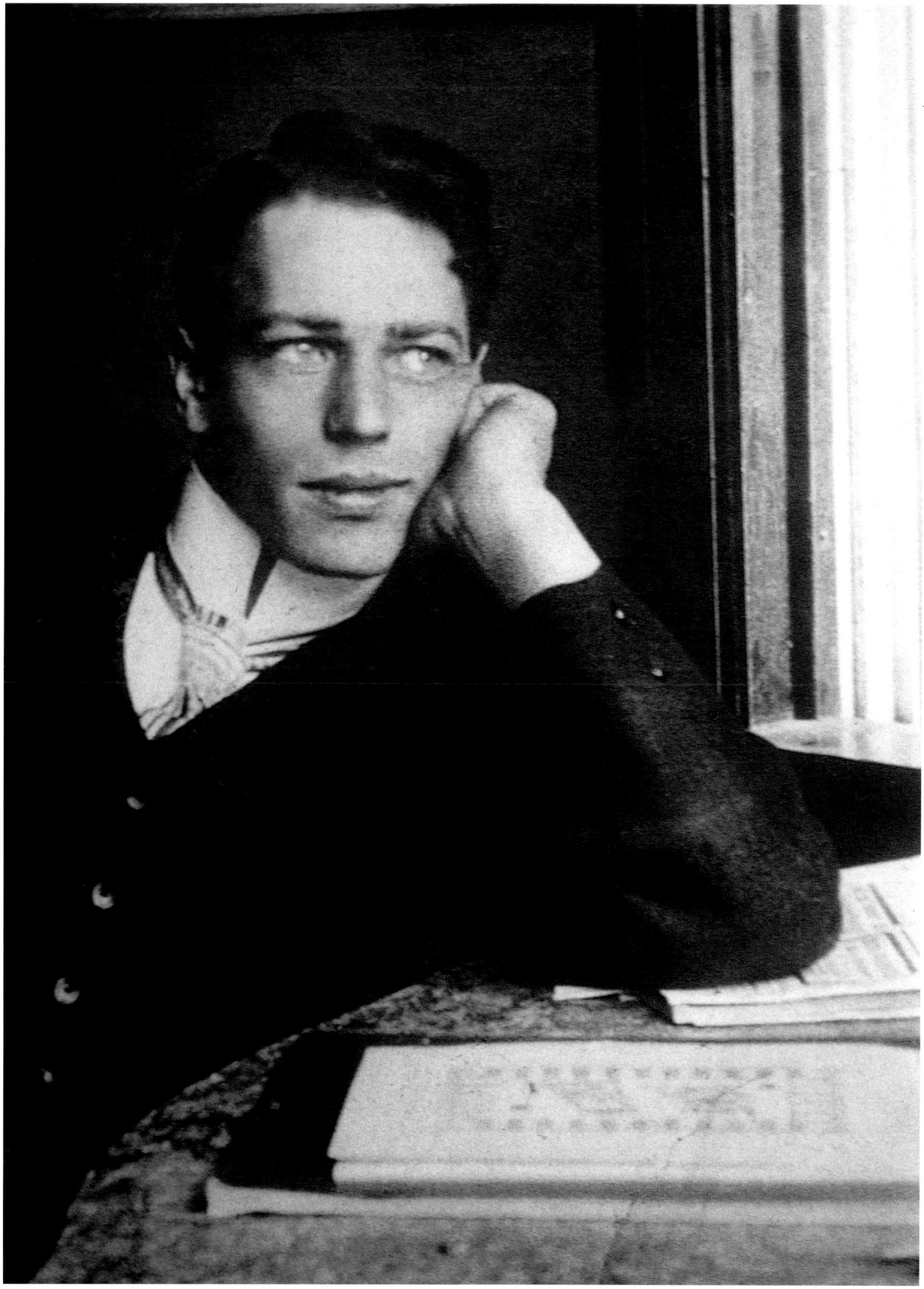

Figure 1 (previous page)
Othmar Ammann (1879–1965), 1904.

Othmar H. Ammann (1879–1965) (fig. 1) learned visual analysis from Wilhelm Ritter largely through graphic statics and exposure to contemporary practice. Unlike Maillart's early career, Ammann's was largely in the United States and postdated Ritter's death. Although his professor could not have a direct influence on him after graduation, Ritter's emphasis on simple analysis and aesthetic form helped to define Ammann's mature work. Ammann's biographer, Fritz Stüssi, emphasized the importance for Ammann of the Culmann-Ritter tradition and of Ritter's early studies of suspension bridges. He further quotes Ammann's testimony of 1955 to the centrality of the educational foundation given him by his alma mater. Stüssi, himself a professor at the ETH (Federal Institute of Technology, Zurich), stated that Wilhelm Ritter was doubtless the most important teacher for Ammann.[1]

Ritter's study of arch and cable bridges (1883) provides one insight into the teaching at Zurich. In these pioneering articles on bridge design, Ritter explored the interaction of cable and deck as well as arch and deck with his flair for presenting the connection between mathematical analysis and physical design. While this seminal writing directly stimulated Maillart in his search for the thinnest possible concrete arches, it had an indirect but powerful influence on Othmar Ammann's search for the thinnest possible deck in a cable suspension bridge. Just as Maillart carried Ritter's emphasis on aesthetics throughout his career, Ammann, too, stressed aesthetics in his designs and writings.

Commentators often treat the works of Maillart and Ammann in isolation and study the engineers disconnected from their formal education. This has led to significant misunderstandings both of the individual engineers' contributions and of the engineering profession. Therefore, it is important to follow the evolution of their works and ideas because these are central to an appreciation of how the austere new art of structural engineering came into being—not by flashy renderings of youth, but by the thoughtful and reflective experience and field observations of a maturing designer. Maillart's evolution began with Stauffacher Bridge, moved through Zuoz and Tavanasa, and reached its full maturity in the bridges of the 1930s. Ammann's evolution began in 1912, when he joined the firm run by America's leading bridge designer, and moved through a series of major steel bridges, culminating with the Verrazano Narrows Bridge of 1964.

While Robert Maillart carried out his important work in Europe, Ammann demonstrated the value of one of Ritter's recommendations—the benefit of international study—by traveling to America. It was planned as a brief visit, but once Ammann applied his engineering and aesthetic talents to the challenges of the New World, he found it increasingly difficult to return to his home country, despite entreaties from family and friends. He battled political as well as technical hurdles and emerged as the designer of a series of large bridges, which are perhaps unrivaled in their beauty and technical splendor among twentieth-century engineering achievements in steel.

Ammann was born in March 1879 in Canton Schaffhausen in Switzerland. In 1898 he entered the ETH in Zurich, where he studied with Ritter and with K. E. Hilgard, among others. Graduating in 1902, he worked as a draftsman in Europe for two years, and in 1904, at Hilgard's suggestion, departed for the United States. He soon took a position with a New York engineering firm, and during the next several years gained valuable field experience, working on several important bridges in Pennsylvania and New York.[2] The possibility of spanning the wide Hudson River, a challenge that would shape his career, also attracted his early attention. As Ammann later recalled, soon after arriving in New York he had traveled to the top of New Jersey's Palisades, on the western side of the Hudson opposite Manhattan: "For the first time I could envisage the bold undertaking, the spanning of the broad waterway with a single leap from shore to shore."[3] This vision was also partly inspired by Ritter, whose lectures were influenced by his visit to America in 1893. In fact, the last illustration in the text of Ritter's book on American bridges (1895) is Gustav Lindenthal's proposal of 1888 for a Hudson River crossing, only a few miles from the site of Ammann's later achievement.

In 1912 Ammann joined Gustav Lindenthal (1850-1935), a railroad-bridge engineer with an international reputation, and Lindenthal soon appointed him as his chief aide in work on the Hell Gate Bridge. Much of Ammann's time during the years 1912–17 was devoted to the Hell Gate, where he was in charge of all office and field operations, supervising a large team of engineers. Here he gained extensive field experience of the kind Ritter had emphasized in his teaching and writings. Ammann also demonstrated that he had learned the importance of clear thinking and writing—central, as we have seen, to Ritter's critique of Cullman and to Ritter's own writings. Once the Hell Gate was completed in 1917, Ammann then undertook the task of describing the alternative designs that were considered, the engineering challenges that had to be met, and the general standards that he believed should be applied in carrying out large engineering projects.

Figure 2
Cantilever design proposal of 1900 for a bridge over the East River at Hell Gate.

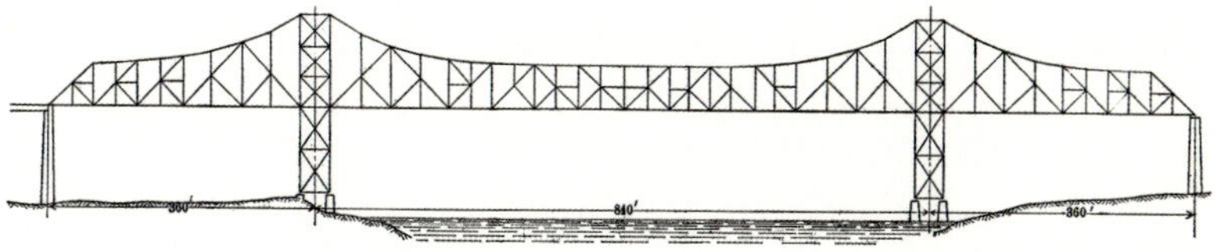

Figure 3
Lindenthal's cable-stiffened suspension design of 1904 for the Hell Gate Bridge.

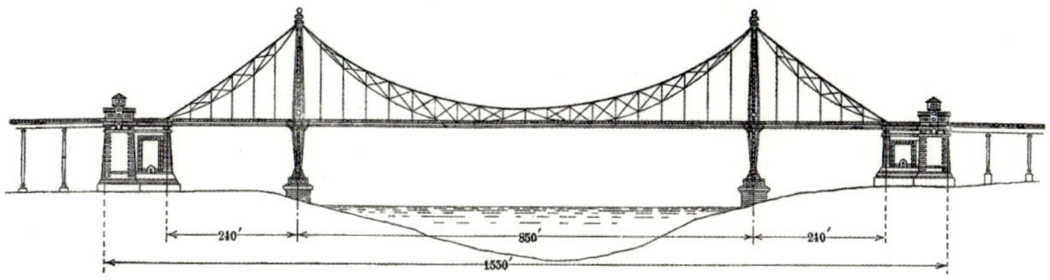

Figure 4
Lindenthal's continuous truss design of 1904 for the Hell Gate Bridge.

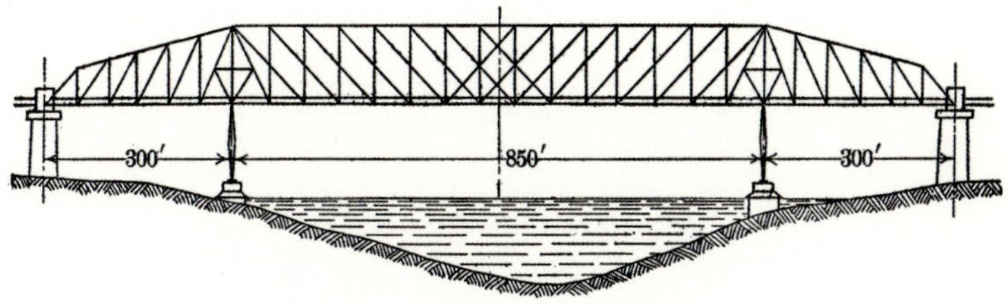

Figure 5
Lindenthal's cantilever design of 1904 for the Hell Gate Bridge.

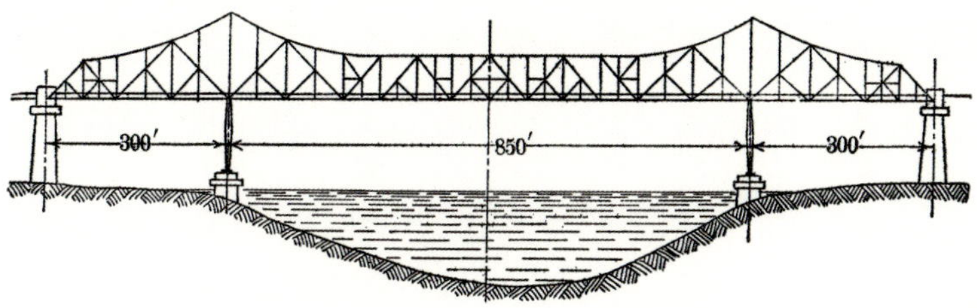

On November 21, 1917, Othmar Ammann presented to a meeting of the American Society of Civil Engineers a paper entitled "The Hell Gate Arch Bridge and Approaches of the New York Connecting Railroad over the East River in New York City." It was a long and detailed, but elegantly clear, presentation of the world's longest-spanning arch bridge. Ammann described the history of the project, including the development of its design, construction, and costs, and he provided a complete explanation of the calculations for the arch. A year later, the full paper appeared in the *Transactions of the Society* (1918), where it occupied 153 pages and stimulated 35 pages of discussion.[4] The following year, the society awarded Ammann the Thomas Fitch Rowland Prize for his cogent analysis. The first winner had been Gustav Lindenthal himself in 1883.[5] What is remarkable about Ammann's paper is its comprehensive sweep from stresses to finances, from aesthetics to construction stages. Ritter's influence is especially notable in Ammann's references to appearance and in his graphic depiction of forces in the arch. He began a discussion of design as follows: "A great bridge in a great city, although primarily utilitarian in its purpose, should nevertheless be a work of art to which Science lends its aid. An elaborate stress sheet, worked out on a purely economic and scientific basis, does not make a great bridge. It is only with a broad sense for beauty and harmony, coupled with wide experience in the scientific and technical field, that a monumental bridge can be created."[6] Later in his paper, Ammann does present elaborate stress sheets that provide evidence of his "wide experience in the scientific field"; but his most striking words claim the bridge to be a work of art. While we may disagree with Ammann's aesthetic judgment of Hell Gate, we cannot emphasize too strongly the role of aesthetics in all his works and especially the way his ideas evolved from the partially imitative tastes of Lindenthal to his own independent sense of sparse but inspiring elegance.

In his paper Ammann describes a series of proposals to bridge the East River at the Hell Gate, starting with a cantilever design in 1900 (fig. 2) and Lindenthal's cable-stiffened suspension bridge of 1904 (fig. 3). Lindenthal made two other designs in 1904, a continuous truss and a cantilever design (figs. 4, 5). Ammann says little about the 1900 design, but he effectively judges it inferior aesthetically by comparing Lindenthal's two truss designs to his suspension one: "In comparison with the bridge types shown in Figs. 4 and 5, the suspension type with its graceful outlines possesses unquestionably the advantage of more pleasing and monumental appearance."[7] He discusses these trusses aesthetically but rejects them as "utilitarian." He also describes how his mentor decided against the "aesthetically superior" suspension type because the arch offered greater rigidity, a major function for a railroad bridge.

Figure 6
Lindenthal's crescent arch design of 1905 for the Hell Gate Bridge.

Figure 7
Lindenthal's spandrel-braced arch design of 1905 for the Hell Gate Bridge, selected to be the final form.

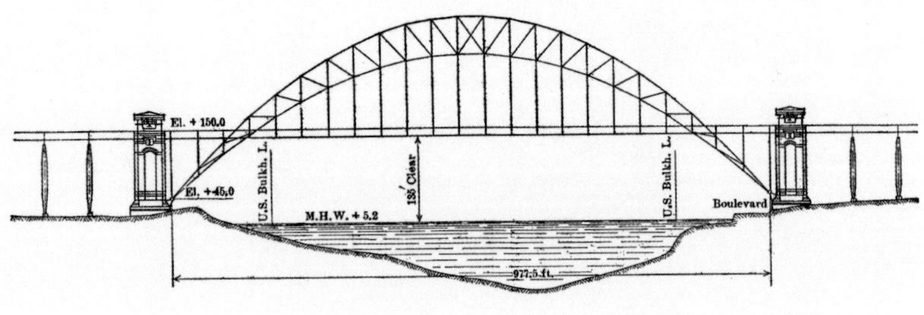

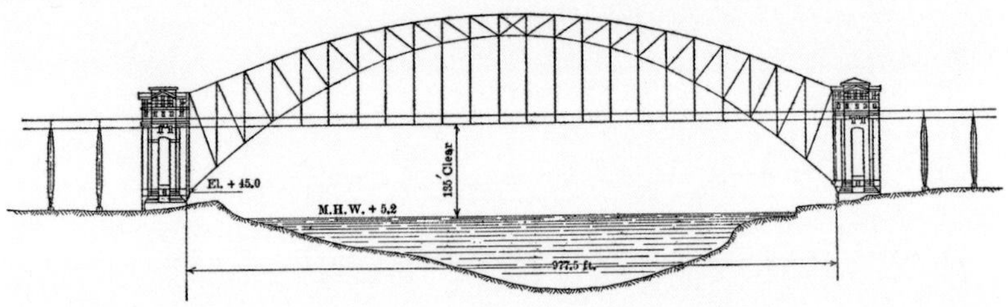

Ammann then turns to the two alternative arch designs that Lindenthal studied in 1905: "[The crescent design] represents the two-hinged arch as used, for instance, for the railroad bridges over the Garabit Valley, in France, and over the Douro River, in Portugal, with spans of 541 and 525 ft., respectively [both designed by Eiffel]. [The spandrel design] represents the two-hinged, spandrel-braced arch similar to bridges over the Rhine at Dusseldorf and Bonn, which have spans of 595 and 614 ft. respectively (figs. 6, 7)."[8] He then gives the justification for choosing the spandrel-arch form of the final design; it is entirely an aesthetic choice since, as he explains, the costs and performance will be about the same: "The estimated weight of steelwork was slightly in favor of the crescent arch, but the spandrel-braced arch offered greater advantages for the cantilever erection. Although both designs are pleasing in appearance, the spandrel arch, owing to its height increasing from the center toward the ends, is more expressive of rigidity than the crescent arch, the ends of which appear to be unnaturally slim in comparison with the great height at the center."

Once Lindenthal had decided in 1906 upon the type and general form, he modified the abutment towers so that there would be a slight reversal of curve in the top chord at its ends and the tower height would be increased (fig. 8). Further detailed changes followed (fig. 9) until construction began in 1912, when Ammann arrived on the scene. Thus the thirty-three-year-old Swiss engineer played no role in the design but served, in his paper, as a reporter on the aesthetic choices and on the detailed calculations. The final structure of 1916 (fig. 10) is essentially the bridge that is still serving in the twenty-first century (fig. 11).

Just as Maillart had made his Zuoz design an imitation of the needless walls at Stauffacher, so Ammann justified Hell Gate's useless end towers and deeper truss ends. The towers carry no loads while all the arch forces at the abutments are concentrated in the lower chords, making the upper chords there unneeded. For both men, the aesthetic of the past was in conflict with that of the future. They were both approaching great structural art.

In a brief conclusion to this monumental paper, Ammann reflected on "some of the broader engineering questions" suggested by the Hell Gate experience. The questions are those raised by Ritter in his teaching, and they would be the ones to guide his pupil throughout a memorable career: How does a great work come into being? What is the nature of teamwork? What are the roles of rules and specifications? How does economics relate to long-term performance? And—his crowning point—how critical is aesthetics? This manifesto for engineering education and practice begins: "A great engineering work cannot be spontaneously created in its final, perfect form, but has to grow and develop gradually, in its entirety as well as in its constituent parts... the engineer must ever be conscious that it is only through extensive and laborious preliminary studies, and untiring efforts to improve, that he can hope to achieve a perfect work."[9]

This gradual improvement requires, as Ammann put it, "collaboration of experts... but a great structure of monumental character must be the product of an individual creative and directive mind." For the Hell Gate, that mind was, as he implied, Lindenthal's. Great works of structural art, like poems, are not made by committees, even though for large structures, collaborations in details are essential. Ammann then observed that rules and specifications, while important, do not lead to fine designs; rather, "such a work requires wide experience and sound judgment," a clear reflection of Ritter's ideas. Moreover, Ammann emphasized that "throughout this paper the importance of rigidity in bridge construction has been pointed out. Large bridges must be built for generations to come. Engineers today cannot afford to build important structures cheaply, to serve their purpose for the time being, and incur the risk of having to replace them after a short period of usefulness."[10]

Figure 8
Drawing of the tower design, 1906, of the Hell Gate Bridge.

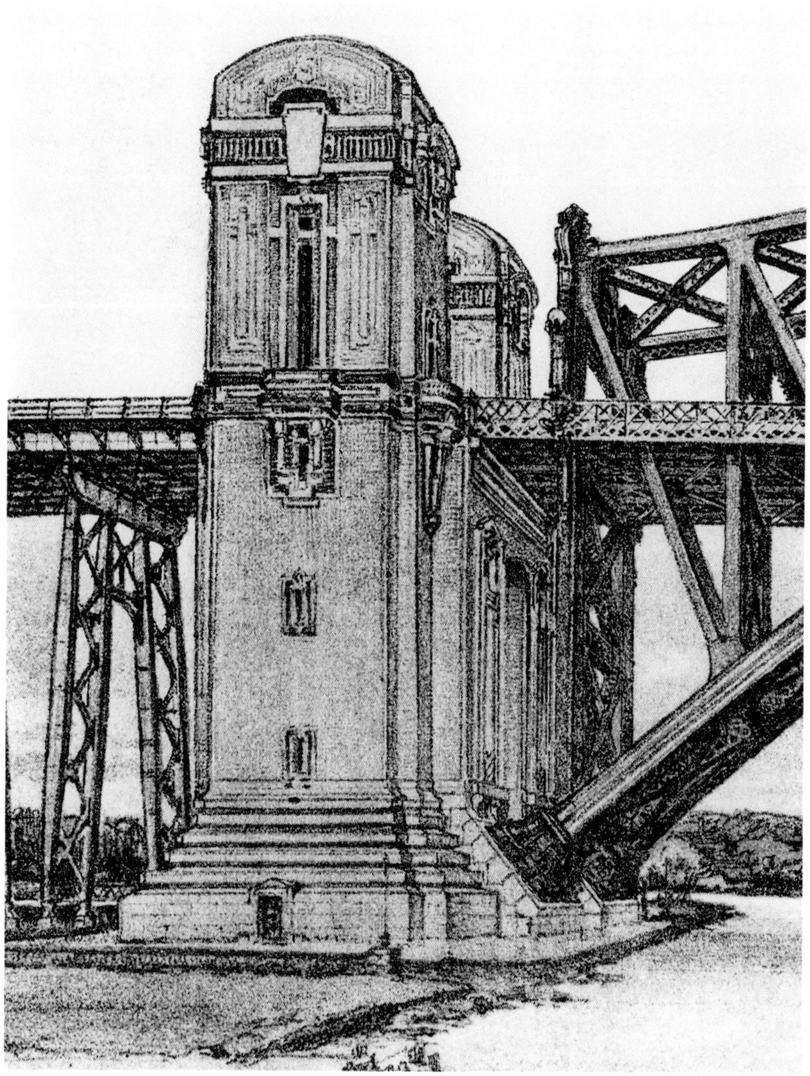

Figure 9
Final tower design of the Hell Gate Bridge.

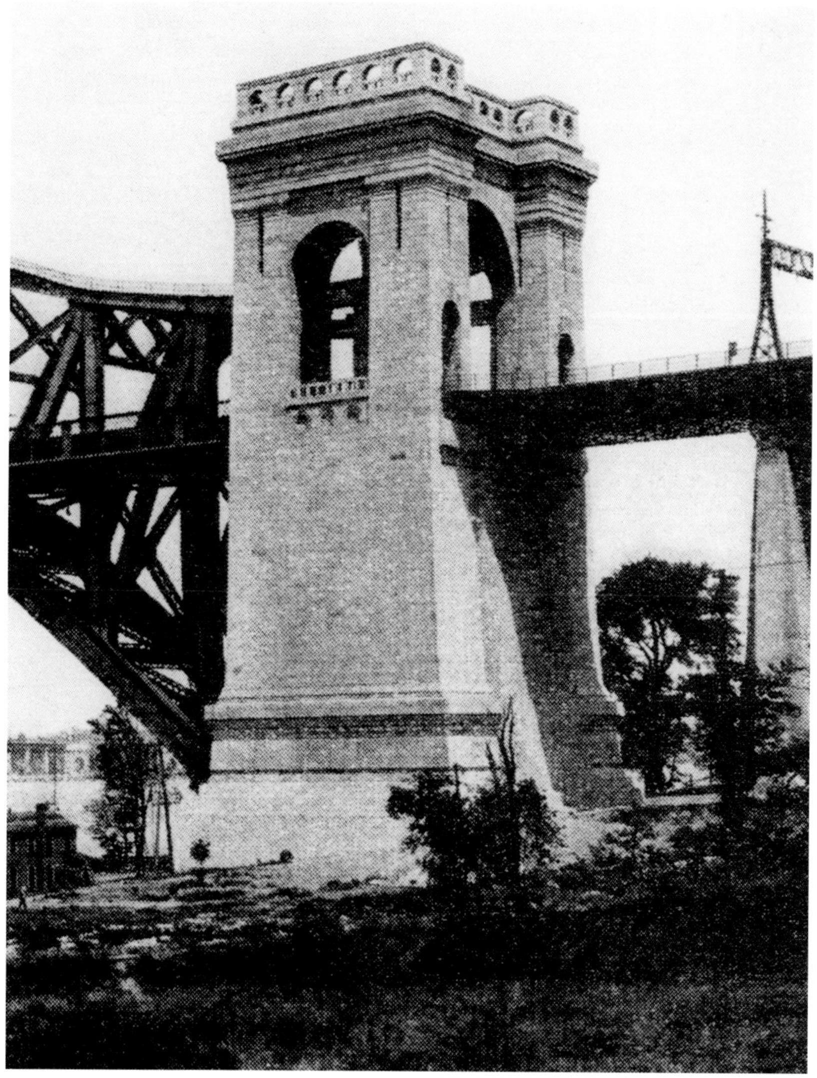

Here the role of deflections is crucial, a test that the Swiss bridges under Ritter's careful direction had to pass by full-scale measurements. Ammann presented such calculations in his paper to show how exceptionally rigid was the Hell Gate. But for him, aesthetics is the true measure of quality, built upon the concerns for safety and economy; as he put it in the paper's concluding climactic paragraph: "Emphasis has been laid on the appearance of the structures described. Engineering structures are still regarded by many engineers as mere works of utility, which deserve no consideration in architectural or artistic treatment. So long as this opinion prevails, the Engineering Profession will not lift itself to a higher plane, and it is even running the risk of being relegated to second place—or after the architect—in the creation of such monumental structures as properly belong in its domain."[11] What Ammann means by "architectural or artistic treatment" is not fully clear, since that treatment at Hell Gate was so clearly nonstructural at the abutments. He is unambiguously clear, however, in his concern that the engineer take aesthetics seriously and by so doing lift the profession "to a higher plane."

Figure 10
Nearly completed Hell Gate Bridge and the leadership design group, with a bearded Lindenthal at the center and Ammann two figures to his right.

Figure 11
Contemporary view of the Hell Gate Bridge, with the Triborough Bridge in the background.

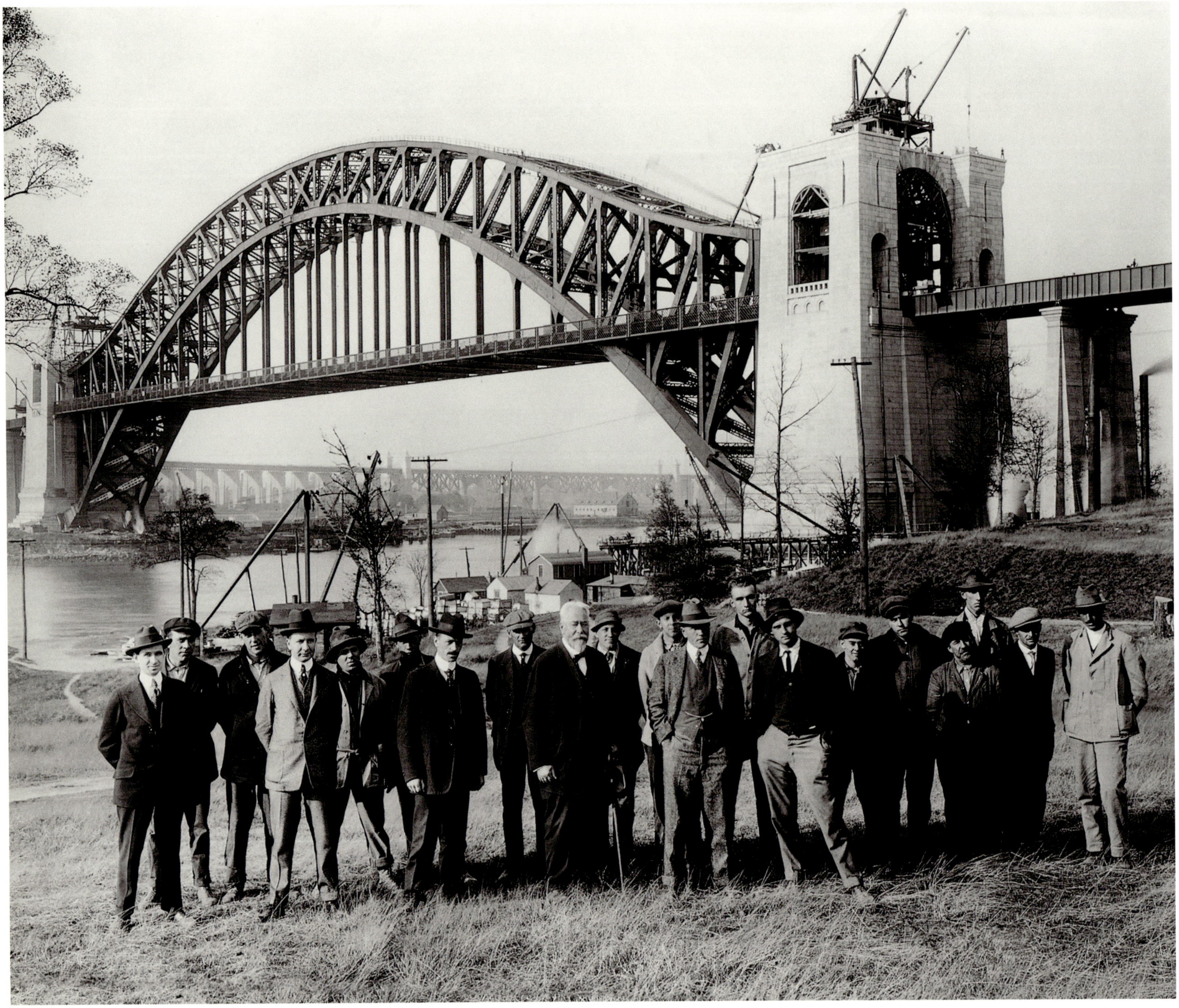

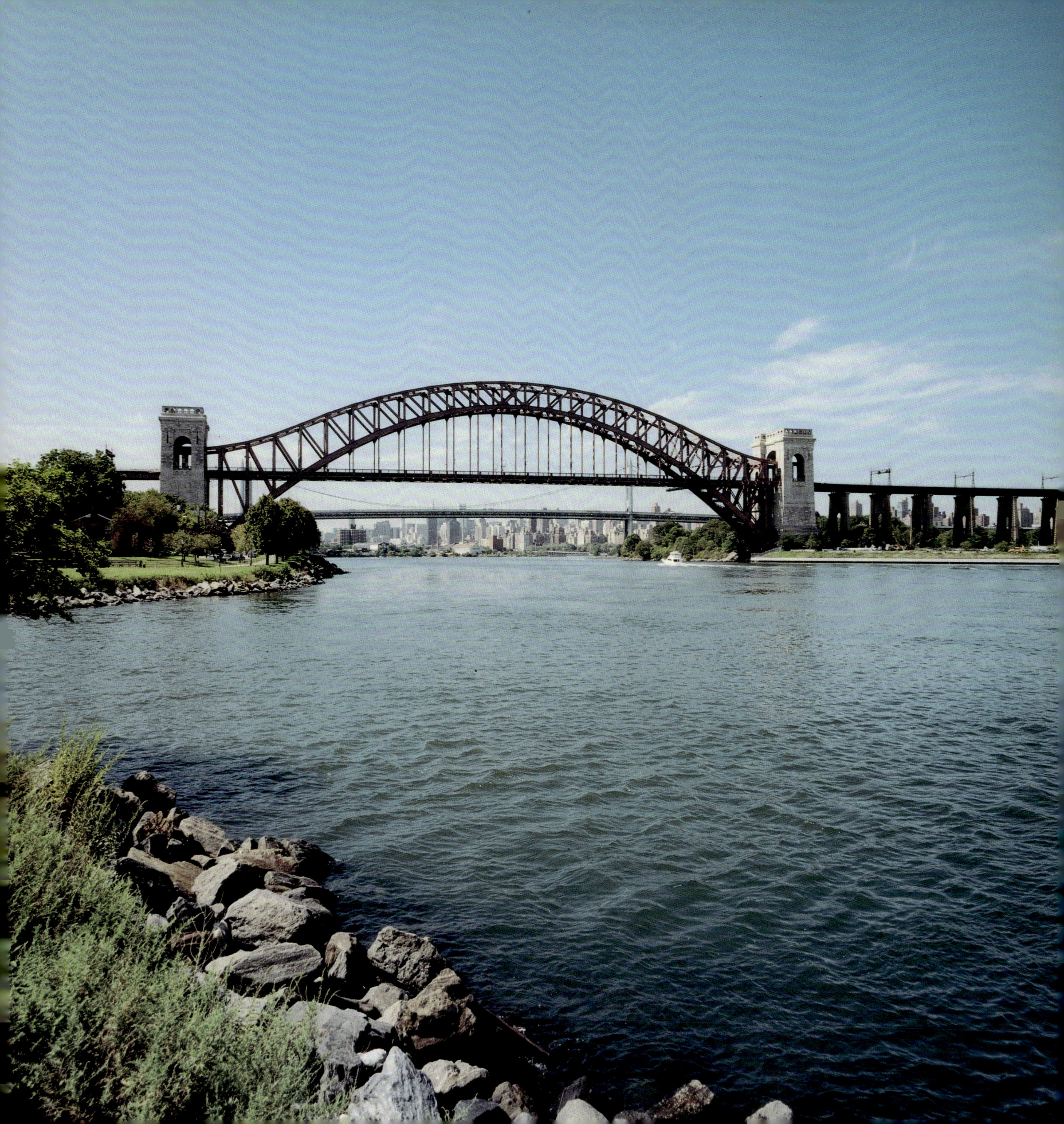

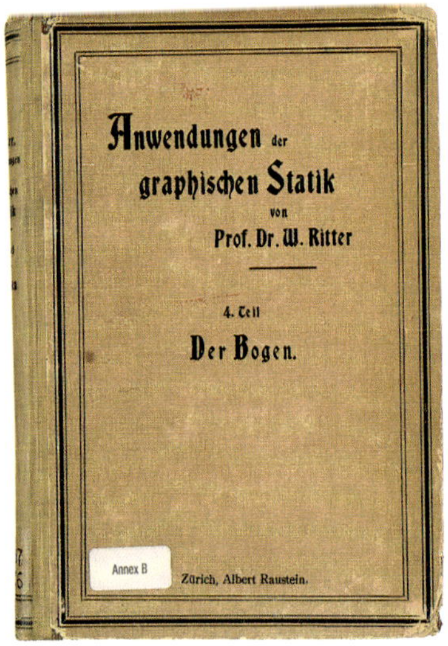

Figure 12
Cover of Wilhelm Ritter's *Anwendungen der Graphischen Statik* part 4, *Der Bogen* (Zurich, 1906), which discusses arches similar to the Hell Gate Bridge.

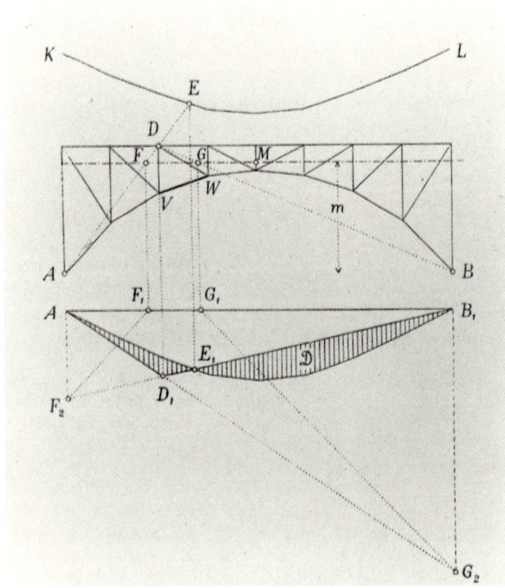

Figure 13
Ritter's design of a trussed arch with a straight top chord.

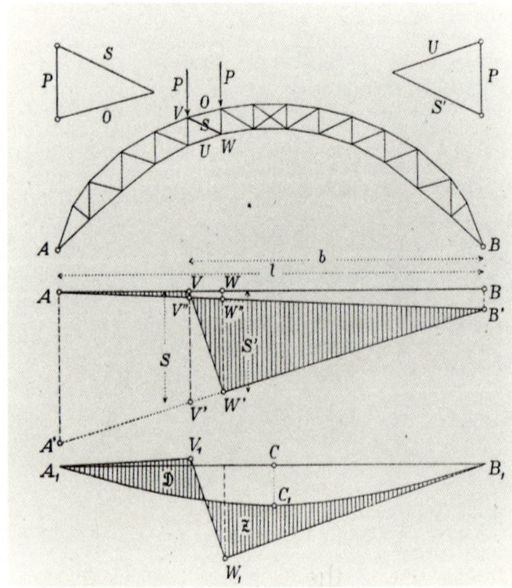

Figure 14
Ritter's design of a trussed arch that clearly expresses the hinged form.

In an appendix to the paper published in 1918, Ammann described the calculations for the arch; here and in the aesthetic discussion, his debt to Ritter is clear. Although he is justifying Lindenthal's decision on form, the description is Ammann's own. In the same way, when Ammann presented the technical justification through calculation of stresses, he made the discussion his own by calling on his educational experience with Wilhelm Ritter.

Ritter's last book, published the year of his death (1906), was devoted entirely to arches (fig. 12),[12] and more specifically to trussed arches similar to the Hell Gate design—except they do not have deeper truss ends. Figure 13 shows one common bridge type and figure 14 illustrates another; both are two-hinged arches and represent the two extremes; the former has a straight top chord (usually the bridge deck) and the latter clearly expresses the hinged form. In Ammann's presentation of the calculations, he resorts to the same graphic type of analysis that he learned in Ritter's lectures of 1901 and which later appeared in the book of 1906. Figures 15 and 16 show typical diagrams found in the appendix to Ammann's "Hell Gate" paper. Although the analyses are not exactly identical to those in figures 13 and 14, the approach is the same.[13]

Once the Hell Gate was finished, Lindenthal had very little engineering work to occupy his staff.[14] He suggested that Ammann take a temporary position in Middlesex County, New Jersey, managing a clay pottery mine in which Lindenthal had invested. Ammann took that job; and with his managerial skills he turned a shaky financial enterprise into a healthy firm. In 1920 Lindenthal called him back to assist in developing plans for a gigantic railroad-vehicular bridge that would cross the Hudson at 57th Street. Ammann worked on this project with Lindenthal from 1920 until 1923. By the middle of 1922, however, Ammann had become concerned that the massive 57th Street span could not be constructed in the near future. Against the project were its great cost, the reluctance of the railroads to commit themselves to using the bridge, and the opposition of Manhattan business and political leaders to an effort that would dump twenty lanes of traffic into the midtown area.[15]

As Ammann pondered what to do in the face of the strong opposition to a bridge at 57th Street, he looked northward to the less crowded precincts around 179th Street, and to the steep Palisades shores of New Jersey, from which a long bridge might be cast. Ammann was convinced that the engineering profession now understood the performance characteristics of long-span suspension bridges sufficiently well to permit such a bridge—even though its span would be nearly twice the length of any bridge yet constructed.[16] The new understanding would make it possible, Ammann thought, to meet high standards of aesthetic as well as technical achievement, a combination that might attract any first-class engineer. Moreover, the expanding use of automobiles and trucks across the New York region, exemplified by long lines of vehicles waiting to use trans-Hudson ferries from Bergen County and northern Manhattan, indicated to Ammann that there was a real need for a bridge located well north of congested midtown Manhattan. He was ready to throw his considerable energies into the dual task of putting into detailed form the

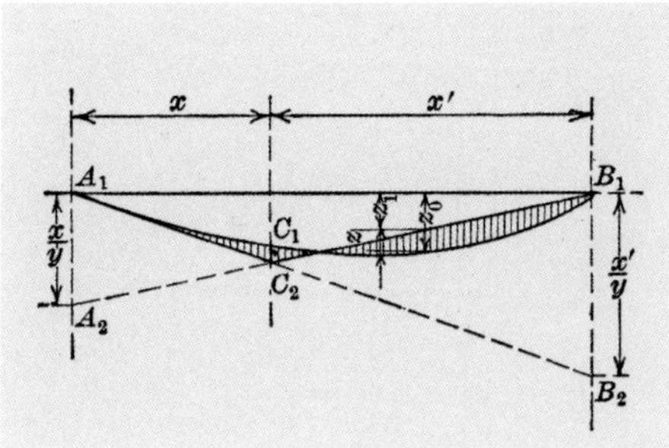

Figure 15
Amman's analytical diagram using Ritter's graphic statics approach.

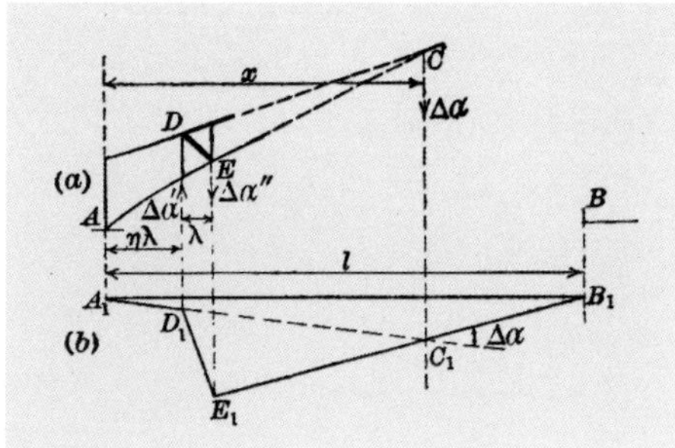

Figure 16
Ammann's analytical diagram using Ritter's graphic statics approach.

pioneering engineering ideas needed to make such a bridge technically feasible and of helping to create the public support that would be required before the great span could be approved and constructed.

Meanwhile, in the fall of 1922, a change in political fortunes offered Ammann a different route to spanning the Hudson—a project that would achieve the goal he had imagined from the top of the Palisades in 1904. In November 1922, George Silzer was elected governor of New Jersey, and Ammann and Silzer soon formed an informal but close alliance that would in time lead to the construction of the George Washington Bridge.[17]

For his part, George Silzer could see the economic advantages to New Jersey that would flow from improved transportation between New York City and the extensive rural areas of northeastern New Jersey; and he could see the advantage to his own political prospects that might follow if the great bridge, with its stimulus to commerce across the northern part of the state, were commenced during his time in office. Elected for a three-year term, Silzer would serve as governor from January 1923 until January 1926, with the state legislature controlled by the Republicans throughout those years. Silzer, a Wilsonian Democrat, was an activist in his philosophy of government. One of his major interests was the improvement of highway transportation; he viewed this goal as crucial to the state's economic growth. Therefore, as governor, he urged the creation of a new state highway agency to devise an efficient road system, and he advocated extensive road-building programs, as well as bridges that would connect his home county, Middlesex, to nearby Staten Island.[18] Silzer might be in a receptive mood if Othmar Ammann could bring him a feasible plan to dissolve the Hudson barrier.

Silzer's victory in the 1922 election provided Ammann with an opportunity to reach out for assistance as he tried to extricate himself from the political and economic obstacles that confronted Lindenthal's hopes to build a giant railroad-highway bridge over the Hudson. Fortuitously, Ammann was already acquainted with Silzer, for Silzer was from Middlesex County—and a director of the clay mine that Ammann had managed for three years. Sometime in the weeks before Silzer took office in mid-January 1923, Ammann talked with him about the importance of a Hudson River crossing. They discussed the economic and political problems that surrounded the Lindenthal bridge and the advantages of constructing a bridge farther north that would join the vast, rural Bergen County to the urbanized New York shore.[19] The bridge could be limited to motor vehicles and light trolleys, so that its cost would be far less than the Lindenthal colossus, which had to be strong enough to carry loaded freight trains. Ammann had studied various sites, and he preferred a bridge that swept from the Palisades cliffs, in the town of Fort Lee, across the river to 179th Street in Manhattan.

Silzer was enthusiastic about Ammann's plan.[20] But now Ammann had to decide how to proceed. While he was certain that a span at 179th Street would face far fewer obstacles than the massive 57th Street bridge, could he persuade Lindenthal to embrace this more modest crossing? Ammann hoped the great man might be persuaded to join forces with him—perhaps take the leading role—and attract the private capital and political support needed for the Fort Lee plan. But if Lindenthal stood firm, the prominent civic leaders and financial men who had joined his board of directors would probably stand with him; and while that combined force might never produce a real bridge at 57th Street, its opposition might doom Ammann's uptown scheme. During the fall and winter of 1922, Ammann urged Lindenthal to embrace a modest span, oriented toward the increasing flow of motor vehicles, north of the midtown area. But that plea was unsuccessful. To Lindenthal, the 57th Street project was the only satisfactory way to solve the region's major freight and passenger problem: how to bring railroad trains directly into Manhattan and thereby integrate New York City's vast economic strength with the economies of New Jersey and the rest of the continent.

Figure 17
Various designs for the crossing of the Hudson River, beginning with Lindenthal's structure of 1888 and culminating with Ammann's proposal of 1923.

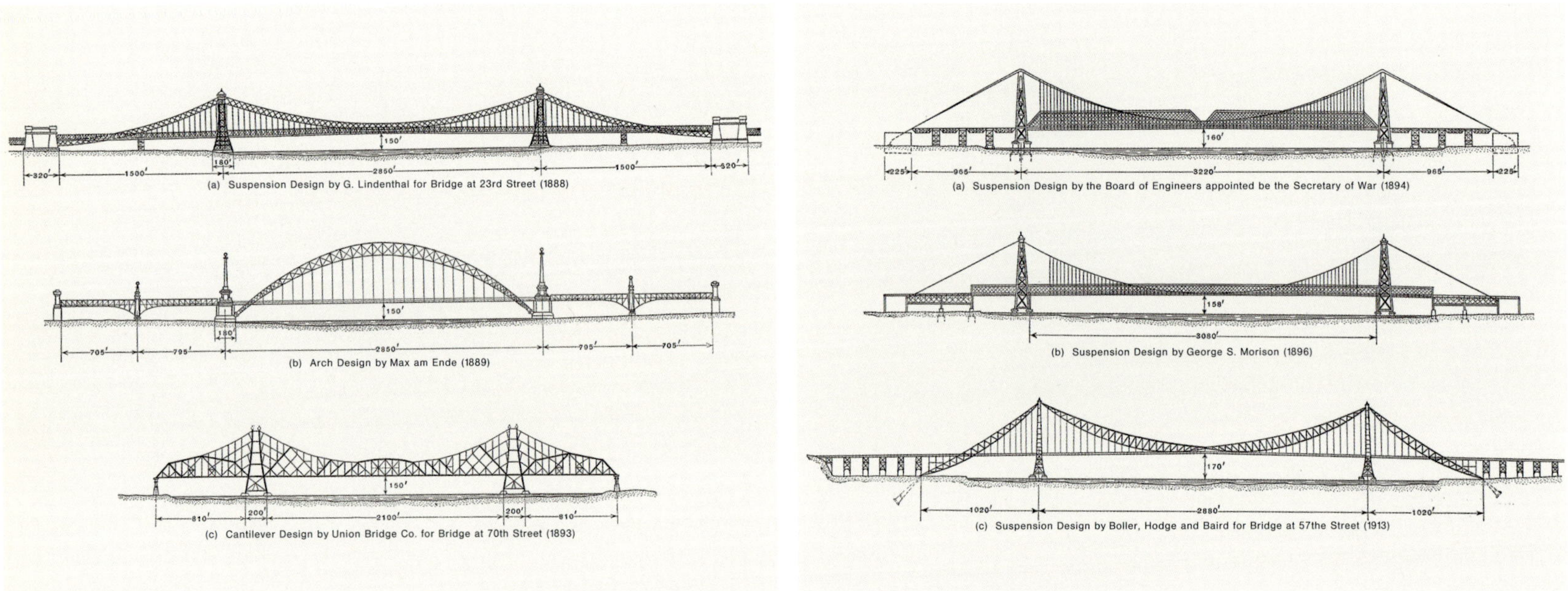

As 1922 drew to a close, it seemed evident to Ammann that he would have to take some initiative on his own. His concerns and hopes during these months were captured in a letter written to his mother a year later:

In order for you to understand my situation for many months, in fact for a whole year, I will no longer conceal from you that the giant project for which I have been sacrificing time and money for the past three years, today lies in ruins. In vain, I as well as others have been fighting against the unlimited ambition of a genius who is obsessed with illusions of grandeur. He has the power in his hands and refuses to bring moderation into his gigantic plans. Instead, his illusions lead him to enlarge his plans more and more, until he has reached the unheard-of sum of half a billion dollars—an impossibility even in America.

However, I have gained a rich experience and have decided to build anew on the ruins with fresh hope and courage—and, at that, on my own initiative and with my own plans, on a more moderate scale. It is a hard battle.[21]

This is a very Swiss attitude in which smaller scale and practicality come to the fore. There are echoes of Ritter's debate with Engesser about the load testing of small bridges and the value of continual observations over the authority of mathematics. What Ammann faced was an authoritarian leader whose success with the huge railroad bridges blinded him to the more modest requirements for motor vehicles. In 1922 the grand days of the railroads were over and those of the automobile had just begun. This new transportation needed fresh ideas for bridging, and those were beginning to germinate in the active mind of the young Swiss engineer. Yet, if Ammann felt compelled to strike out on a new course at the end of 1922, joining with Governor Silzer to press for approval for a bridge crossing far north of mid-Manhattan, he would need to meet challenging engineering questions, and at the same time convert his own enthusiasm into wide public support for such a project. In the spring of 1923, he formally left Lindenthal and began a period of unemployment.

How best to proceed? For Ammann, the answer came in three parts: aesthetic vision, political realities, and engineering calculations. He would need to sketch out a bridge design so dramatic, so arresting, that it would claim the attention and approval of opinion leaders and broad groups of citizens in New Jersey and New York. Then he would need to take the design, and his ideas on how the bridge would benefit the region, directly to public officials and local groups on both sides of the Hudson. But he could not base his design on the structural theories of his mentor Lindenthal or on others that

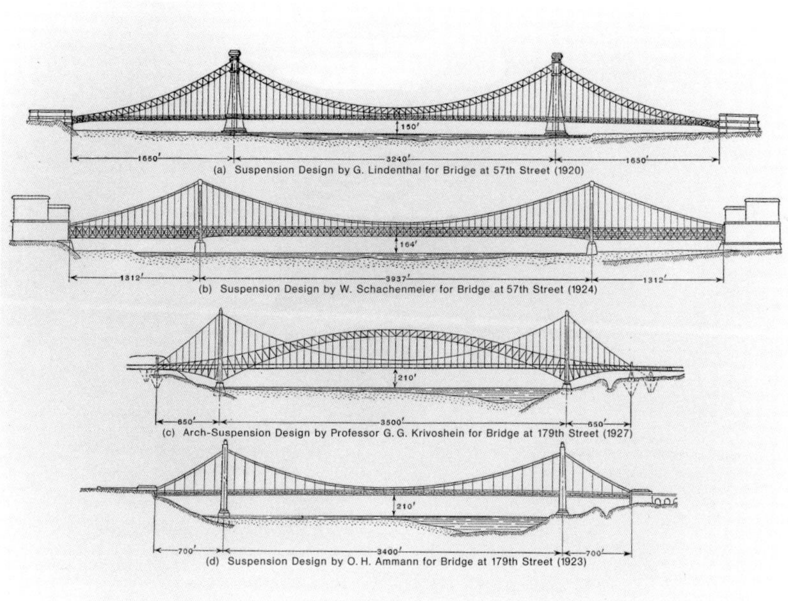

used standard analyses of bridges. He needed engineering ideas as striking as his visual sketches. Thus Ammann would need to think deeply, and perhaps creatively, about the technical issues involved and how they could be solved—at reasonable cost. Even so, this immense span would doubtless be the most expensive bridge project ever built. There had been many proposals for this crossing, beginning with Lindenthal's design in 1888 and concluding with Ammann's in 1923 (fig. 17).[22]

In order to achieve his goal of a record length, combined with a cost that would render the span politically feasible, Ammann needed to rethink basic issues of bridge design. A bridge, together with its towers and anchorages, needed to be strong enough to carry its own weight (the dead load), together with the heaviest convoy of vehicles that might travel across the span at one time (the live load). The dead-load component of total weight had a straightforward technical solution, in mathematical terms, and agreement on how to compute the effects of the dead load had been reached by 1823.[23] These effects consist essentially of the force of gravity, as the deck pulls down on the suspenders, which transfer that vertical weight nearly uniformly to the cable. The cable carries these loads to the towers by assuming a curved shape and by stretching slightly. The towers easily resist the vertical force from the cables but the horizontal force, which could pull the tower tops inward, needs to be held by cables leading from the towers to anchorages on either shore. When traffic crosses a bridge, the additional weight (the live load) will force the deck downward, and all components (deck, suspenders, and cable) will be stressed and alter shape. The effect of these live loads is especially difficult to predict for a suspension bridge because the cable will tend to distort in a variety of ways as heavy traffic moves across the bridge; the result can be large stresses in the bridge deck. To prevent excessive movement of the deck and cables, engineers add stiffening trusses, which extend throughout the entire length of the suspended deck and are usually placed below the roadway.

During the nineteenth century, heavy loads meant railroad trains; by the end of the century, the distribution of loads due to moving trains was well understood, and bridge engineers could design railroad bridges to prevent excessive distortion.[24] The advent of motorized vehicles in the first years of the twentieth century, however, meant that new calculations were needed to take account of the weight—and uneven distribution—of heavy trucks and a variety of lighter vehicles as they moved across the bridge. Ammann recognized that his design would have to be based on a careful analysis of this quite different pattern of live loading and on new ideas for determining the kinds of stiffening trusses needed.

In addition, because Ammann's proposed bridge would have a span twice as long as any previously constructed suspension bridge, his reach into the unknown was far more complicated than if he had been designing a highway bridge in the standard range for long-span bridges (then about 800–1600 feet). It was not reasonable to copy older models, and bold ideas seemed essential to meet the central challenge: how to design a bridge that would endure, and at a cost low enough that the financing hurdle could be overcome. In moving into untried territory, however, Ammann would find it difficult to anticipate all the problems that can arise in a pioneering work. A century earlier, Telford had designed and built the Menai Bridge, which, when completed in 1826, was the longest span bridge in the world; but he was not able to evaluate the effect of wind on a structure of that size, and after severe wind gusts had damaged the deck, funds had to be allotted to strengthen the span. Forty years later, John Roebling used his system of cable stays in designing the Brooklyn Bridge, providing stiffness against wind oscillations of the deck. In solving Telford's problem, however, Roebling unwittingly created an ambiguity in the distribution of vertical loads, and to help overcome that problem, deck trusses had to be added years later.[25]

Figure 18
Ammann's design of December 1923 for a Hudson River crossing at Fort Lee, New Jersey.

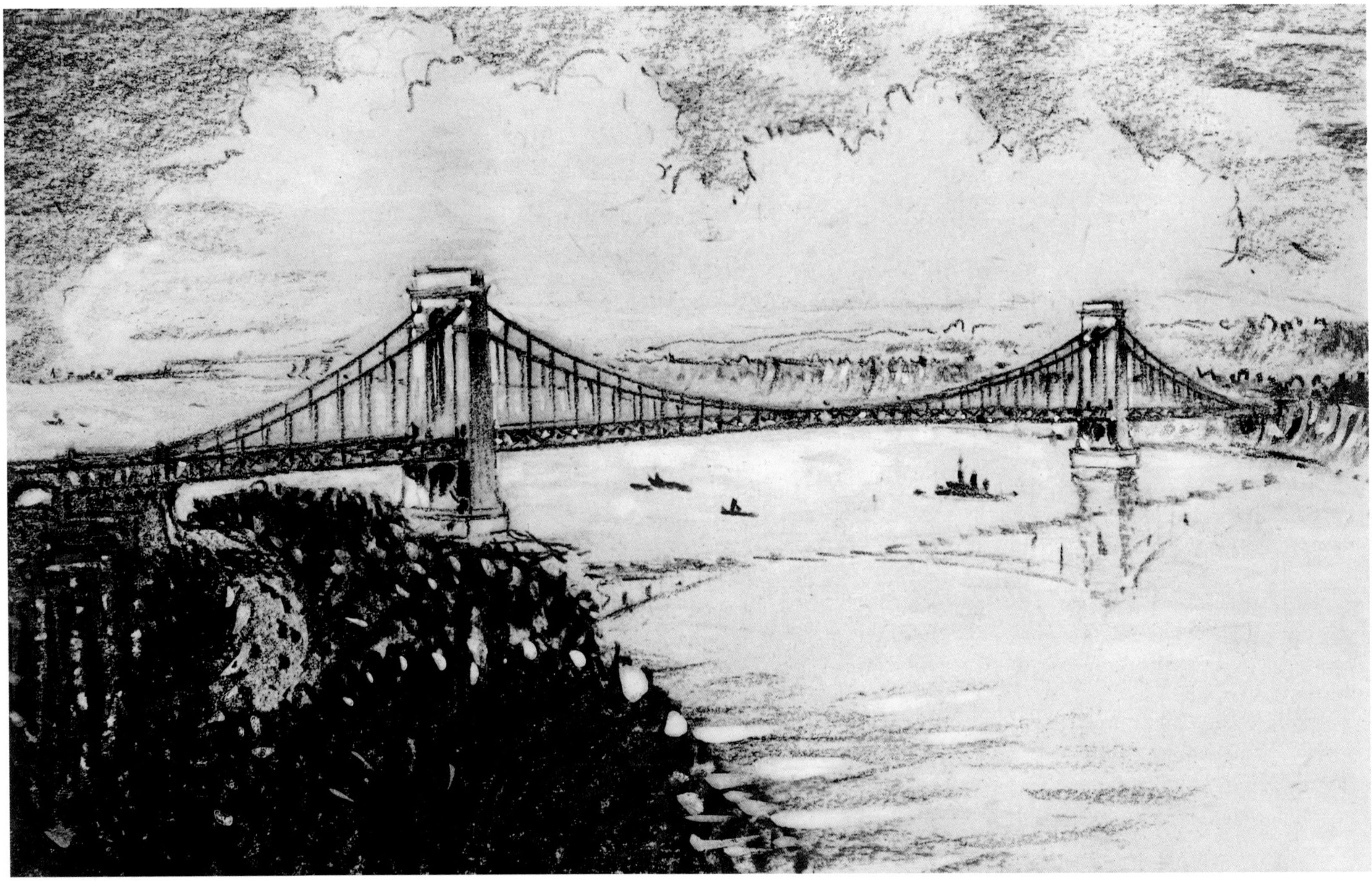

Two major decisions by Ammann permitted his design to be strikingly economical—but also daringly thin and hence elegant. Ammann worked out these issues in 1923 and 1924 while he also met daily with civic and political leaders, attempting to persuade them of the need for this bridge—and of its advantages over competing proposals for a set of vehicular tunnels under the Hudson. Much of his analytical work was relegated to late evening hours. By December 1923, Ammann had developed the design sufficiently to send Silzer a detailed report on the bridge's feasibility, including a detailed description of the design (fig. 18).[26]

The first of these two decisions focused on traffic patterns, which could cause uneven movement of the deck and cables. Engineers knew that short bridge spans should be strong enough to hold a series of heavily loaded trucks, close together end-to-end. As the span length increased, however, that degree of truck congestion seemed very unlikely, since a mix of automobiles and lighter trucks would probably be interspersed with the heaviest trucks. If there were no limitation on cost, an engineer might decide to design even long-span bridges to hold lines of traffic consisting only of heavy trucks; but engineering projects almost always must be justified within budget constraints, and the best bridges invariably have been designed by engineers whose experience derived from strict economy. Therefore, the central engineering question, then as now, was how to balance economy and safety.

Ammann's first step was to determine the maximum conceivable load (also called the "congested load"); this he calculated as 46,000 pounds per foot. Had he followed the convention of the early 1920s, he would then have taken half the congested load, or 23,000 pounds per foot, and designed his bridge for that total.[27] His sad experience with Lindenthal's extravagant plans had underscored the importance of finding ways to reduce costs. As Ammann commented in a speech in 1924 on his proposed Hudson River crossing, the "financial feasibility of the undertaking depends largely upon the design, by which millions of dollars may be saved or wasted."[28] Now the question he confronted was whether a reduction of more than fifty percent from the congested load could be justified.

As he proceeded in his dual role of bridge designer and political advocate, Ammann embarked on a new approach to justify reductions from the congested load. He first considered the likely proportion of heavily loaded trucks on a span of 3500 feet, assuming that only one lane were available. Ammann recognized such conservatism as costly, especially for his much longer span; so he decided on a ratio (which he called K) of one-quarter. This ratio (K = 0.25) represents an estimate that heavy trucks will be only a small fraction of the total number of vehicles on the bridge at any one time. Thus, Ammann took a much greater reduction than the precedents of the Manhattan and Ben Franklin Bridges would admit.[29]

Ammann's second step was to ask if all the lanes would simultaneously be carrying a load corresponding to K = 0.25. If the bridge had only one lane, the answer would be yes; if there were two or more lanes, probably not. The more lanes, the lower would be the probability of all lanes reaching the reasonable one-lane maximum. Ammann then introduced another ratio, C, which would be equal to 1.0 for a one-lane bridge and would decline toward 0.5 for a bridge with a large number of lanes. Applied to a single-deck Hudson River crossing, this formula yields a value of slightly under 0.7 for eight lanes.[30]

When Ammann combined these two ratios (0.25 x 0.682), the result was 0.17, and this is the estimate he used in designing the bridge. Abandoning the traditional one-half reduction, Ammann concluded that the bridge would be adequately safe if it were built to withstand only seventeen percent, or about *one-sixth*, of the maximum conceivable load! Was Ammann justified in reducing the traffic loads so far below the standard ratio? When the various studies carried out in the 1920s had been completed, and the first *Standard Specifications for Highway Bridges* was finally published in 1931, loadings for one lane and number of lanes were both included in the calculations, and the reduction factors were almost the same as those used by Ammann.[31] In the early 1920s, however, he was decidedly pioneering in choosing such a low estimated live load. What was the cost impact of Ammann's analysis? The cost of steel for the George Washington Bridge was $23,000,000; if the standard reduction had been used, it is estimated that the bridge would have cost at least $7,400,000 more.[32]

While it appears that Ammann took risks with such a low live load, his reasoning included an estimate of safety based upon cable stresses. For the wire Ammann planned to use, the maximum stress (called yield point) was about 150,000 psi (pounds per square inch, like a pressure), and Ammann calculated about 82,000 psi under dead load plus the live load. With the dead load plus the maximum conceivable live load, the total cable stress would be 148,000 psi, just under the maximum stress. Thus Ammann's bridge would not have collapsed even had the maximum conceivable live load appeared.[33]

In a speech devoted to gaining political support early in 1924, Ammann argued that by designing a very light stiffening truss to run the length of the suspended deck, he would be able to save "about 25 percent of the entire cost of the bridge."[34] In his extensive reflections on the project after it was completed, he explained how such a drastic saving was possible: "Some formulas would increase the weight of stiffening roughly in proportion to span and live load. On that basis, when compared with the stiffening of [the Manhattan, Bear Mountain, and Delaware River Bridges], the system of the George Washington Bridge… would weigh roughly from 13,000 to 14,000 lbs. per ft." Yet, as Ammann then noted, the truss for his bridge—with both decks in place—would weigh only about *one-sixth* of that total, and the cost savings would be $10,000,000.[35]

The savings were immense for a bridge whose total initial estimated cost for construction (after omitting the $10,000,000) was only $30,000,000 in 1924; it was all but irresistible, providing that a technical justification existed. Ammann found that justification in the "deflection theory" of suspension bridges, which indicated that *the greater the dead weight of the span*, the less deck stiffening was required. Others had already used that theory to analyze both the Manhattan and the Ben Franklin spans, but no one had carried the theory's implications as far as did Ammann, whose design would permit the George Washington Bridge to be constructed with practically no deck stiffness at all.

Once his bridge was built, its dramatic appearance and the cost savings associated with the theory had a significant impact on bridge design. Ammann and other leading designers proceeded to use the theory to justify bridges with decks that had very little stiffness, until the dramatic collapse of the Tacoma Narrows Bridge in 1940 demonstrated that the theory was incomplete: it did not take into account the dynamic character of the wind. The George Washington Bridge never suffered any significant wind-induced damage, even when it had no second deck (from 1931 to 1962), but it did lead designers to think that deck stiffness (apart from dead weight) could be omitted in long-span bridges.[36]

By the 1880s, Wilhelm Ritter and Josef Melan in Austria had developed a theory that considered truss deflections but assumed the cable to remain in its dead-load shape (a parabola) even under live loads.[37] Melan expanded his theory to include changes (or deflections) in cable shape due to live loads, which was used to analyze the Manhattan Bridge (1909) and the Ben Franklin Bridge (1926). Called the deflection theory, it gained favor with specialists, thanks to publications by prominent engineers, and it worked well for spans far shorter than the Hudson River bridge.[38]

When Ammann began serious study of a bridge over the Hudson, in 1923–24, this new theory was waiting to be of service. It told Ammann a remarkable, almost incredible story: the heavier the bridge, the lighter it could be. Stated otherwise, the longer a bridge span was, the more its dead weight alone could provide the needed stiffness against traffic loads and wind. Melan's theory provided apparent proof via a long and complex mathematical elaboration. Once that general proof was accepted, the design implications for his bridge were dramatic, and the cost savings equally significant.[39]

The deflection theory encouraged Ammann to see ways of making the George Washington Bridge economically attractive and therefore more politically appealing. By early 1924, he was able to report that the deck of his proposed span would have minimal trusses and be unusually thin. "The far-reaching economy of this design," he wrote, "may be realized by the fact that the weight of these stiffening trusses amounts to only about one-eighth of that of the chains," while in many existing suspension bridges "the stiffening system is heavier than the cables or chains themselves." Even at that early stage, he estimated that "the savings effected in the trusses," together with resulting savings in the suspenders and other elements, would amount to "about 25 percent of the entire cost of the bridge."[40]

The deflection theory shows that nearly the entire live load is carried directly up through the truss and the suspenders to the cable; the deck carries almost none of the live load. It was for these modest forces that Ammann designed the truss. But it tells an even more remarkable tale. When Ammann studied the effect of a single deck, he found that the required truss size approached zero. The deflection theory results showed that the more flexible the truss, the less the forces in it. The cable, in short, adjusts to carry all the live load *without any need for a truss.*[41]

Here Ammann justified his aesthetic urge to create as thin a deck as possible; this, as he wrote in 1933, led him to design a "very shallow and flexible truss, which not only resulted in far-reaching economy, but also effected a light and graceful appearance." Ammann expressed both a logical conclusion on economy and a personal vision on aesthetics in his preference for a very light span. He clearly identified his preferences: "Finally, whatever influence these various considerations may have had on the general conception of the design, the writer has admittedly been influenced by his personal conceptions and taste. He has always been an admirer of the early English suspension bridges with their general simple appearance, their flat catenary, light, graceful, suspended structure, and their plain, massive and, therefore, monumental towers."[42]

While his design for the bridge led to no difficulties, it did set an unfortunate precedent because of its neglect of the dynamic effect of wind loads. Driven by his aesthetic of thinness, Ammann, like Maillart, was entering into a new world of form that would define structural art in modern bridges. As such, there were risks. Ammann led the way and the profession followed. The Golden Gate, the Deer Isle, and the Bronx-Whitestone Bridges all followed Ammann's lead, and because they were all substantially lighter than the huge George Washington Bridge, they all suffered more motion than predicted. Finally, the incompleteness of the deflection theory became catastrophically clear with the collapse of the Tacoma Narrows Bridge in November 1940. Nevertheless, Ammann's Hudson River structure never had any such problems.[43]

Ammann's perspective was quite different from that of Lindenthal. Lindenthal's great bridge, in its original dimensions, would be wonderful. To Ammann, however, the substantive arguments and the political strength of the opponents deserved the same steely-eyed analysis that a good engineer devoted to understanding the stresses in bridge cables and the stability of the ground under proposed bridge towers. If influential opponents thought the bridge too large, or badly located, Ammann could draw on his engineering experience and perhaps find ways to modify the plan, rather than let it go down to defeat. On the subordinate issues of location, type of bridge, sponsorship, and financing, Ammann was a pragmatist. His main goal was to span the Hudson. Since the original design was not *his* plan, he had the psychological advantage in thinking about alternatives. Lindenthal had been working on a railroad bridge across the Hudson to the middle of Manhattan since 1888—for more than thirty years. And Ammann was a mere forty-two, nearly thirty years younger than his boss; possibly youth permitted greater flexibility.

Perhaps most important, however, Ammann—in contrast to Lindenthal—could encompass political obstacles, and strategies to overcome them, within the analytic framework of his engineering mind. Any good engineer knew, for example, that you had to design your bridge in relation to the character of the terrain where the towers would sit. Therefore, if preliminary studies suggested the tower footing would be solid rock, and closer exploration revealed softer ground, adjustments and even major redesigns would be necessary; and sometimes long weeks and months of arduous work would be needed to solve the problem and ensure that the tower and the bridge would hold. Moreover, bridge engineering was not an armchair activity; you had to go into the field continuously, marshal and motivate your workers, and modify your abstract designs as required by the land, the weather, and the impact of human mischance. Close exploration of the *political* terrain associated with any large project was essential; and the exploration might require meeting with local politicians, business people, and interested citizens in towns and county courthouses across the region in order to work through the proper combination of engineering, aesthetic, and political designs. This was not easy for Ammann, who was a modest man inclined to reticence.[44]

Ammann and Silzer had agreed that Ammann would prepare an extensive report on the Fort Lee project—covering technical engineering issues, traffic projections, financing questions, and probable impact of the bridge on regional development—and on December 17, 1923, Ammann's twenty-two page analysis reached Silzer's desk. That afternoon, the governor forwarded the report to the Port of New York Authority, with a letter from Ammann that concluded that the Fort Lee bridge could be paid for in twenty years. Silzer also released a public statement on his actions, noting that the Ammann plan was consistent with his own 1923 inaugural statement on the need for more Hudson crossings, and suggesting that the bi-state Port Authority could finance the Fort Lee bridge by issuing tax-exempt bonds—with "ample security" provided by tolls on the bridge. The governor's efforts for the day had not yet ended. He also wrote a private letter to Commissioner Julian Gregory at the Port Authority, suggesting that, in carrying out its studies of the various bridge plans, the Authority might want to secure the services of "such a man as Mr. Ammann, who is thoroughly skilled in this kind of work."[45]

The Port Authority's staff reviewed Ammann's report, and on December 21, the commissioners reported to the two governors that they would carry out a detailed study of the Fort Lee plan. Ammann's efforts were beginning to bear fruit. But the events thus far brought a measure of pain as well as pleasure. Most hurtful was the behavior of Lindenthal. Silzer had sent him a personal copy of Ammann's detailed report on the Fort Lee plan, and on December 20 the great engineer responded with a letter to the governor, alleging that Ammann had stolen Lindenthal's own ideas, and condemning his former associate for unethical behavior.[46] This unfair criticism did not affect Silzer's continued enthusiasm for Ammann's project.

The Port Authority would study his Fort Lee plan and would find, Ammann hoped, that his engineering design and his analyses of costs, traffic flows, and financing were sound. But he knew that the bi-state agency—created in 1921 to solve railroad problems, and staffed by railway engineers and statisticians—would be far more likely to take the next step and agree to *build* the great bridge if it found a groundswell of popular support for Ammann's 3500-foot span. Moreover, legislative approval and probably some initial state funding would be required to get the project underway. Here again, Trenton and Albany would be much more willing to commit their funds and the Port Authority's efforts to this project if local groups on both sides of the Hudson demanded action for a bridge they felt was sorely needed.

So Ammann once again threw his energies into the effort to organize public support for the Fort Lee enterprise. Between late December 1923 and April 1924, he held dozens of meetings with chambers of commerce and other groups in Bergen and nearby Passaic and Morris Counties in New Jersey. He wrote to and visited similar associations in the Bronx, Harlem, Washington Heights, Westchester, and Yonkers in New York State. He traveled into Connecticut and explained his arguments for the Fort Lee Bridge before the Civil Engineering Society of that state. By December 1923, Ammann had developed working sketches of the proposed bridge—with its thin, graceful roadway, and its great towers, which would be vast metal structures, sheathed in monumental stone.[47] And now, when he spoke, Ammann could show his audiences some visual hint of his own deep motivations, which lay beyond engineering technique, beyond matters of practicality (see fig. 18).

With increasing evidence of local support for the project, Ammann met with the state senator from Bergen County, William Mackay, and found him ready to press for legislation authorizing the Port Authority to construct a bridge at Fort Lee and smaller spans between Staten Island and New Jersey. During the spring of 1924, the New Jersey legislature took the first bite, endorsing the Port Authority study and construction of two Staten Island bridges, and New York State approved similar legislation. In late May, Governor Silzer sent a brief note to the Port Authority's general counsel, Julius Henry Cohen: "It has just occurred to me, in connection with the two bridges over Staten Island and your other bridge work, that the Port Authority ought to avail itself of the services of Mr. O. H. Ammann… I understand that just at the moment he is available."[48] "Just at the moment" was an exaggeration, since he had been unemployed for more than a year now. But still the Port Authority did not call.

Ammann met with a committee of North Jersey mayors that had been formed to press for action on the bridge, and he drafted a strategic plan that the committee could use in gathering further support. But the committee failed to mount the needed campaign. So Ammann once again found himself taking the lead, engaged in a series of meetings with local groups and state legislators.[49] Finally his political efforts produced significant results. In January 1925, the New Jersey Senate passed a bill authorizing the Port Authority to construct a bridge across the Hudson at Fort Lee, and the State Assembly soon followed suit. Ammann then crossed the Hudson to New York, where companion legislation had been introduced, and met with local and state officials to urge favorable action; and in late March, New York State approved the bill.[50]

The Port Authority now moved forward to construct a bridge from Fort Lee to 179th Street, and to build two spans that had been authorized between New Jersey and Staten Island. In a letter to the governor on March 27, 1925, Ammann expressed the hope that he would be asked to "take charge of the working out of the preliminary plans" for the Fort Lee bridge, but he thought there would be opposition, and that an engineer "with long practice and wide reputation" might be selected instead. Reviewing his many activities on behalf of the bridge project, Ammann concluded that he would appreciate "anything you may be able to do to help" achieve a favorable outcome.

Two weeks later, with the Port Authority bills now signed in Trenton, Governor Silzer once again wrote to Julian Gregory, who was now chairman at the Port Authority. Noting that the Authority would soon be proceeding with the Fort Lee bridge, Silzer suggested that "you take into consideration for the doing of this work the name of O. H. Ammann." Silzer continued: "Mr. Ammann was one of the pioneers in this project, has spent two years of his time in advising the public of its advantages, has drawn freely upon his own ability as engineer, and in every way has probably done more than any other one man to bring this bridge into being."[51] Silzer also sent a copy of this letter to Chief Engineer William Drinker at the Port Authority; and he sent a copy to Ammann, with a note: "I have it in the back of my mind somewhere that Mr. Drinker had an impression that you were an able assistant, but that you had not had the experience to independently undertake work of this kind." The governor suggested Ammann talk with Drinker about this impression.

A few days later, Ammann met with Drinker. He thought it was an "encouraging interview," though it contained disappointing news: Drinker told him that the Port Authority had concluded that their first projects—the two Staten Island Bridges—should not be awarded to Ammann but rather to "an engineer of long established reputation."[52] The job went to Waddell & Hardesty, a firm led by the prominent engineer J. A. L. Waddell.

In its short and thus far uneventful life, however, the Port Authority had already begun to develop a few traditions. One of these was a preference for hiring its own engineers and other experts as regular members of the staff, rather than relying heavily on outside consultants. By late April, the chief engineer (himself a railroad man) had concluded that the Authority ought to hire a staff engineer with bridge-building experience, and—Chairman Gregory wrote to Governor Silzer—Drinker had recommended Othmar Ammann for the post. The commissioners soon concurred with Drinker's recommendation, and on July 3, 1925, Ammann sent a letter to Silzer noting that he had assumed his duties as "bridge engineer on the Port Authority staff" on July 1, and thanking the governor for his "goodwill and efforts on my behalf." The long and active campaign ended at last. Ammann had finally landed a job; he would now have to show that he had the capacity not only to fight for—but also to build—a great bridge.

At midyear 1925 Othmar Ammann began a six-year epoch unmatched by any bridge engineer in history for its dramatic sweep of planning, design, and construction of astoundingly original and gargantuan structures. Before July 1, 1925, the forty-five-year-old engineer had never been in charge of designing any

Figure 19
Ammann surveying the construction of the George Washington Bridge from the cliffs on Washington Heights.

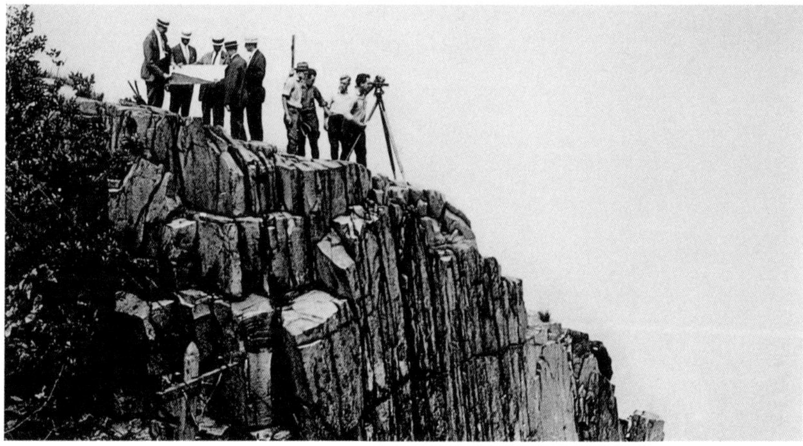

Figure 20
Construction of the steel towers of the George Washington Bridge.

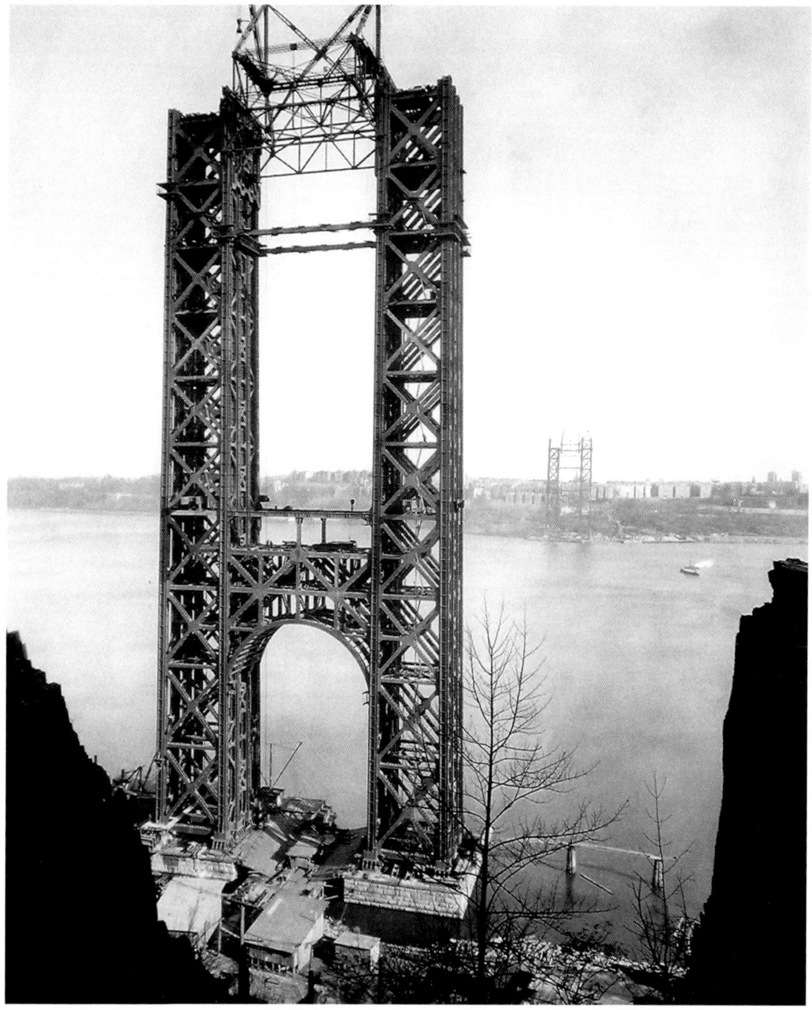

structure at all; by October 1931 he would be acclaimed the greatest bridge builder alive. Yet by 1925, his education was the finest of anyone who ever designed steel bridges. Not only did he go through the last classes of Wilhelm Ritter in Zurich, but also he went through the last bridges of Gustave Lindenthal in New York. In retrospect, these were the best pair of teachers he could have found anywhere. His apprenticeship had been lengthy, but it placed him at precisely the right place at precisely the right time to enter the profession as a leader. He "entered into a sequence" of bridge events that for steel bridges would never occur again, and by the 1930s he would stand as one of the masters of bridge design—joining Roebling, Eiffel, and Maillart. As Kubler had put it, the gifted artist needed in addition to fortunate timing also an "allotment of its temperament."[53] Ammann, like Maillart, had the steady, focused temperament with natural self-confidence but little bravado.

With the Port Authority now committed to his dream, the engineer poured his energies into the tasks before him, completing the detailed design for the Hudson River bridge, organizing the staff and consultants to carry out that large effort, and at the same time supervising the constructions of two smaller bridges—the Goethals and Outerbridge—authorized by the two states in 1924. In 1925, the two states approved a third Staten Island span, from Bayonne, and so Ammann, while building the longest suspension bridge in the world at Fort Lee, also designed what would turn out to be the longest arch bridge in the world, for the Bayonne crossing. While resisting political pressures that threatened to slow progress and to undermine his engineering decisions, Ammann used his considerable organizing talents to press ahead with this array of projects. The Goethals and Outerbridge were completed in 1928.

Of all these works, the Hudson River crossing was Ammann's major concern. He was constantly on the site, beginning with the survey (fig. 19). The construction proceeded rapidly: first the foundations; then the steel towers (fig. 20), the cables, and finally the extraordinarily thin deck (fig. 21). The expected date of completion was mid-1932, but the bridge was finished eight months ahead of schedule at a cost below the 1925 estimate and opened in October 1931. In 1930, Ammann had returned to his alma mater in triumph, where he received an honorary degree at the celebration of the seventy-fifth year for the ETH. He had left in 1904 an unknown apprentice and returned to receive the accolades of his country and its great engineering school.[54]

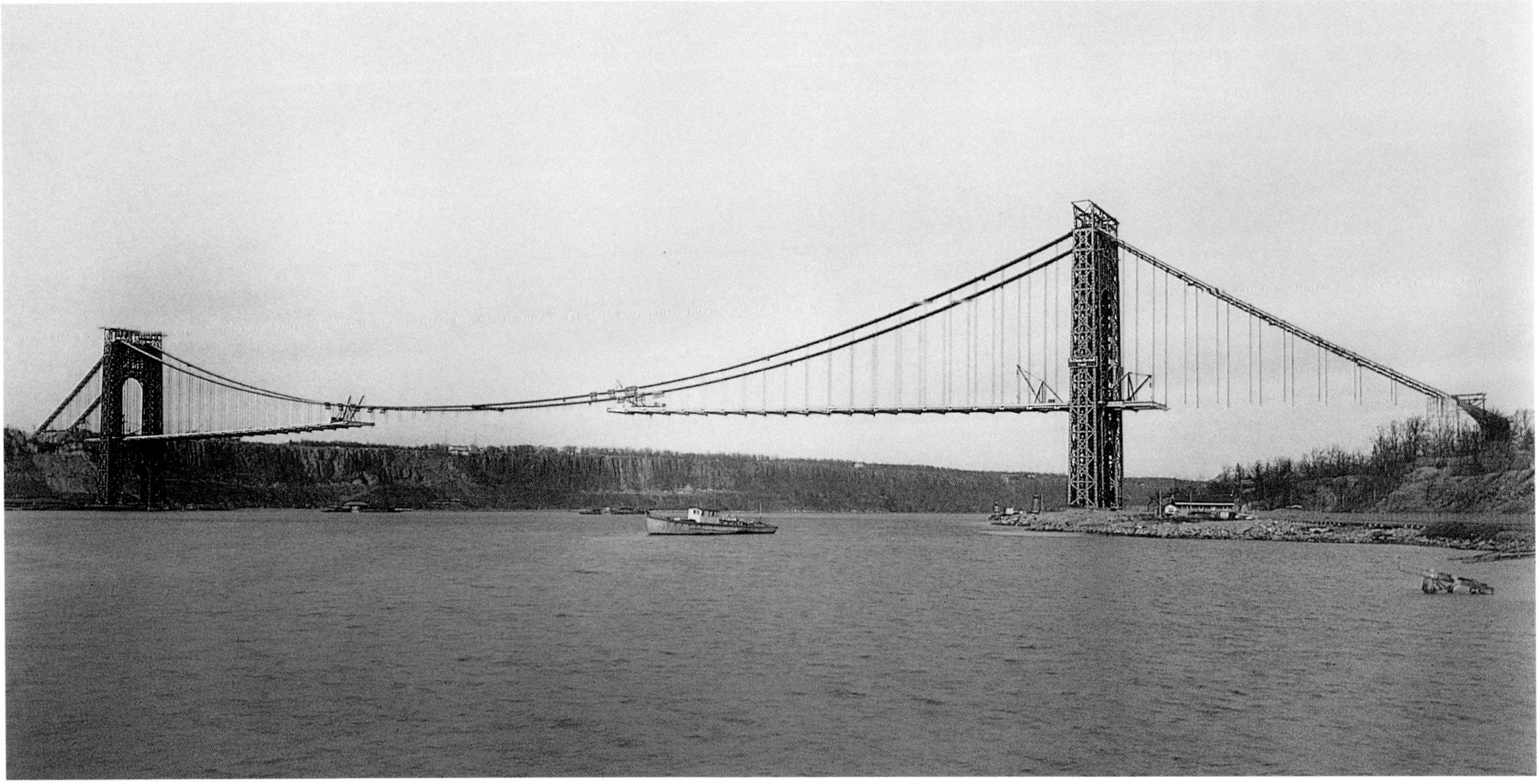

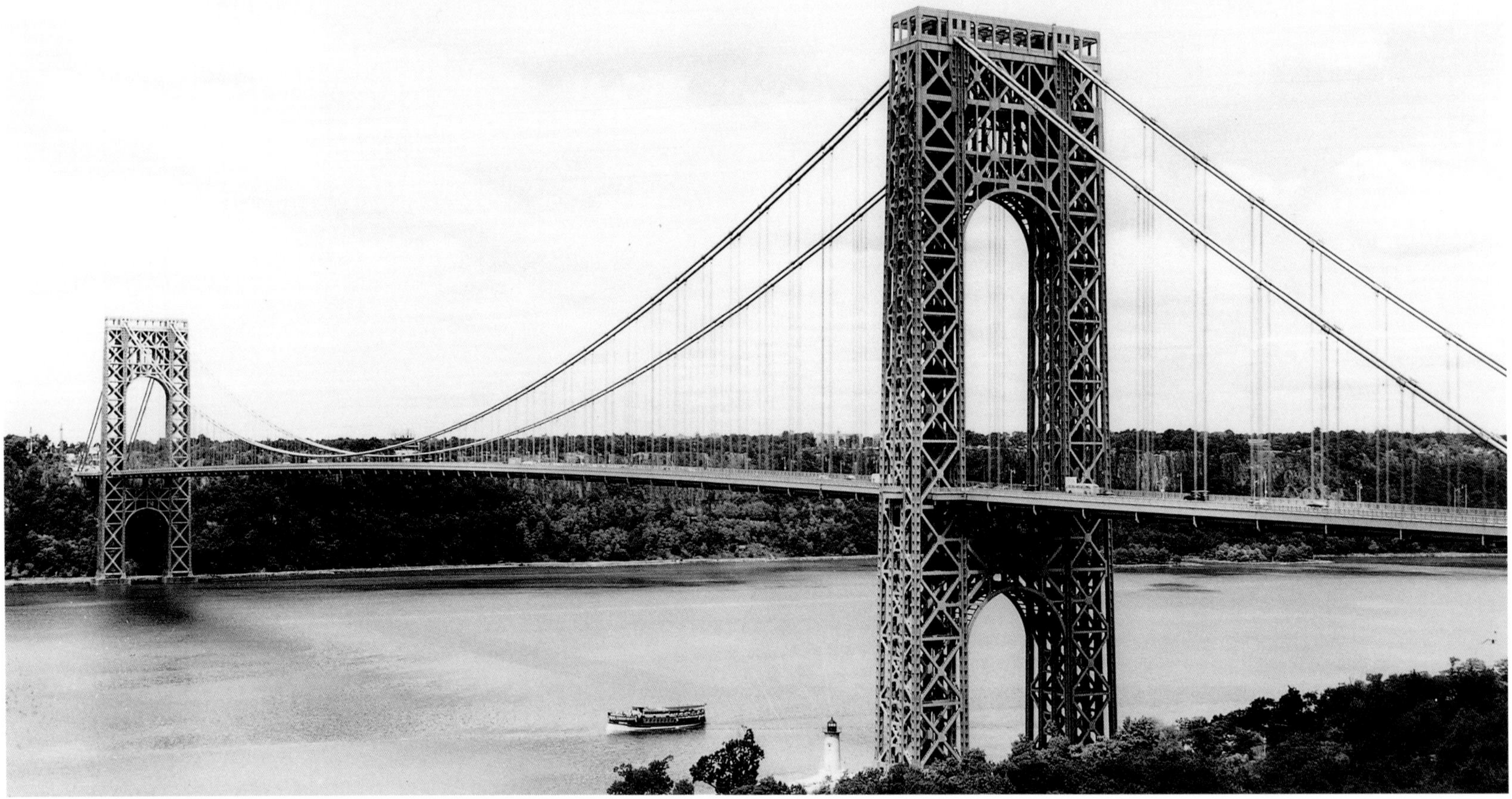

Figure 21
Construction of the George Washington Bridge's single deck. This thinness was made possible by Ammann's application of Ritter's ideas and Melan's deflection theory.

Figure 22
The George Washington Bridge, which opened in October 1931, ahead of schedule and below cost estimates.

In 1936, the famous Swiss-born architect Le Corbusier came to New York and wrote about the visit in his book *When the Cathedrals Were White*. Earlier, in 1923, he had produced his most famous book, *Towards a New Architecture*, in which he had singled out Gustave Eiffel's Garabit Viaduct as the leading image of spare elegance for his chapter "The Engineer's Aesthetic." There he stated that the engineer "achieves harmony" whereas the architect "realizes… a pure creation of his spirit… it is then that we experience the sense of beauty."[55] During his New York visit in the mid-1930s, Le Corbusier came to a deeper understanding of the engineering profession, which he expressed dramatically in a chapter entitled "A Place of Radiant Grace":

> The George Washington Bridge over the Hudson is the most beautiful bridge in the world. Made of cables and steel beams, it gleams in the sky like a reversed arch. It is blessed. It is the only seat of grace in the disordered city. It is painted an aluminum color and, between water and sky, you see nothing but the bent cord supported by two steel towers. When your car moves up the ramp the two towers rise so that it brings you happiness; their structure is so pure, so resolute, so regular that here, finally, steel architecture seems to laugh. The car reaches an unexpectedly wide apron; the second tower is very far away; innumerable vertical cables, gleaming against the sky, are suspended from the magisterial curve which swings down and then up. The rose-colored towers of New York appear, a vision whose harshness is mitigated by distance.[56]

This is not steel architecture. It is structural art of the engineer, and it has since been recognized by the profession and the general public (fig. 22). But Le Corbusier responded to Ammann's great bridge in the light of Eiffel and of John Roebling, whose Brooklyn Bridge he described in his next chapter. Not only did Ammann more than double Roebling's span, but he would also design an arch three times the span of Eiffel's longest.

On November 14, 1931, Othmar Ammann stood before a modest crowd (fig. 23) to speak at the opening of the new Kill van Kull Bridge connection from Bayonne, New Jersey, to Staten Island, New York.[57] This would be the last of four interstate bridges that the Port Authority would build, and all were carried out under the direction of Ammann. His speech, typically short and modest, took place on the New Jersey shore and was followed by the reading of a cablegram from the Lord Mayor of Sydney, Australia, which read: "You win by five feet. On behalf of the citizens of Sydney, extend hearty greetings, felicitations and good wishes and hope the Bayonne Bridge will further enhance the position Bayonne occupies in the family of cities."[58] The Lord Mayor's concession recognized that the Sydney Harbor Bridge soon to be completed would not hold the world's record, now securely held by Ammann's arch at 1652 feet and 1 inch. In a progress report of 1930, Ammann had listed the Sydney Bridge with a span of 1650 feet, so the Lord Mayor seems to have been a bit generous, but perhaps his implied comparison of Bayonne to Sydney was less so.[59] The race for the longest span does tell us something about the so-called rationality of engineers. There was no technical reason to make the George Washington Bridge exactly twice the span of the previous suspension-bridge record holder (the Ben Franklin at 1750 ft.), nor could any engineering reason be found for providing the Bayonne with a span 2 feet and 1 inch greater than that at Sydney.

Indeed Ammann had studied several designs of suspension bridges for the Bayonne span but abandoned them with technical arguments that are not fully convincing.[60] Why build another 1600-foot suspension bridge when you can build the longest span arch in the world? Later on, he would surpass the Golden Gate Bridge span by a mere 60 feet for his Verrazano structure. These motivations do not detract from the achievements; they only make their exact numbers seem more human. Apart from this harmless competition for span, the Bayonne Bridge was a remarkable work for its efficiency of materials, economy of construction, and elegance. Ammann stressed this last requirement for successful bridges in his remarks: "Finally, in this as well as all of its projects, the Port Authority has recognized the fact that its structures must not only be useful, but they must also conform to the aesthetic sense. This was one of the motives for the selection of an arch spanning the entire river in one sweeping graceful curve." He followed this expression with one further aesthetic comment, showing his debt to Lindenthal, and indeed, his own sense of monumentality: "The huge abutments of the arch which are yet exposed in their crude construction are eventually to be marked by massive pilons, and will thus further enhance the appearance of the structure in its setting in the landscape."[61]

The renderings for the bridge (fig. 24) all showed heavy stone blocks at either abutment, which, like the masonry facing for the towers of the George Washington Bridge, have no purpose except decorative. Like Maillart with his Romanesque approaches to the bridges of the 1920s, Ammann was shedding the imitative forms of the past slowly, and with his last two major bridges, all such excretions would disappear. Even so, the Bayonne blocks are far less obtrusive than the useless towers on the Hell Gate or on the Sydney Bridge. Today, the legacy of the Bayonne plan is a light steel framing over the abutments (fig. 25) that does not harm the distant profile view but adds some minor visual disturbance close up. Of all the steel arches on that scale (there is only

Figure 23
Ammann at the opening of the Kill van Kull Bridge, which connected Bayonne, New Jersey, to Staten Island, New York, November 14, 1931.

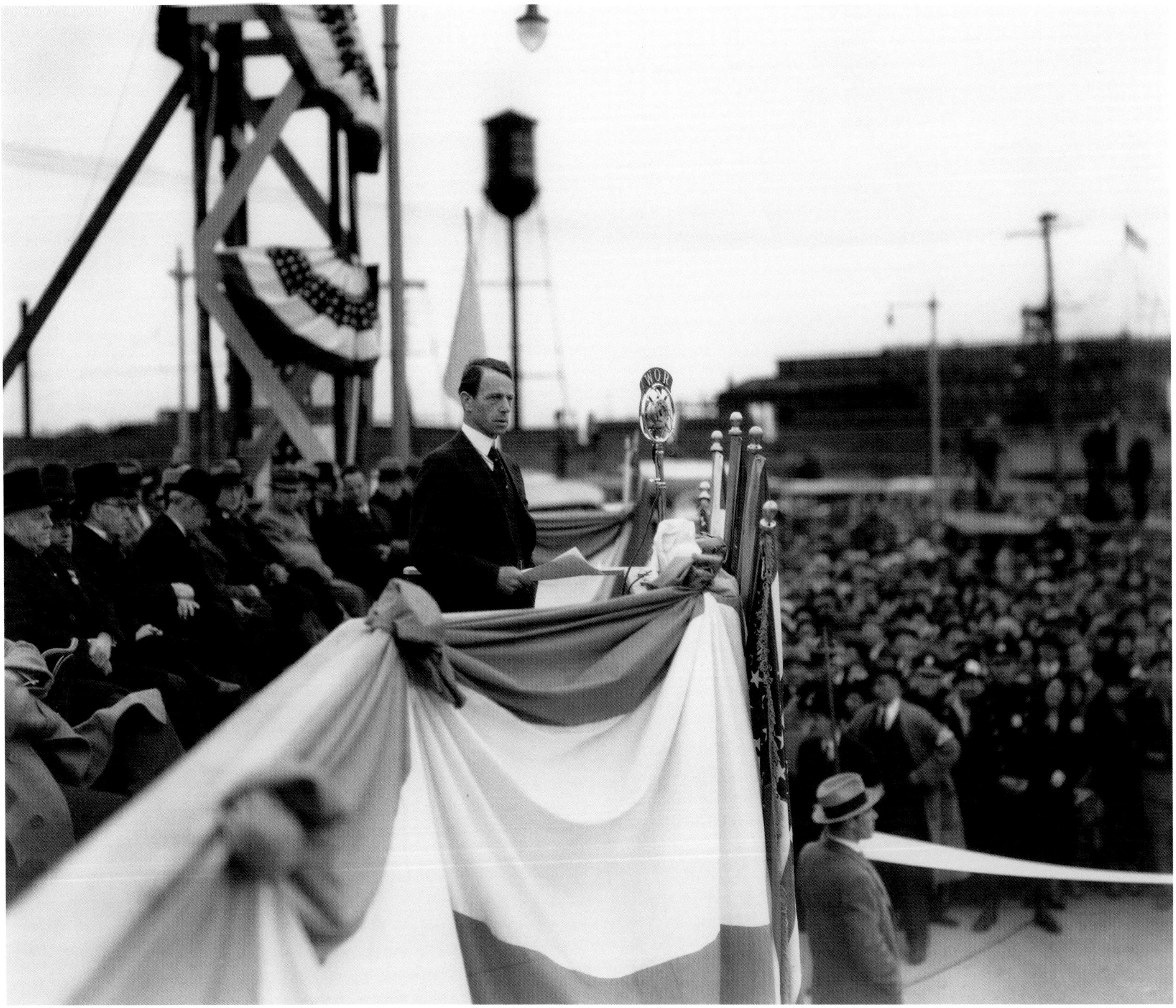

Figure 24
Rendering showing the decorative, heavy stone abutments of the Bayonne Bridge, from the Port of New York Authority Progress Report, 1930.

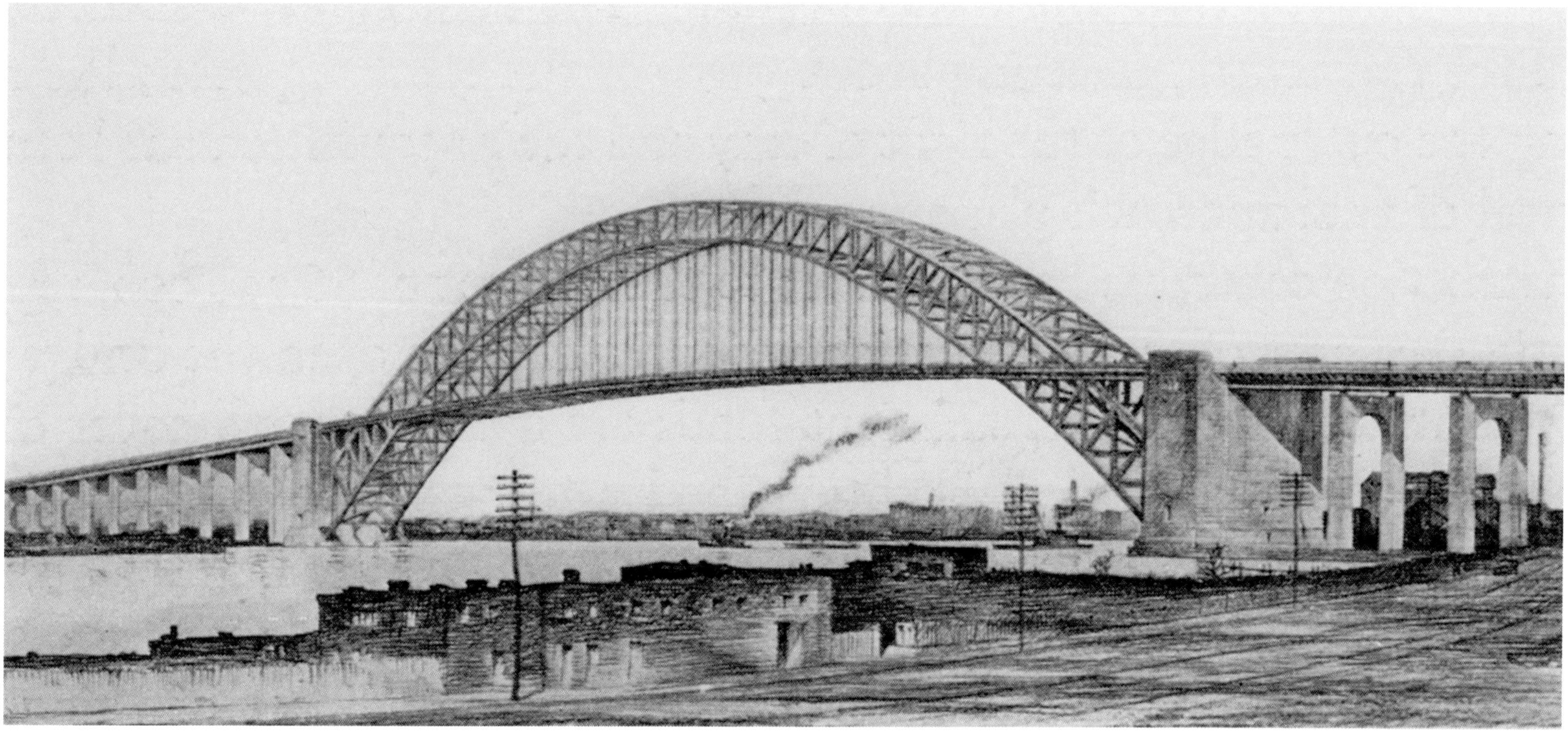

one other besides Sydney and that is over the New River Gorge, with a 1700-foot span, completed in 1974), Bayonne is the most elegant.

When the American Institute of Steel Construction inaugurated an award in 1931 for the most beautiful bridge, it chose Bayonne and gave second place to the George Washington. The architect on the jury rejected the Hudson River crossing because "the setting of the Washington Bridge with the lofty Palisades on one side and the low Manhattan shore on the other did not call for symmetrical treatment and might be filled better by an unsymmetrical structure." A good cartoonist might have fun imagining how an unsymmetrical suspension bridge with that span would look. The architect, Raymond Hood, thought one tower should have been higher than the other; Ammann's response was simply that such a structure would be a monstrosity.[62]

Although the aesthetics of the two great bridges completed in 1931 were judged of similar merit, the way each came into being was radically different. Ammann had much less to do with getting the Port Authority to take on the Staten Island bridges. The local people agitated for three bridges (the Outerbridge, the Goethals, and the Bayonne) in 1924, but not until 1925 had both state legislatures authorized the Port Authority to design and build the Bayonne Bridge. In April 1926 Ammann began to study the project, he reported on its feasibility in early 1927 and settled on the arch design.[63] Ammann and his engineers began the design in earnest in 1927 and by June 1, 1928, produced the complete plans. The first construction contract was awarded in July, ground breaking took place on September 18, 1928, and the foundation work ended in May 1929. During the design stage, Ammann decided that a scale model, nine feet in span, should be built and load tested to compare with mathematical analyses. He was especially concerned about the lateral bracing under transverse loading since the ratio of span to width is substantially greater than the Hell Gate (22.5 compared to 16.3).[64] The construction of the Bayonne required considerable design skill because of the many different structures temporarily created from the arch during its partial forming (fig. 26). The bridge was completed well before its scheduled date of 1932 and, like the Hudson River span, for less cost than the estimate.[65]

Figure 25
Contemporary view of the Bayonne Bridge at sunset, selected in 1931 by the American Institute of Steel Construction as the most beautiful bridge for that year.

Figure 26
Construction of the Bayonne Bridge, creating several temporary structures.

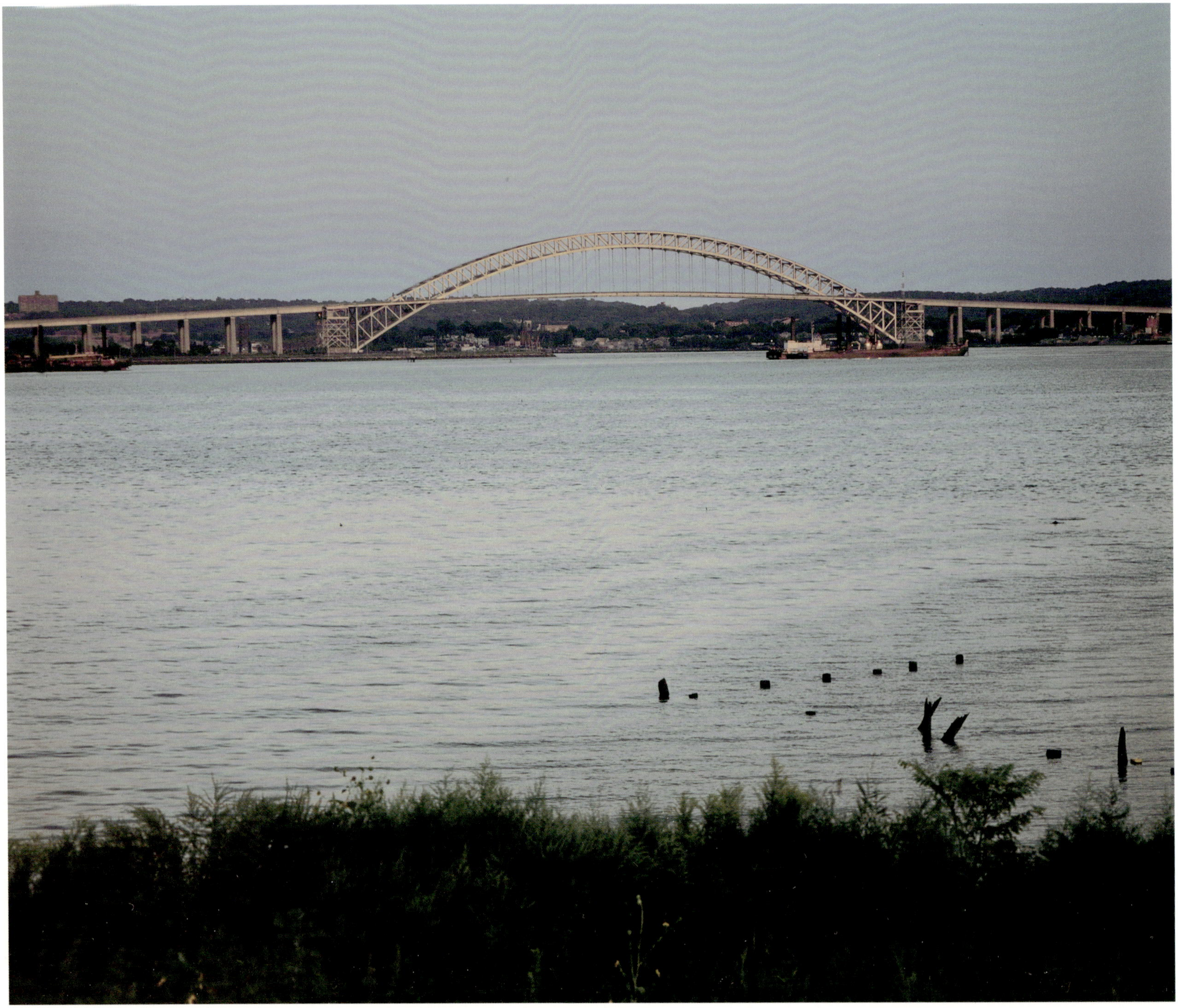

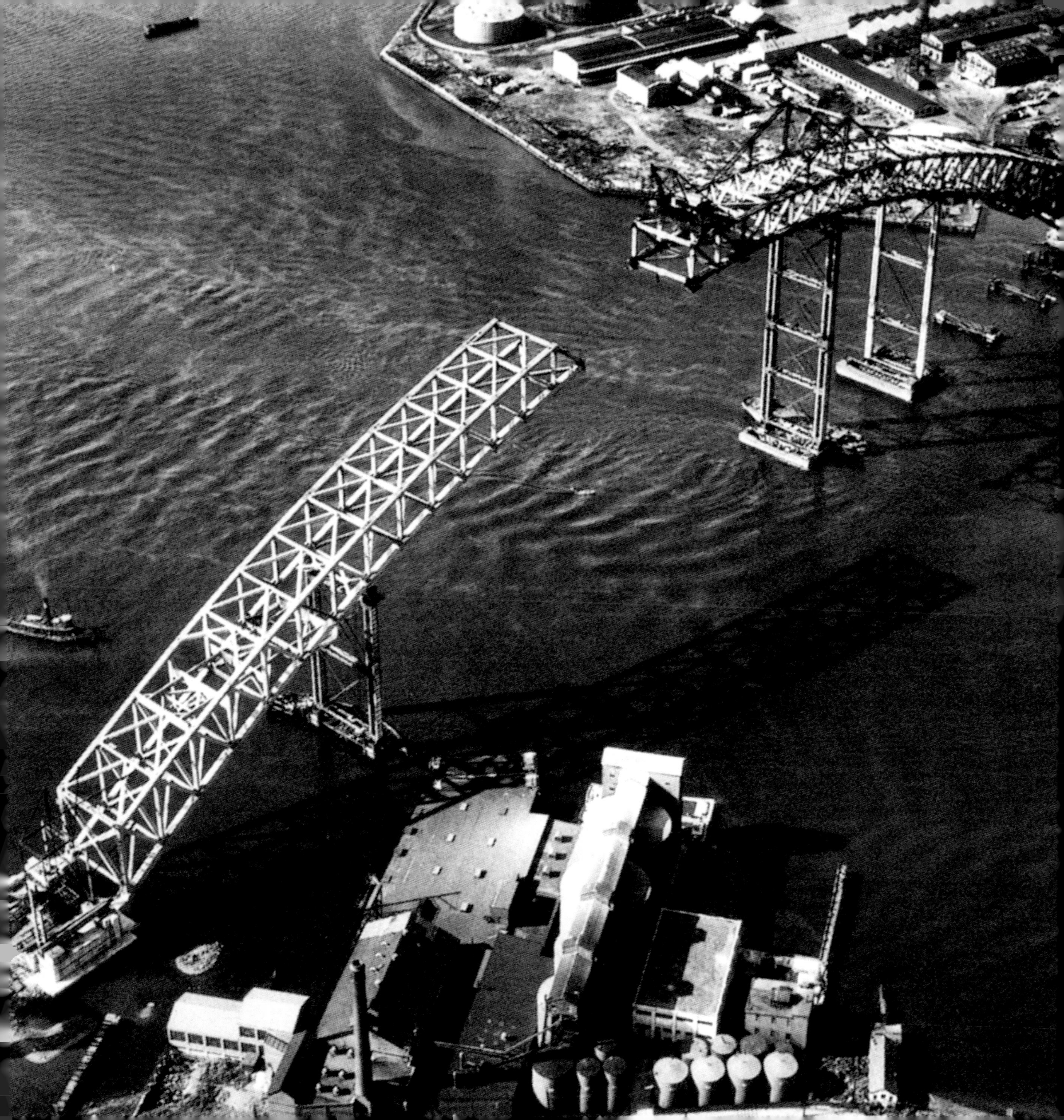

Figure 27
Profile of the Bronx-Whitestone Bridge, opened in 1939.

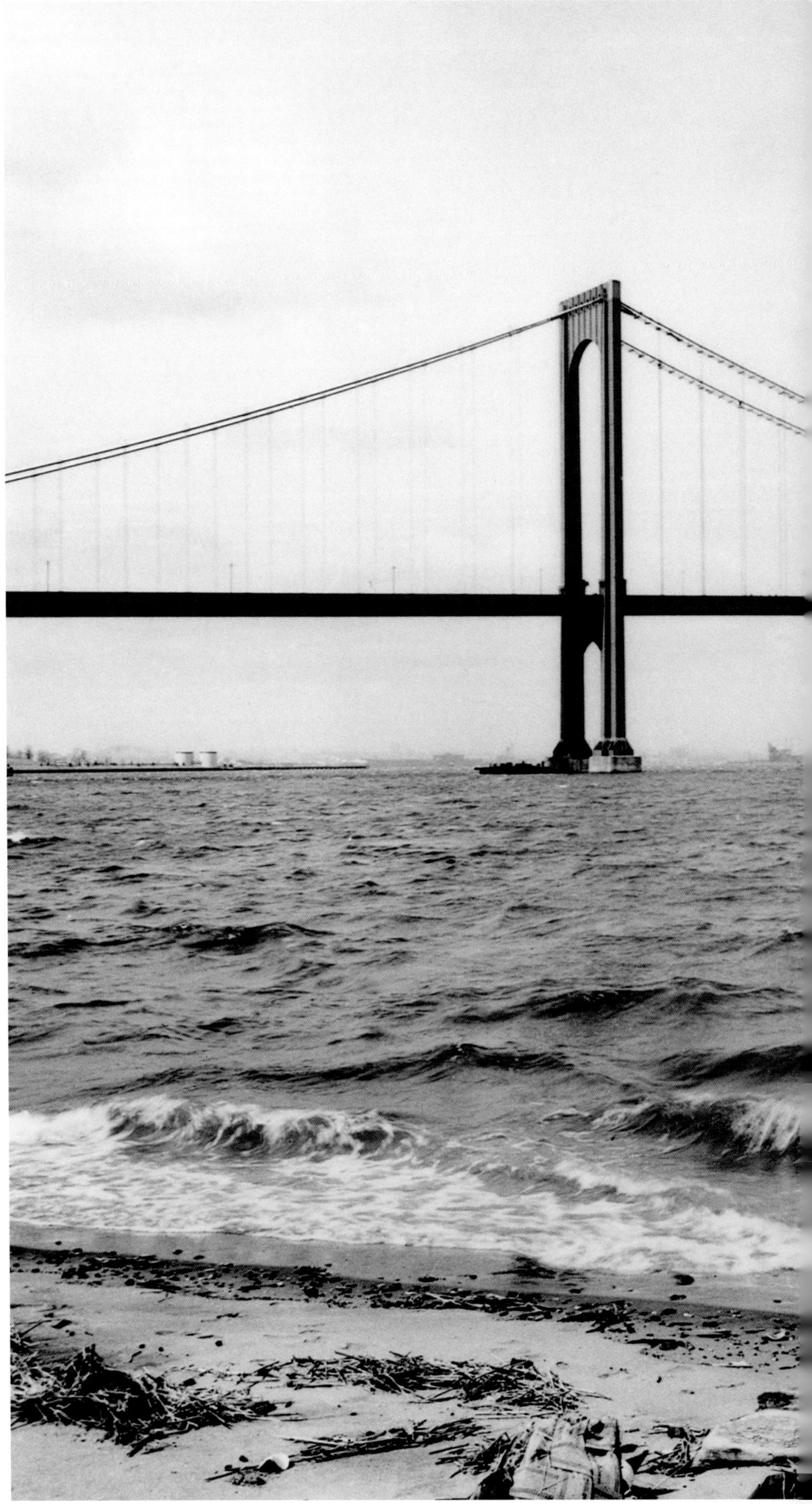

By 1931, Ammann had risen from bridge engineer at the Port of New York Authority to the agency's chief engineer. He was in charge of construction for the Lincoln Tunnel's first tube, which emerged from under the Hudson River at 39th Street in Manhattan; that project was begun in 1932 and completed in 1937. Between 1929 and 1937, he also served on the board of engineers that redesigned the Golden Gate Bridge. And in 1934, when the Triborough Bridge Authority was created and Robert Moses named as its head, Ammann served as that agency's chief engineer, directing the planning and construction of the Triborough Bridge and then the Bronx-Whitestone span. Moses had enlisted Ammann to clean up the managerial and design chaos of the Triborough Bridge. This bridge does not reflect Ammann's design ideals; the towers, having been devised by his architectural consultant, were especially foreign to him, and its decoration marks it as someone else's visual concept.[66]

In 1935, Moses presented Ammann with a new task: to connect the boroughs of Bronx and Queens with a bridge that would provide a link to the World's Fair planned for Flushing Meadows. The new project, the Bronx-Whitestone Bridge, gave Ammann the most powerful visual challenge of his career. Here was a site where the entire bridge from anchor to towers to anchor would be fully visible from nearly every viewpoint (fig. 27). It was to be the third longest spanning bridge in the world (when the designing phase began, only the George Washington Bridge had a longer, completed span). Ammann had strong opinions regarding the new design and, as his consulting architect, Aymar Embury, stated: "We were in a sense his instruments, and were guided by his desire as to the lines along which we should proceed."[67]

Ammann's ideals came directly from his experience with the Hudson River crossing, which led him to three basic visual goals: the most slender deck possible; the strong but spare steel towers, and anchorages expressive of their function. The earlier bridge without any deck stiffening was performing excellently by 1938, but the uncovered steel towers had a look of complexity that argued for a far simpler steel design. There would never again be a proposal for a major suspension bridge tower to be covered over by masonry. Embury described his chief's desires: "Mr. Ammann indicated his desire for a design of extreme simplicity, in which plate girders would take the place of trusses in stiffening the roadway, in which no lattice bracing would be used (steel webs being substituted), and in which no wind bracing would be required between the legs of the towers near the roadway level, the sole connection being at the top, with the bracing in the form of an arch."[68]

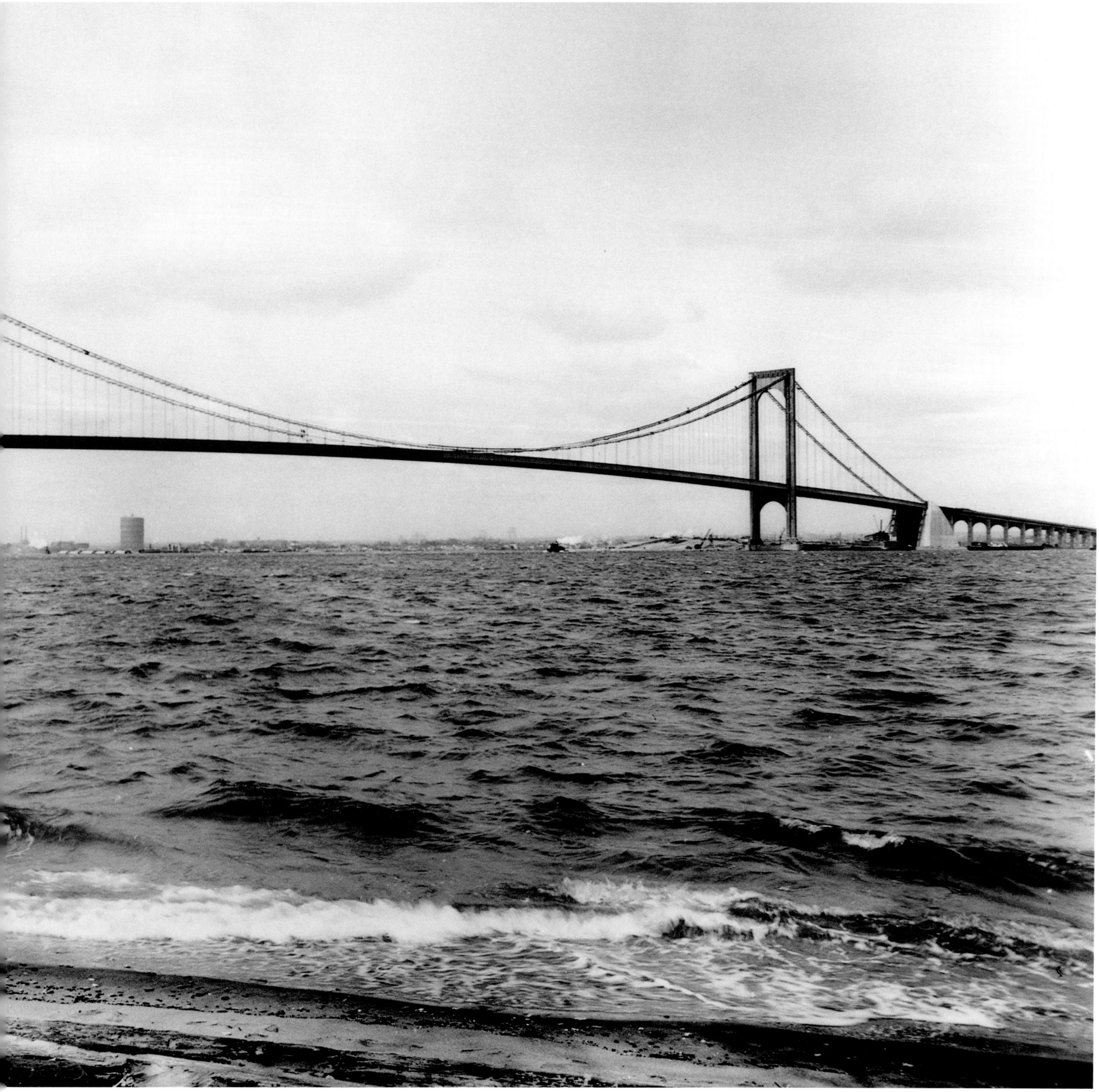

Figure 28

Anchorage of the Bronx-Whitestone Bridge, reflecting the cables' horizontal pull, vertical pull, and changing slope.

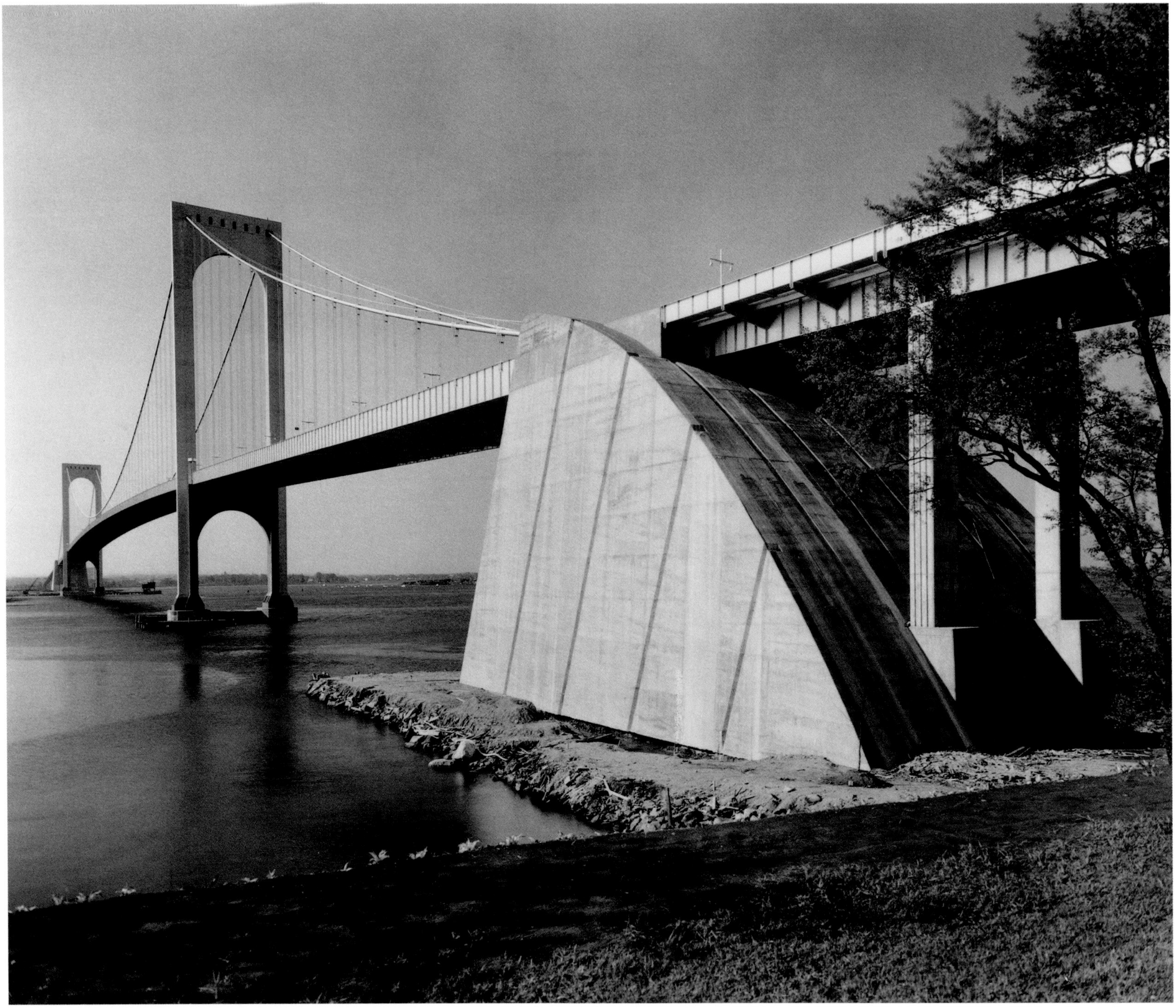

Figure 29
Throgs Neck Bridge, completed in 1961.

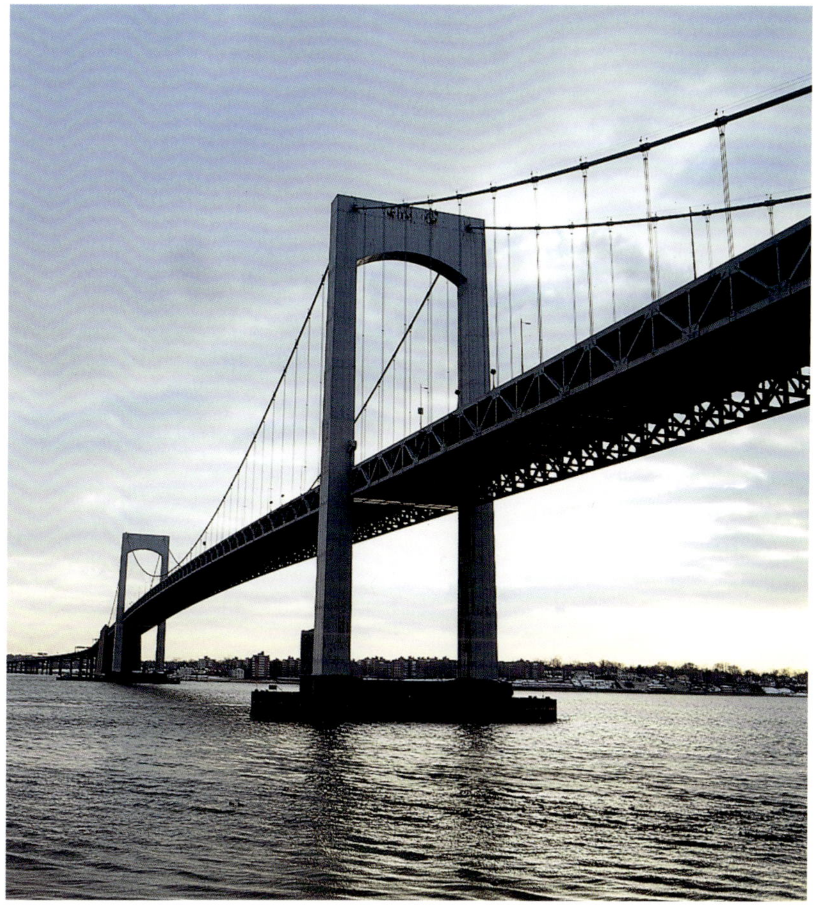

The anchorages presented Ammann with a new problem, which he directed Embury to study along the lines of structural expression rather than the decorated masonry blocks that Cass Gilbert had proposed for the George Washington design or especially those actually built for the Philadelphia-Camden Bridge (Ben Franklin, 1926). There the architect Paul P. Cret used huge blocks of ashlar as facing and the "structure, no matter how one feels about it as a separate entity, is a failure in that it appears trivial compared to the terrific scale of the towers and steelwork... It might be a magnificent warehouse or a superb old fortress, but it certainly is not an anchorage."[69] What Ammann and Embury intended to create was an anchorage that reflected three separate structural aspects: the horizontal pull of the cables inward toward the towers, the vertical upward pull of the cables, and finally the change in slope of the cables within the anchorage to bring them to the foundations without extending far out toward the shore (see fig. 28). As Embury stated, "We hope that its very boldness and stark simplicity will mark it as belonging to a bridge and to nothing else in the world."[70] The entire bridge represented a bold adventure in the stark simplicity that made this bridge the most acclaimed suspension bridge of structural art. Nevertheless, the bridge did exhibit some of the same wind-induced motion as the Tacoma span and has been modified several times to reduce that effect. The Bronx-Whitestone Bridge was held up as the greatest example of structural art in long-span bridges for students at the ETH in the late 1940s.[71] But Ammann's best-known work lay in the future at the gateway to New York harbor.

In 1939, Ammann retired from the Port Authority and opened his own consulting firm. During the next several years, he was involved in the design of the Delaware Memorial Bridge, in the investigation of the collapse of the Tacoma Narrows Bridge, and in other projects, including a detailed study of the Brooklyn Bridge.[72] In 1946, he formed a partnership with Charles S. Whitney, a specialist in concrete structures, and the new firm, Ammann & Whitney, worked on the Walt Whitman Bridge (between Philadelphia and New Jersey) and a range of other engineering works. In 1955, when the Port Authority and Robert Moses joined forces to propose a series of arterial highways and bridges in the New York region, Ammann agreed to take a central role—designing the Verrazano Narrows and Throgs Neck bridges (fig. 29), as well as a second deck for the George Washington Bridge. The Narrows Bridge set a new standard for long-span bridges, exceeding the George Washington in span length by 760 feet (fig. 30). On February 21, 1979, the Swiss confederation issued a stamp commemorating the centennial of

Figure 30

The Verrazano Narrows Bridge, completed in 1964, set a record for span length at 4260 feet.

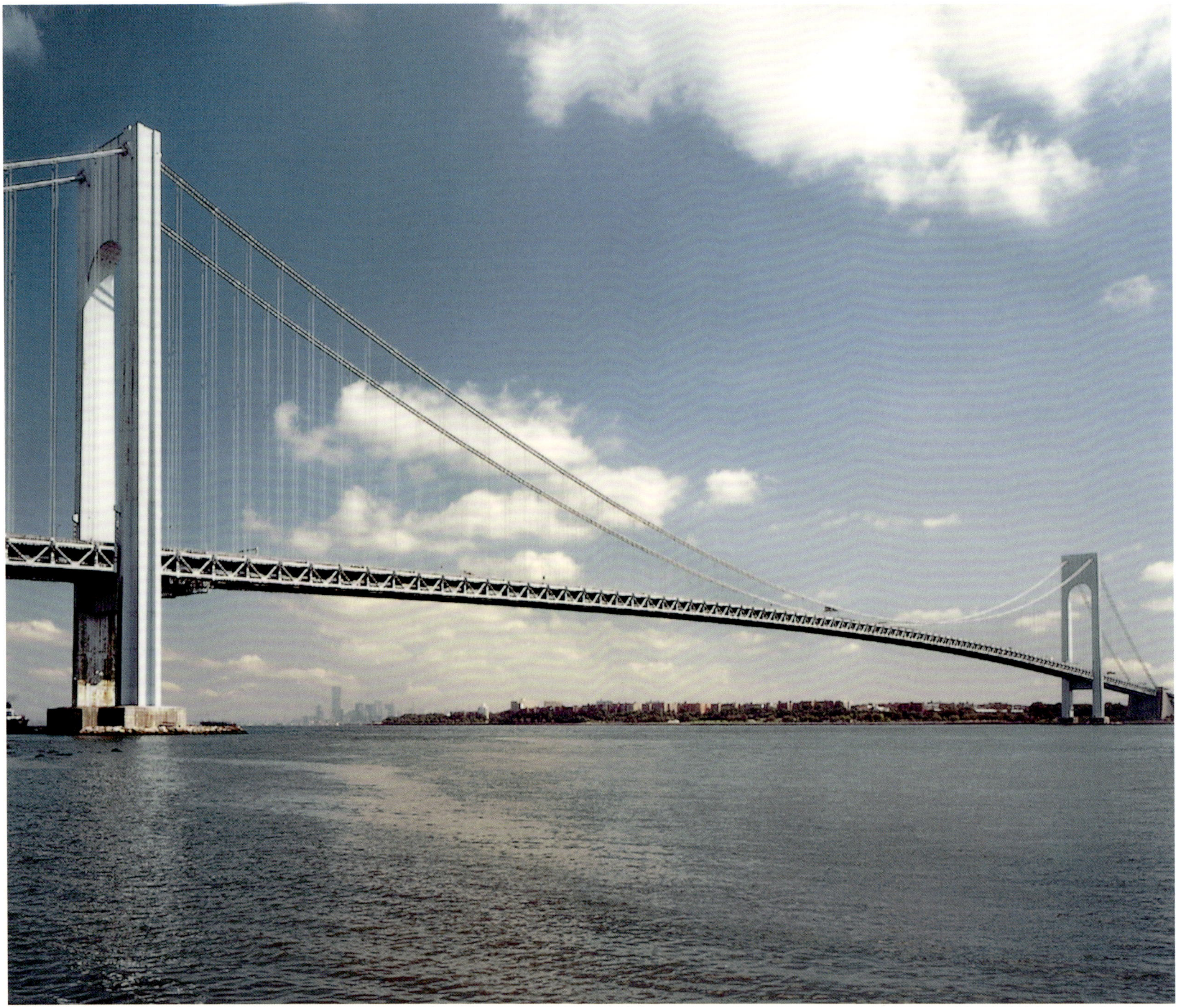

Figure 31
First-day cover of the stamp issued by the Swiss confederation to commemorate the centennial of Ammann's birth.

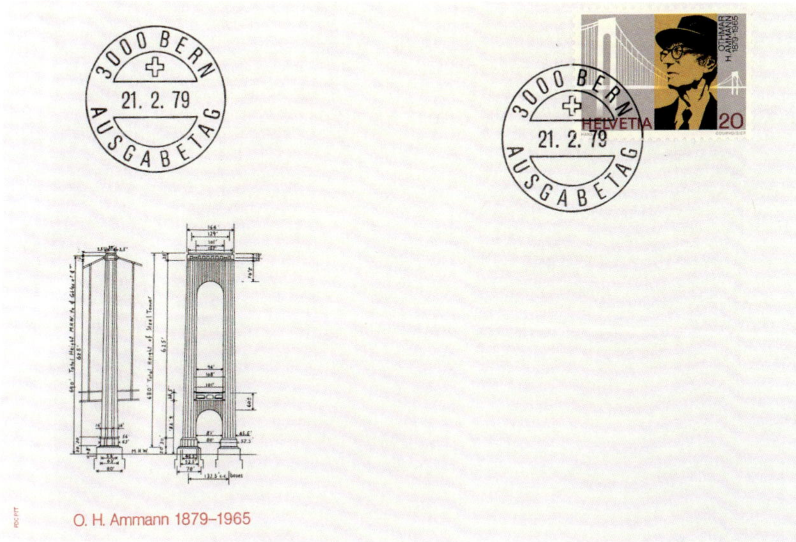

Figure 32
Tower of the Verrazano Narrows Bridge, a two-legged frame rising almost 700 feet.

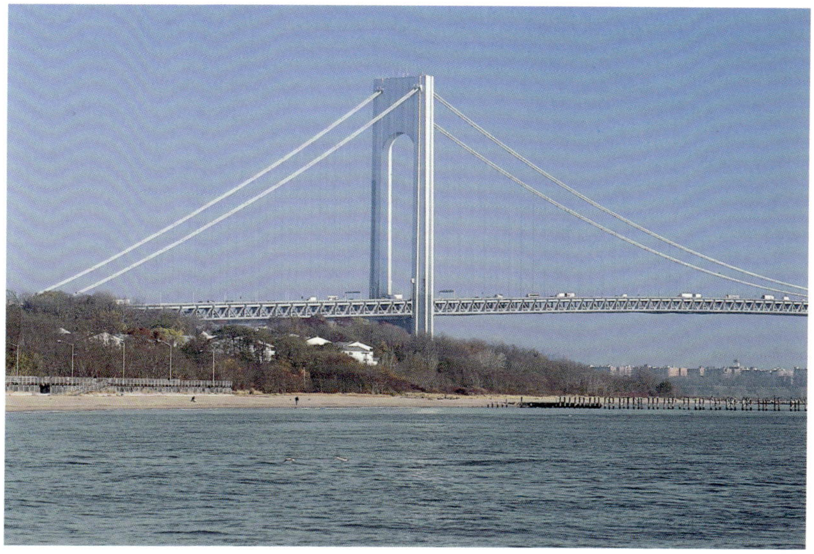

Ammann's birth (more than a month early, since he was born on March 29, 1879). The stamp had a photo of Ammann taken late in life with a drawing of the Verrazano Bridge behind, and the first-day cover showed drawings for one of the great towers (fig. 31). Completed in 1964, the span of 4260 feet exceeded the Golden Gate by a mere 60 feet to regain for Ammann the record span title.

On March 13, 1963, Ammann presented a paper to the New York Academy of Sciences entitled "Planning and Design of the Verrazano-Narrows Bridge." He first stated that such a project "is not a scientific subject" even though an essential aspect of science is required in the sense of "forces and stresses, knowledge of the behavior of materials, and soil mechanics."[73] The completed bridge represents a summary of Ammann's education to guide future designers. It was to be the last major New York bridge and part of a comprehensive study begun in 1954 for metropolitan traffic facilities. That study led to those major projects: the Throgs Neck Bridge near the Bronx-Whitestone structure, the second deck of the George Washington Bridge, and the Narrows Bridge. All three designs were in Ammann's hands, with the first bridge completed in 1961, the second in 1962, and the big bridge in 1964.

There was little discussion of the form for the last bridge; no type other than a suspension bridge would be considered for the 4260-foot main span and the two 1215-foot side spans. Ammann oversaw each of the main visual features so that they conformed to his aesthetic ideals: thin deck, strong towers, and expressive anchorages. Ammann had to design the deck in the light of the failed Tacoma Narrows Bridge of 1940 when modest wind speeds caused that bridge to vibrate up and down and finally twist itself to destruction. Ammann's solution consisted of a rigidly connected trussed box about 25 feet deep and 100 feet wide. The top supported six traffic lanes and the bottom, within the box, also carried six lanes. Thus the ratio of span to depth is 170, bolder than the George Washington with two decks of 120 and of the same ratio as the Golden Gate and Deer Isle, neither of which uses a trussed box and both of which were found to be unstable in wind and have been considerably stiffened since their completion in the 1930s. Ammann thus achieved a visually slender deck but with none of the wind-induced difficulties characteristic of pre-Tacoma structures.

Ammann succeeded also with the towers, which show his desire to avoid decoration and fussy forms. The single deep cross-brace at the tower top contrasts starkly with the facaded multibraced towers at Golden Gate. Ammann's towers follow directly from those at Bronx-Whitestone and Throgs Neck, and the slight increase in width at the base reflects the higher loads there due to tower weight. There is a second bracing member that Ammann placed below the roadway so that the image of each tower is one of a single two-legged frame rising almost 700 feet from the concrete pedestals (fig. 32). A set of vertical flutes serves to increase visually the sense of height without impairing the spare elegance of the single tower form.

Figure 33
Edward Steichen, American 1879-1973. *George Washington Bridge*. Gelatin silver print. Museum of the City of New York, gift of Sidney Singer, 92.6.1.

Finally, the anchorages swallow the four cables with their concrete bulk, which is sloped into a triangular form that reflects the studies for the Bronx-Whitestone. Again, the sloping top of the anchorages emphasizes the incline of the cables. Ammann described it this way: "As in the case of the towers, aesthetic considerations were essential in view of the prominent exposure of the anchorage blocks which reach a height of a ten-story building. A number of studies were made in collaboration with the board of architects. The finally developed design is believed to meet the aesthetic requirements very satisfactorily. The triangular shape of the major part of the concrete block is simple in design and expresses clearly the function of the anchorage in resisting the inclined cable pull."[74] Although there was a board of architects, it is clear that Ammann directed the design, as Embury had said, so that the entire project expressed his aesthetic sense as a structural artist. As he put it in the published article (1963): "This span and clearance arrangement… resulted in a well-balanced design of light and graceful appearance, in harmony with the magnificent landscape. The bridge will form a monumental portal at the entrance to New York Harbor."[75]

The Verrazano Narrows Bridge is Ammann's climax. It demonstrates the aesthetic vision that grew from his studies with Ritter, expanded through the experiences with Lindenthal, and matured in the George Washington, Bayonne, and Bronx-Whitestone structures. But in all those earlier bridges, great exemplars of structural art as they are, there were either technical or aesthetic flaws. All of these were overcome in the masterpiece of 1964. Yet, for Ammann and for the art of structural engineering, still the most significant design advance was the Hudson River bridge of 1931. Suitably, the most poetical evocations of Ammann's bridges are the photographs of Margaret Bourke-White and Edward Steichen (fig. 33) and the writing of Le Corbusier, whose brief essay captures the aesthetic vision of Othmar Ammann and the aesthetic hopes of the engineering profession.

The highest standards of the engineering profession are efficiency, economy, and grace, captured in a structure that is actually built, and that works. The goal of efficiency entails a desire for minimum materials, which result in less weight and less visual mass. The discipline of economy means a desire for less cost, usually gained by simplicity in construction. The search for grace, or engineering elegance, involves the visual expression of efficiency and economy through thinness and a final integrated form, and through contrast with the surrounding environment.[76] All of Ammann's engineering achievements—from his designs for the George Washington and Bayonne Bridges in the 1920s through his Verrazano Narrows span in 1964—emphasize these values at a very high level of distinction. Moreover, his writing—which also strikes a high level in clarity and detail—underscores the importance of these values in his own work.[77] And, like the engineers of the earlier period, Ammann was "independent, austere," and "self-confident."[78]

From the early 1930s onward, Ammann was viewed by members of his profession and by the wider public as one of the immortals of bridge design; and his quiet manner and self-assurance reinforced the perception that this was an engineer's engineer. When the historian or engineer asks why Ammann was chosen to design the George Washington Bridge, the answer may seem self-evident: the Port Authority chose Ammann because he was the best man for the job. And so, in exploring why the Port Authority shifted direction to take on this task, and how public support for building the bridge was obtained, the researcher easily passes over Ammann, and looks for the answers in the activities of real-estate developers and other interest groups in the region, and in the imperial visions of an ambitious Port Authority.[79] Not that these factors are irrelevant; but it now seems clear that they provide an inadequate explanation. The ambition, the vision, and the sustained political efforts of this engineer as entrepreneur were also essential.

In addition, an understanding of political strategies necessary to gain support for major engineering projects will be enriched by exploring technical dimensions. Ammann's detailed studies of traffic loads and structural performance were essential to his case for a bridge of unprecedented size. In a larger sense, moreover, these studies are a part of the conversion from an industrialized society dependent upon railroads to one entering the era of large trucks, private automobiles, and suburban decentralization. More than any other American structure, perhaps, the George Washington Bridge symbolizes that conversion (fig. 34), and the engineer who created this span characterizes the entrepreneurial, political, and engineering activity that lay behind that achievement. At the end of his career Ammann challenged his profession of civil engineers to think beyond the immediacy of its works. "The engineers… must judge the significance of such a structure [the Verrazano Narrows Bridge] by the problems they have to solve and by the extent to which they must utilize progress in science and technology to produce superior design, combined with greatest economy and expediency of construction; they must be conscious that the public will judge such an imposing and conspicuously located bridge by its aesthetic appearance."[80] This is Ammann's message to the future as well as his final implicit tribute to his bridge professor in Zurich.

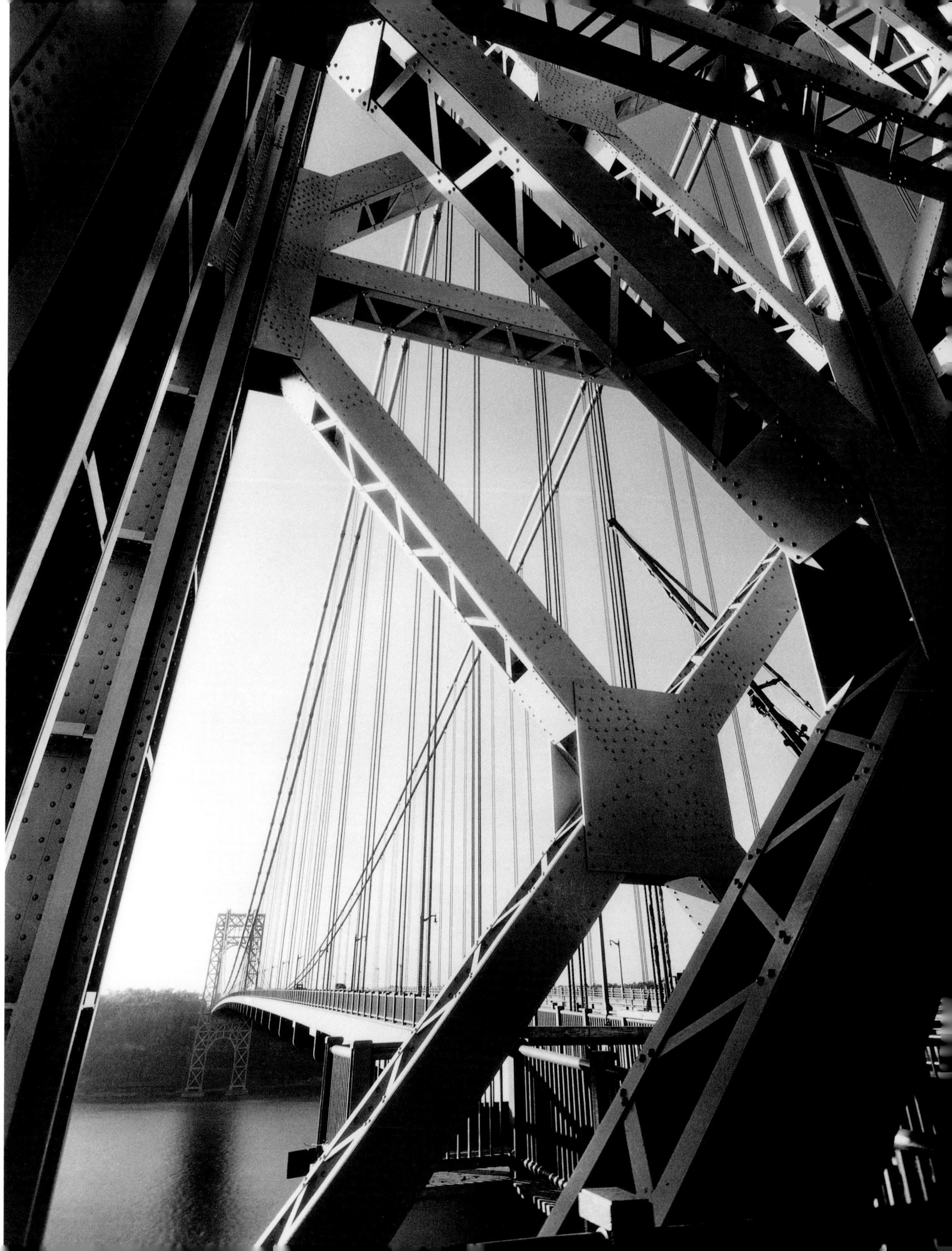

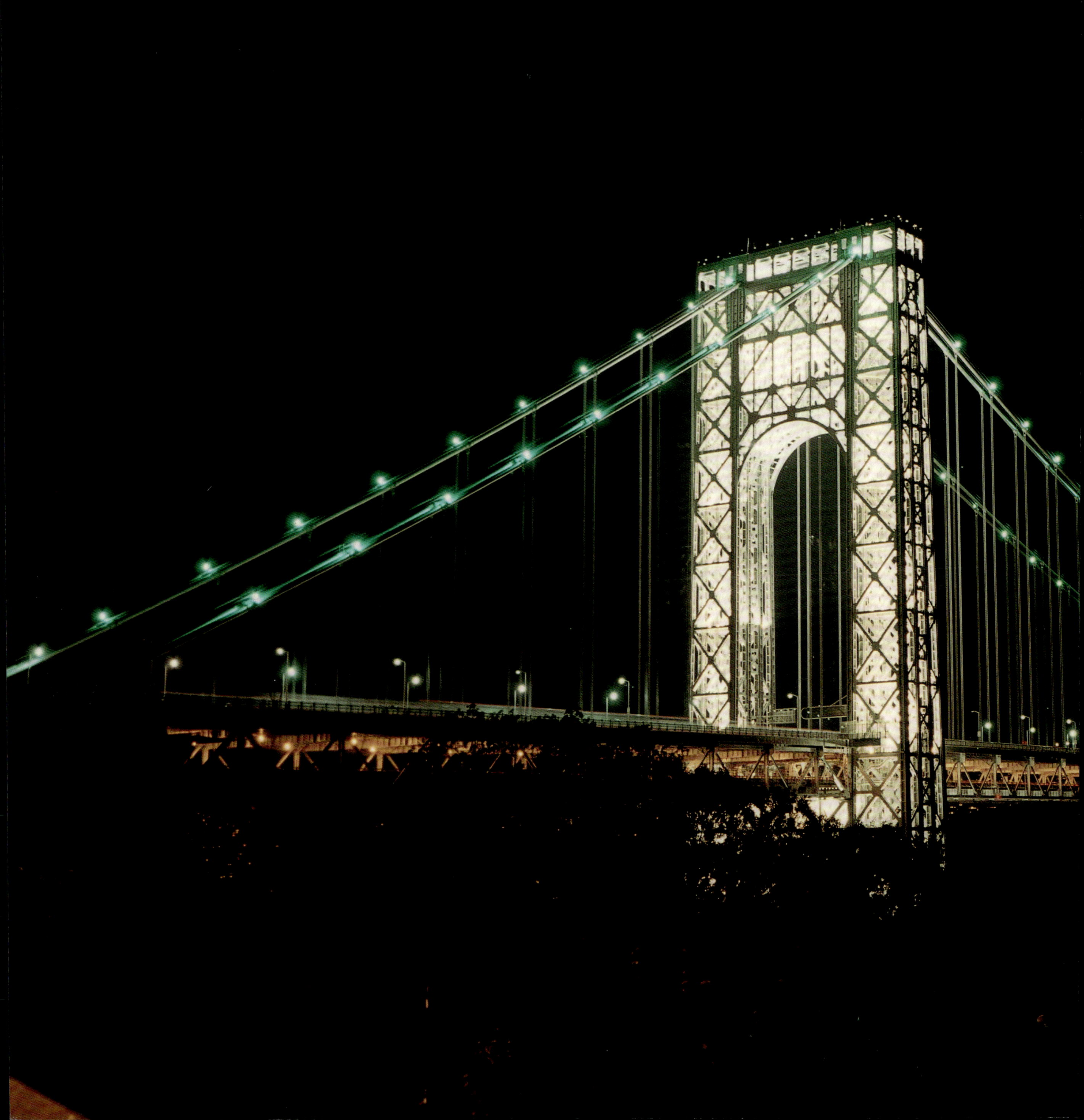

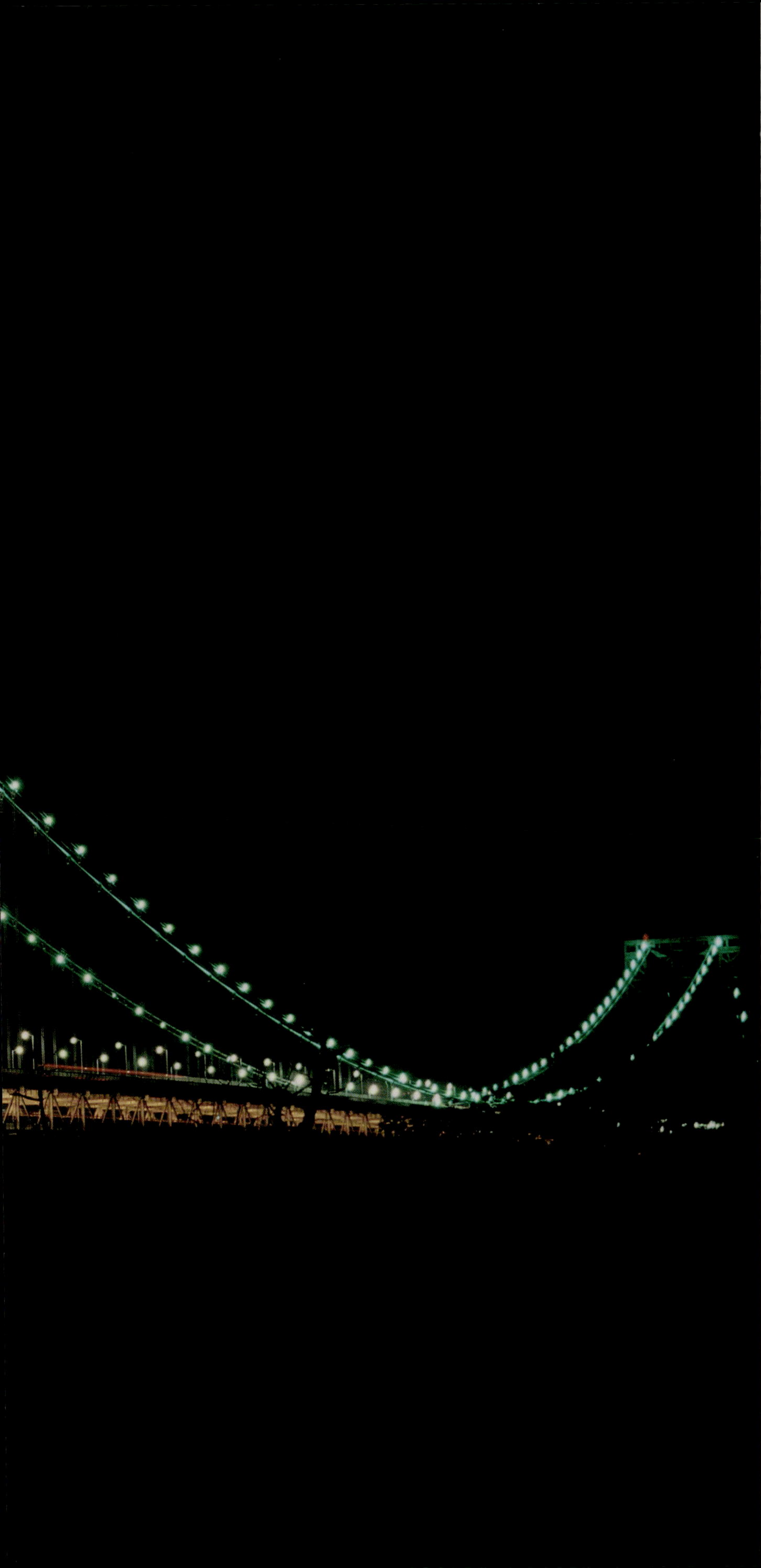

Figure 34
Contemporary view of the George Washington Bridge.

Notes

1. F. Stüssi, *Othmar H. Ammann: Sein Beitrag zur Entwicklung des Brückenbaus* (Basel, 1974), 11–12.

2. Ibid., 12–15. For an important presentation in English of Ammann's career in the United States, see D. Rastorfer, *Six Bridges: The Legacy of Othmar H. Ammann* (New Haven, 2000), 1–38.

3. U. Widmer, "Othmar Hermann Ammann, 1879–1965," *Swiss American Historical Society Newsletter* 15 (1979): 5–6.

4. O. H. Ammann, "The Hell Gate Arch Bridge and Approaches of the New York Connecting Railroad over the East River in New York City," *Transactions of the American Society of Civil Engineers* 82 (December 1918): 852–1004.

5. G. Lindenthal, "Rebuilding of the Monongahela Bridge, at Pittsburgh, Pa.," *Transactions of the ASCE* 12 (September 1883): 353–85.

6. Ammann, "Hell Gate," 863.

7. Ibid., 867.

8. Ibid., 871; both this quote and the next one are from the same page.

9. Ammann, "Hell Gate," 986.

10. Ibid.

11. Ibid.

12. W. Ritter, *Anwendungen der Graphischen Statik nach Professor Dr. C. Culmann*, part 4, *Der Bogen* (Zurich, 1906).

13. Ibid., 72, 79; and Ammann, "Hell Gate," 993, 989.

14. Lindenthal had another huge project, the Sciotoville Railroad Bridge over the Ohio River, on which design had begun when Ammann arrived in 1912. Construction had started in 1914 and ended in 1917. Ammann served as Lindenthal's principal assistant on this continuous two-span truss just as he had on the Hell Gate. Like Hell Gate, Sciotoville was the longest spanning bridge of its type in the world. Lindenthal himself wrote the major paper on this bridge and also won the Howard Award in 1922; see G. Lindenthal, "The Continuous Truss Bridge over the Ohio River at Sciotoville, Ohio, of the Chesapeake and Ohio Northern Railway," *Transactions of the ASCE* 85 (1922): 910–53.

15. For a more extensive treatment of this case, see J. W. Doig and D. P. Billington, "Ammann's First Bridge: A Study in Engineering, Politics, and Entrepreneurial Behavior," *Technology and Culture* 1994, 537–70.

16. The main span of a bridge at 179th Street would be 3500 feet; the longest suspension-bridge spans then in existence or under construction were the Brooklyn Bridge (1595 feet), the Manhattan Bridge (1470 feet) and the Williamsburg Bridge (1600 feet), all across the East River to Manhattan; the Bear Mountain Bridge over the narrower Hudson farther north (1630 feet); the Delaware River [Benjamin Franklin] Bridge to Philadelphia (1750 feet); and the Ambassador Bridge over the Detroit River to Canada (1850 feet). The engineering advances that, in Ammann's opinion, would make possible this giant step in span length included the creation of new alloy steels, development of more accurate methods of shop fabrication and shop assembling of bridge parts, better methods of calculating stresses and of model experimentation, and an improved conception of how to evaluate the forces that stabilize (or "stiffen") a massive bridge. Ammann's views on these issues are set forth in several speeches and papers during the 1920s and 1930s; for a summary discussion, see his paper "Brobdingnagian Bridges," *Technology Review* 33 (July 1931): 441ff. Whether the information available in 1922–23 was *technically* convincing, however, is an issue we consider below (see notes 27, 29, and 30). Ammann had surely "convinced himself" that the technology was avail able to permit doubling the length of a bridge span, but in view of the uncertainties regarding the impact of wind and other factors, a leap of faith—based on Ammann's strong desire to create such a bridge—was almost certainly an element in his conviction.

17. Doig and Billington, *Ammann's First Bridge*, 544–49. For a detailed treatment of the political context of these works, see J. W. Doig, *Empire on the Hudson* (New York, 2001), 120–77.

18. On Silzer's values and programs, see his inaugural address (January 1923) and his first and second annual messages to the legislature (January 1924 and January 1925): I. S. Kull et al., *New Jersey: A History*, vol. 3 (New York, 1930), 1080–83, and *The Governors of New Jersey, 1664–1974*, ed. P. A. Stellhorn and M. J. Birkner (Trenton, 1982), 194–96. Silzer had served in the state legislature when Woodrow Wilson was governor (1911–1913), and he had been a strong supporter of Wilson's reform program.

19. Lindenthal's letter of Jan. 30, 1923, to Silzer suggests that earlier discussions of financial details of the 57th Street span had taken place between the two men. In addition to its great cost, the 57th Street Bridge faced political hurdles, especially in Manhattan, where business and governmental leaders argued that it would cause massive traffic congestion.

20. We have no direct account of the views that Silzer expressed in this meeting, but later evidence clearly suggests the opinions summarized in the text.

21. Othmar Ammann to Rosa Labhardt Ammann, Dec. 14, 1923 (trans. Margot Ammann, 1988); from the files of Margot Ammann.

22. O. H. Ammann, "General Conception and Development of Design, George Washington Bridge," *Transactions of the ASCE* 97 (1933): 34–36.

23. C. Navier, *Memoire sur les ponts suspendus* (Paris, 1823), 80–81. For a review of these theories, see S. G. Buonopane and D. P. Billington, "Theory and History in Suspension-Bridge Design from 1823 to 1940," *Journal of Structural Engineering* 119, no. 3 (March 1993): 954–77.

24. See, for example, T. Cooper, "Train Loadings for Railroad Bridges," *Transactions of the ASCE* 31 (February 1894): 174–84.

25. See D. P. Billington, *The Tower and the Bridge* (Princeton, 1983), chaps. 2 and 5.

26. O. H. Ammann, "Study of a Highway Bridge across the Hudson River at New York between Washington Heights and Fort Lee," Dec. 14, 1923. This report was sent to Governor Silzer, who sent it on to the Port Authority. The report included several appendices. It is reproduced in its entirety in Stüssi, *Othmar H. Ammann*, 92–114.

27. The components of the 46,000 pounds per foot were an 80-foot roadway filled with heavy trucks (20,000 lbs./ft.), four electric railway tracks filled with trolley cars (24,000 lbs./ft.) and two 10-foot sidewalks crowded with pedestrians (2,000 lbs./ft.). See A. Dana, A. Andersen, and G. M. Rapp, "George Washington Bridge: Design of Superstructure," *Transactions of the ASCE* 97 (1933): 104. For the congested load see "Engineering Studies for Philadelphia-Camden Bridge," *Engineering News-Record* 86 (June 23, 1921): 1086–90; H. D. Hussey, "Proposed Loading for Highway Bridges," *Transactions of the ASCE* 87 (1924): 414–20 with discussion, and 421–37 (where Hussey endorsed a fifty percent reduction for spans of 200 feet and greater); and two specifications published in 1924 for the American Association of State Highway Officials [hereafter AASHO]: *Standard Specifications for Concrete Highway Bridges, Division V—Design* (February 1924), 7, and *Standard Specifications for Steel Highway Bridges*, U. S. Department of Agriculture, Dept. Bulletin no. 1259 (Oct. 9, 1924), 21–23 (both endorsing a fifty percent reduction for spans of 200 feet and longer).

28. O. H. Ammann, "The Problem of Bridging the Hudson River at New York with Particular Reference to the Proposed Bridge between Fort Washington Point and Fort Lee," an address given February 19, 1924, and reprinted in the *Annual Report of the Connecticut Society of Civil Engineers* (1924); the quotation is on page 18.

29. Ammann used a simple formula to express the likelihood of full loading: K = 0.2 + 160/(200+L) where L is the span length. The results of this calculation for different lengths, starting with L = 0 and going to infinity, give K = 1.0 down to 0.2. His formula actually gives K = 0.24 for L = 3500 ft., which he rounded up to K = 0.25. The published version of his reduction formula appears in O. H. Ammann, *Specifications for Design of Bridges Carrying Highway and Electric Rail Passenger Traffic*, Port of New York Authority (July 1, 1929). See also Dana et al., "George Washington Bridge," 103–4.

30. The formula is C = 0.5 + 2/(n+3) where n is the number of loaded lanes. In Dana et. al., "George Washington Bridge," we find n = 8 and hence C = 0.682, which is the value used in the design. This formula also appears in Ammann's *Specifications*, 2. A similar formula appeared in the AASHO preliminary specifications of July 1, 1927. For loaded widths above 18 feet, the formula gave $C = \frac{W+18}{18n}$ (where W is the loaded width and not the number of traffic lanes). Where W = 80 ft. and n = 8, then C = 0.68, which is the number Ammann reached by a quite different formula.

31. AASHO, *Standard Specifications for Highway Bridges* (Washington, D.C., 1931), 176. In this early document, reduction factors are based on roadway width in feet; later editions put the reduction in terms of loaded lanes.

32. We have used the following reasoning to reach this conclusion. If the standard reduction (C x K = 0.5) had been applied to the maximum conceivable load of 46,000 pounds per foot, Ammann would have designed the bridge to withstand a live load of 23,000 pounds per foot. This large loading would have increased the total main-span load to 39,000 + 23,000 = 62,000 pounds per foot, or an increase of 62/47 = 1.32 over the actual design loads. Since the total steel cost was about $23,000,000, we estimate that such a high live load would have increased the cost by 0.32 x $23,000,000 = $7,400,000, plus the additional cost caused by the dead weight of the additional steel.

33. The cable stress is directly proportional to the magnitude of the uniformly distributed loads. Ammann used 39,000 pounds per foot for dead load and 8,000 pounds per foot for live load; the total load of 47,000 pounds per foot gave a cable tension stress of 82,000 psi. Thus, for a maximum conceivable load of 39,000 + 46,000 = 85,000 pounds per foot, the cable tension stress would be (85/47) 82,000 = 148,000 psi.

34. Ammann, "Bridging the Hudson," 22.

35. O. H. Ammann, "General Conception and Development of Design, George Washington Bridge," *Transactions of the ASCE* 97 (1933): 43, 62.

36. D. P. Billington, "History and Aesthetics of Suspension Bridges," *Journal of the Structural Division of the ASCE* 103, no. 478 (August 1977): 671–687, with discussions.

37. W. Ritter, "Statische Berechnung der Versteifungsfachwerke der Hängebrücken," *Schweizerische Bauzeitung* 1, nos. 1–6 (1883), and Josef Melan, "Theorie der eisernen Bogenbrücken und der Hängebrücken," in *Handbuch der Ingenieurwissenschaften* (Leipzig, 1888), vol. 2, part 4, 1–144.

38. See R. Modjeski, *Manhattan Bridge: Report 1909*. F. E. Turneaure made these calculations and then included a full presentation of the theory in his text: J. B. Johnson, C. W. Bryan, and F. E. Turneaure, *The Theory and Practice of Modern Framed Structures*, 9th ed. (New York, 1911), part 2, 276–321.

39. Ammann, "General Conception"; Ammann's analysis shows that the stiffening truss designed for the two-deck George Washington Bridge would be able to carry the live load of 8,000 pounds per foot (placed in one critical position around midspan) by forces that are far less than the value of those found from the standard method developed in the nineteenth century. The forces anticipated by using that earlier method would have required a huge stiffening truss, a great increase in weight, and an immense rise in cost.

40. Ammann, "Bridging the Hudson," 22. See especially Ammann, "General Conception," 43. Since Ammann's design used about 2400 pounds per foot, there would be an addition of about 11,000 pounds per foot of steel if he had followed that precedent. Thus 11,000 x 4760 = 52,500,000 pounds of steel would be required for those stiffening trusses. This additional weight would require more material in the suspenders, cables, towers, and anchorages, which weigh 152,600,000 pounds, and would increase by (47 + 11)/47 = 1.23, or an increase of 0.23 x 152,600,000 = 35,000,000 pounds, for a total increase of 87,500,000 pounds. The bid price for steel averaged 12.25 cents per pound in 1927; thus, the increase in cost would have been $10,700,000.

41. This surprising result comes from the fact that the truss deflections must be about the same as the cable deflections. In a long-span bridge with a normally light truss deck, the cable deflections are hardly influenced at all by the truss stiffness. If two trusses have the same deflection curve, then their bending moments will be proportional to their stiffnesses, much like springs. If two springs of different stiffness are compressed by the same distance, the forces in the springs will be proportional to their stiffness. For a detailed discussion of this analysis and its influence on engineers, see Buonopane and Billington, "Theory and History in Suspension-Bridge Design."

42. Ammann, "General Conception," 39.

43. Billington, "History and Aesthetics of Suspension Bridges"; see especially discussion and closure.

44. As he wrote to his mother in 1921: "Toward strangers one is always covered with a veil... The human soul must not expose itself to the profanities of the world" (Feb. 12, 1921; from the files of Margot Ammann). Our views on Ammann's personality also benefited from discussions with Margot Ammann and Sylva Brumer, wife of Milton Brumer, partner of O. H. Ammann.

45. Ammann to Silzer, Dec. 12 and Dec. 13, 1923. Ammann wrote that he would be glad to accept a position with the Port Authority to carry out the studies for the bridge. (Having left Lindenthal's employ early in the spring of 1923, Ammann had continued to work full-time on the Fort Lee Bridge Project without pay, using his savings to support his family.) For the report, see Ammann to Silzer, Dec. 17, 1923 (with attachments); Silzer to the Port Authority Commissioners, Dec. 17, 1923; Silzer to Gregory, Dec. 17, 1923.

46. For Ammann's report and the commissioners' actions, see Gregory to Silzer, Dec. 19, 1923; Port Authority Commissioners, Minutes, Dec. 19, 1923; Port Authority Commissioners, letters to Governors Silzer and Smith, Dec. 21, 1923. For Lindenthal's criticism: "Mr. A. had been my trusted assistant and friend for ten years," Lindenthal wrote, "trained up in my office and acquainted with all my papers and methods. But I know his limitations. He never was necessary or indispensable to me... Now it appears that A. used his position of trust, the knowledge acquired in my service and the data and records in my office, to compete with me in plans for a bridge over the Hudson and to discredit my work... He does not seem to see that his action is unethical and dishonorable..." (Lindenthal to Silzer, Dec. 10, 1923). We have no direct evidence that Governor Silzer showed this letter to Ammann, but Silzer's general habit was to send letters he received to other interested parties, and Ammann's letters to Silzer in later months make it clear that he knew Lindenthal had criticized his behavior on professional and personal grounds. As earlier sections of this article indicate, Lindenthal's characterization of Ammann's actions was quite unwarranted.

47. Ammann's sketch of 1923 shows the stone towers that he favored.

48. Silzer to Cohen, May 22, 1924.

49. These events are described in Englewood Board of Trade, letter to Silzer, April 3, 1924, and Ammann to Silzer, May 7 and May 27, 1924. For newspaper reports of Ammann's efforts during these months, see "Fort Lee Bridge is Advocated by Engineer Ammann," *Bergen Evening Record*, March 5, 1924; "Bridge at Fort Lee Sure, Says Ammann," *Palisadian*, April 25, 1924; and "Bridge the Hudson Meeting Monday: Engineer Ammann Will Tell of Proposed Structure," *Bergen Evening Record*, May 3, 1924. See also Ammann to Silzer, July 23, Nov. 23, Nov. 29, 1924; Silzer to Ammann, Nov. 24, 1924; *Palisadian*, Nov. 14, 1924; *Boonton Times*, Nov. 28, 1924.

50. See *Bergen Evening Record*, Jan. 13, 1925; Ammann to Silzer, Jan. 29, Feb. 25, and March 27, 1925. See also J. W. Binder letter to Silzer, March 4, 1925, informing the governor that the "Mackay Hudson River Bridge Association" had been formed to urge that the Port Authority build the Fort Lee Bridge.

51. Silzer to Gregory, April 14, 1925.

52. Ammann to Silzer, April 17, 1925.

53. G. Kubler, *The Shape of Time* (New Haven, 1962), 7.

54. O. H. Ammann, "Die Ersten interstaatlichen Brücken zwischen New York und New Jersey," *SBZ* 95, no. 22 (May 31, 1930): 285–88.

55. Le Corbusier, *Towards a New Architecture* (Paris, 1923), trans. F. Etchells (London, 1927), 15–16.

56. Le Corbusier, *When the Cathedrals Were White*, trans. by F. E. Hyslop Jr. (New York, 1947), 75.

57. "Bayonne Bridge: Addressses Delivered at Dedication," Nov. 14, 1931, 22–25 for Ammann's address.

58. Ibid., 26, for the cablegram.

59. O. H. Ammann, "First Progress Report, Bridge over the Kill von Kull," New York, April 1930.

60. *Proceedings of the ASCE*, March 1930.

61. "Bayonne Bridge: Addresses," both quotes in this paragraph come from page 24.

62. Widmer, "Amman: 1879–1965," 19, and M. MacKaye, "Profile: Poet in Steel," *The New Yorker*, June 2, 1934.

63. Ammann, "Die Ersten interstaatlichen Brücken."

64. Ibid.

65. O. H. Ammann, "Second Progress Report," 1931.

66. A. Embury II, "Esthetic Design of Steel Structures," *Civil Engineering* 8, no. 4 (April 1938): 261–65.

67. A. Embury II, "Esthetics of Bridge Anchorages," *Civil Engineering* 8, no. 2 (Feb. 1938), 85.

68. Embury, "Steel Structures," 265.

69. Embury, "Bridge Anchorages," 88.

70. Ibid., 87.

71. Personal conversation with Christian Menn. For a comprehensive study of the wind problems and structural modifications, see M. Barelli, "Bronx-Whitestone Bridge," senior thesis, Department of Civil and Environmental Engineering, Princeton University, 2002.

72. O. H. Ammann, T. Von Karman, and G. B. Woodruff, "The Failure of the Tacoma Narrows Bridge," Report to the Federal Works Administrator, Washington, D.C., March 28, 1941; and O. H. Ammann and H. D. Robinson, "Report of Board of Consulting Engineers on Technical Survey of Brooklyn Bridge," New York, Feb. 5, 1945, 174 pages.

73. O. H. Ammann, "Planning and Design of the Verrazano-Narrows Bridge," paper presented to the New York Academy of Sciences, March 13, 1963, and published in *Transactions of the New York Academy of Sciences*, April 1963, 598–620.

74. Ibid., 617.

75. Ibid., 604.

76. See J. W. Doig, *Empire on the Hudson* (New York, 2001), chapters 6 and 7. Discussions of engineering ideals are found in E. E. Morison, *From Know-How to Nowhere: The Development of American Technology* (Cambridge, Mass., 1974), esp. 6–8, 88–96, 127–29, and Billington, *The Tower and the Bridge*, esp. chap. 1 and 266ff. Morison's comments capture a value which would apply to engineers more generally: "a daring elegance... the ultimate morality of the engineer—if it works" (p. 8).

77. See, for example, his paper of 1918 on the Hell Gate Bridge (note 4 above), which won the Civil Engineering prize that year for its quality of analysis and exposition, and his extensive report of 1933 on the George Washington Bridge (note 22 above). When teachers of English at an engineering school prepared a book of essays to demonstrate to their students how to write clearly, one of the examples they chose was another Ammann paper, "Tentative Report on the Hudson River Bridge," written in March 1926; see *Engineers as Writers*, ed. W. J. Miller and L. E. A. Saidla (New York, 1953), 237–51.

78. The references to nineteenth-century engineers are taken from Morison, *From Know-How to Nowhere*, 68, 93.

79. For example, the discussion in Binder's letter to Silzer of March 24, 1925 (see note 50 above) and in M. N. Danielson and J. W. Doig, *New York: the Politics of Urban Regional Development* (Berkeley, 1982), 186–94.

80. Ammann, "Planning and Design of the Verrazano-Narrows Bridge," 598.

CHAPTER 4

PIERRE LARDY: FROM MATHEMATICS TO AESTHETICS

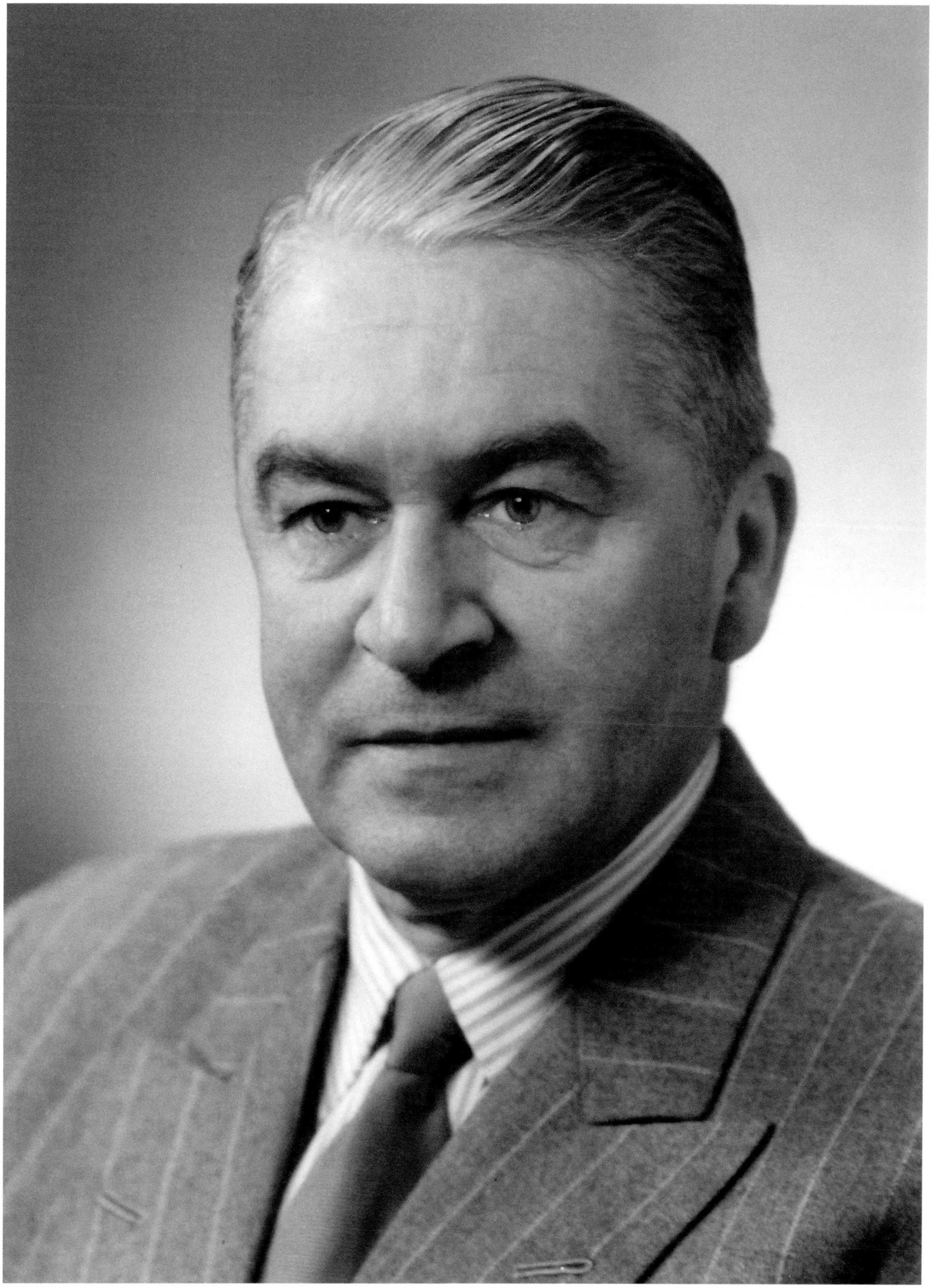

Figure 1 (previous page)
Pierre Lardy (1903–1958).

Figure 2
Diagram showing the prestressing of concrete.

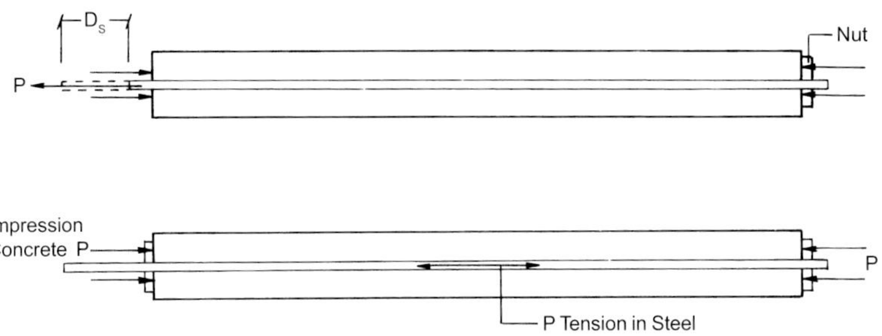

Figure 3
Cracks appearing in the concrete when low-strength steel was used for prestressing.

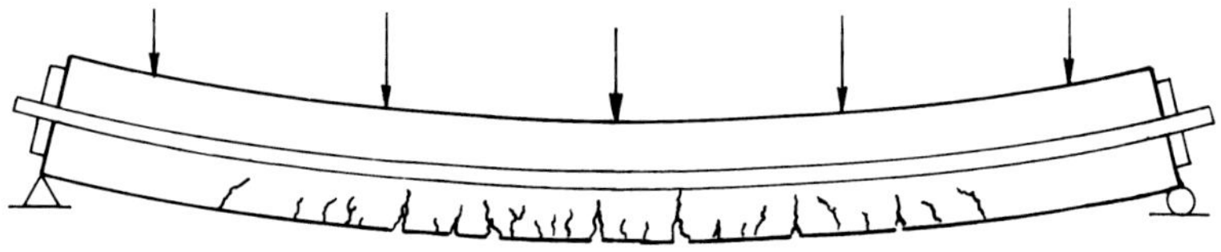

On November 8, 1930, following a two-day symposium commemorating the seventy-fifth Anniversary of the ETH, Othmar Ammann and Robert Maillart sat together at a celebratory luncheon high above Zurich in the Grand Hotel Dolder. There Wilhelm Ritter's two greatest students conversed for a long time.[1] It was apparently their first such conversation, and it occurred just after the inauguration of the Salginatobel Bridge and a short time before completion of the George Washington and Bayonne Bridges. As these two veteran engineers met, there was in the mathematics department of their school a young doctoral candidate beginning his second year as an assistant. Unknown to anyone, this mathematician would one day pick up the tradition of Wilhelm Ritter and facilitate the early careers of the next generation of structural artists.

Following Ritter, the next Swiss professor of structures in Zurich was Arthur Rohn (1878–1956), who was appointed in 1908 and who eventually became president of the ETH. His technical abilities were not on the same level as those of Wilhelm Ritter, and his view of aesthetics emphasized the central role of architects in bridge design. He had bitter academic disputes with Maillart; these debates, published in the *Schweizerische Bauzeitung*, showed his incomplete understanding of some important structural issues.[2] Rohn was followed at the ETH by Max Ritter (1884–1946), no relation to Wilhelm Ritter, who carried on Rohn's attacks on Maillart, although not in the open press.[3] Thus, the tradition of Culmann and Wilhelm Ritter was not carried forward with the same emphasis on design and aesthetics until Max Ritter's death in 1946, when his assistant Pierre Lardy (1903–1958) took over the professorship of concrete structures.

Pierre Lardy (fig. 1) grew up in German-speaking Bern within a French-speaking family from Neuchâtel. He was completely bilingual and was known to have an unusual mastery of the two languages and cultures, even for a Swiss. When he began to publish technical papers, he would sometimes write in German and sometimes in French. After graduation from secondary school in the early 1920s, Lardy had to choose a career. He had many talents, the two most prominent being music and mathematics. Like Wilhelm Ritter, he was a gifted pianist, and he never abandoned the piano. He decided, finally, to pursue mathematics, entering the ETH department of mathematics and physics in the fall of 1923. Lardy graduated from the ETH in 1928 and took a position as a mathematics teacher in the technical high school in Schiers. He arrived there just as the competition for the new Salginatobel Bridge was under consideration and much discussed in the town. In mid-1929, with the design completed and construction about to begin, Lardy decided to leave Schiers, return to Zurich, accept an assistantship with a mathematics professor, and study for an advanced degree in mathematics, which he received in September 1933. But structures had begun to take hold so that immediately after finishing mathematics, he launched himself into the study of structural engineering, receiving his degree at the ETH in 1936 with unusually high grades, much like Ritter's performance there almost seventy years earlier.[4]

After graduation he opened a consulting practice, but the approaching war disrupted his engineering work, and in 1942 he accepted a position as an assistant to Professor Max Ritter, who had been a strong opponent of Maillart's, severely criticizing his work and taking an applied-science view of engineering. He emphasized a more mathematical approach to teaching and to analysis and was offended by Maillart's much simpler design-oriented approach.[5] It is easy to see why a mathematically trained engineer like Lardy would appeal to Ritter. In spite of his views on Maillart, Max Ritter had begun to study the most revolutionary idea in structures at that time—prestressed concrete. In 1938 the Swiss Society of Engineers and Architects had formed a special committee with Ritter as its chairman to study prestressed concrete, and starting in 1942 Pierre Lardy became secretary and a researcher for that group.[6]

The idea of prestressing concrete had arisen almost at the same time in the 1880s as reinforced concrete entered construction practice, but the original concept was based on a misunderstanding of concrete. Imagine a concrete beam with a hole running its length into which is inserted a steel rod, anchored at one end by

Figure 4
Prestressing concrete using low-strength (A) and high-strength (B) steel.

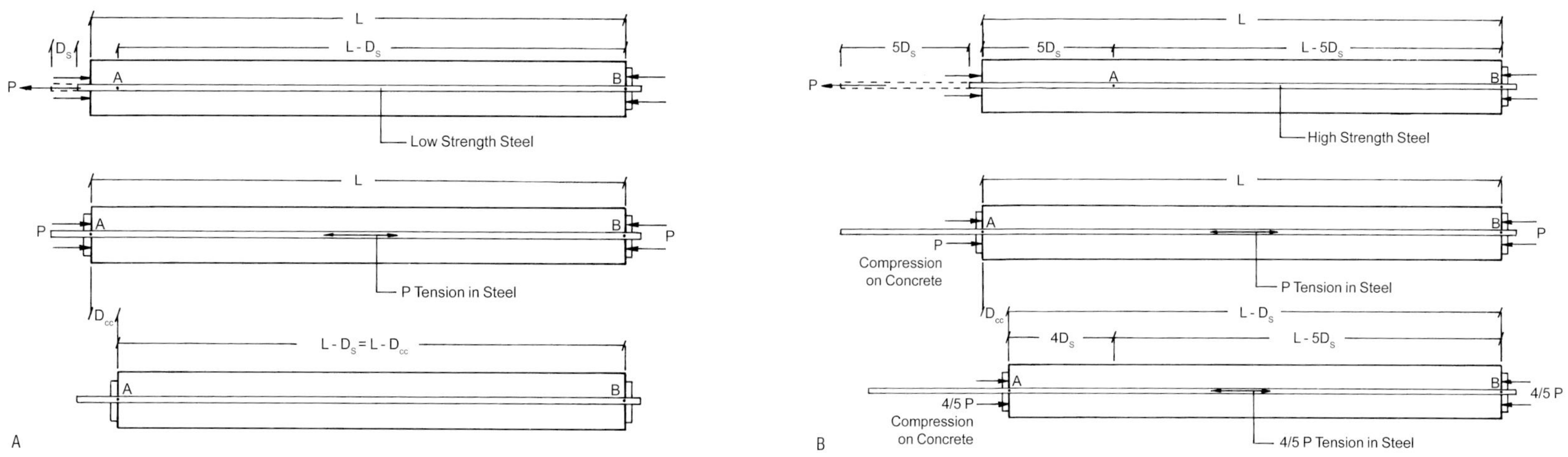

a nut (fig. 2). The rod is pulled with a force (P) at its other end so that it elongates a distance (D_S), after which it is anchored there again by a nut. In this way, the elongated rod, which is thus in tension like a violin string, puts its force (P) onto the concrete beam at both ends as if a rubber band had been stretched around the beam lengthwise. The force (P) compresses the beam with the same force that it has itself in tension. The original idea for prestressing claimed that this precompression (or prestress), applied before load came onto the beam, would counteract any tension caused by the loading. This idea never worked because engineers found that after a few weeks of loading, tension arose in the concrete; this was clearly observed since large cracks appeared just as if no prestressing force (P) had been applied (fig. 3). From 1884 (the first patent) to 1928, no one convincingly identified the problem and its solution. In 1928 the French engineer Eugène Freyssinet (1879–1962) took out a French patent that explained the difficulty and proposed a way to make prestressing work.[7] A concrete beam or column under permanent compression shortens when that force is applied (initial shortening), as all engineers knew; but what many failed to understand was that concrete structures under compression continue to shorten without any additional compression force. This effect gained the slightly animated title of "creep," and its result could be two or three times greater than the initial shortening. Thus, when the beam was prestressed, it shortened over time, thereby also shortening the steel rod; but the prestressing force (P) depended upon the rod's elongation, which the creeping concrete reduced. Early attempts to provide permanent prestressing failed because the concrete creep made the rods lose their force.

Like many great engineering innovations, Freyssinet's solution is obvious in hindsight, it was to propose high-strength steel to overcome the problem of concrete creep. Reinforcing steel then and now was of a relatively low grade (low strength) whose design stress was around $f_S = 20{,}000$ psi (pounds per square inch). The tension force $P = f_S A_S$ where A_S is the cross-sectional area of the steel. Figure 4A shows a beam in which such a steel rod is pulled to an elongation of $D_S = 0.8$ inches (point A on the steel moves to the end of the beam [the distance on figure 4A is greatly enlarged]) to give that steel stress (20,000 psi). When anchored at both ends, the rod transfers a large (P) onto the concrete. Freyssinet proposed using steel that was stronger, for example, with $f_S = 100{,}000$ psi, and therefore using one-fifth the amount of steel so that the total force (P) is the same. Such steel, while stronger, has almost the same value of stiffness, i.e., its elongation $D_S = f_S L / E_S$, where L is the beam length and E_S is the material stiffness (modulus of elasticity), which, when large, leads to small values of D_S. Now if E_S is the same for low-strength steel as it is for high-strength steel, then D_S will vary directly with f_S for a given beam length.

Freyssinet recognized that the concrete beam under the permanent load (P) will creep; for example, in this case D_{CC} could easily equal minus 0.8 inches (inward) so that the steel rod will also shorten by 0.8 inches, leaving a net elongation for the low-strength steel rod of D = 0.8 minus 0.8 = 0 (the distance AB returns to its original length). All prestressing is lost because of creep. The Frenchman saw the solution in high-strength steel, which to reach 100,000 psi requires an elongation $D_S = 4.0$ inches (see fig. 4B), or five times that for the low-strength steel (100,000/20,000 = 5). Thus, for the same force P on the concrete, the net steel elongation after creep would now be $D_S = 4.0$ minus $0.8 = 3.2$ inches (the distance AB no longer returns to its original length), so that the stress remaining in the rod would now be 80,000 psi ($\frac{3.2}{4.0}$ x 100,000), enough to keep the beam from cracking.[8]

Figure 5

Illustrations from Isler's notebooks on Lardy's lectures of beam theory and beams of variable depths.

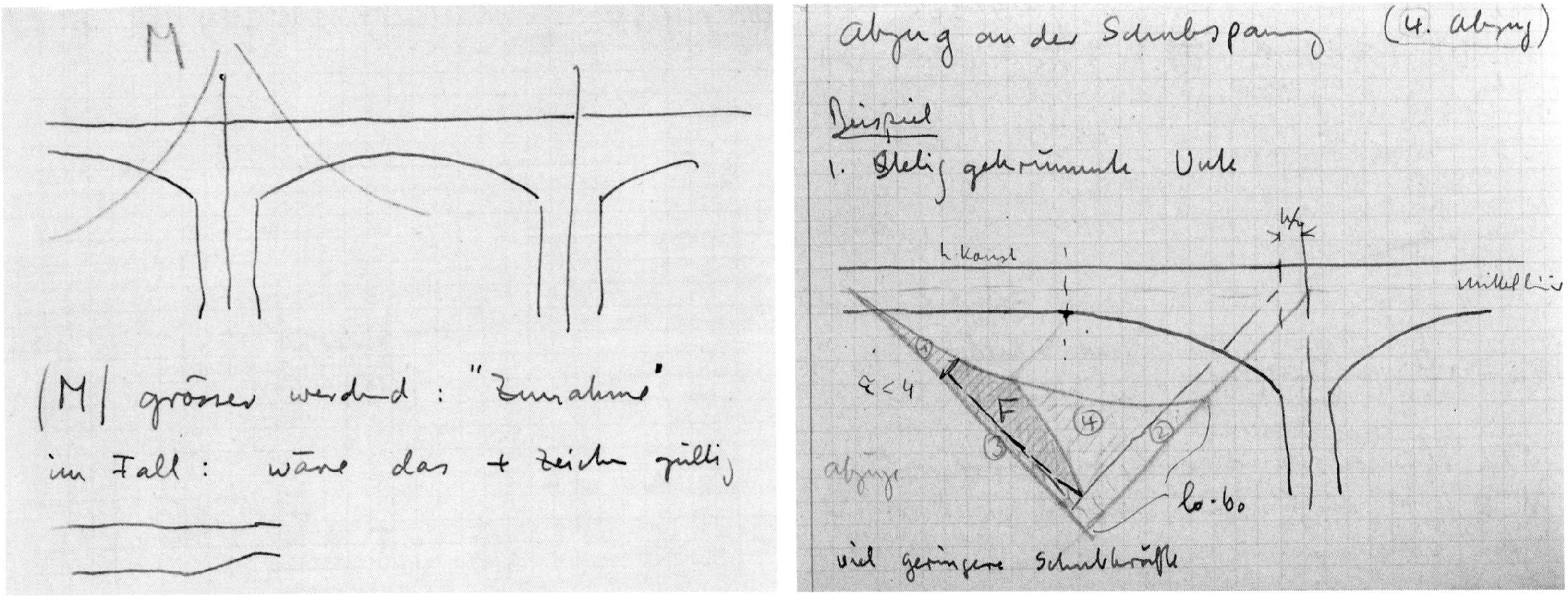

The added virtues of prestressing are that less steel (one-fifth in this case) is required for the same force (P), it is economical to use high-strength concrete (because the entire section is in compression), and the resulting designs can be substantially lighter than with beams in reinforced concrete. By the late twentieth century, hollow-box beams of prestressed concrete have spanned 1,000 feet. Lardy grasped this idea and began to put the principles of prestressing in writing for Swiss engineers in 1943 and 1944.[9] He was building upon systematic testing of beams in Lausanne and Zurich that Ritter had promoted beginning in 1940. Events had conspired to put Lardy on the forefront of this revolutionary design and construction approach to concrete structures. Along with the Belgian Professor Gustave Magnel, Lardy played a significant role in bringing prestressed concrete to the European engineers of the immediate postwar era. Freyssinet was a brilliant engineer and an even more brilliant builder, but he was not a gifted communicator, either in speaking or in writing. So it was that a small group of professors led the way. In 1946 Ritter and Lardy published their book on prestressed concrete, largely directed to Swiss engineers and partly the outcome of Ritter's committee.[10] Surely Ritter stimulated its publication, but the systematic development of design and analysis was entirely the work of Lardy; it was in fact a slightly revised version of his previously published articles in the *Schweizerische Bautzeitung.*

Max Ritter's sudden death in 1946 resulted almost immediately in the appointment of Lardy as his successor. There was some bitterness over his appointment because Lardy had so little practical experience, but his personality quickly made him popular.[11] He took on the teaching of advanced structural analysis (*Baustatik II*), concrete structures (*Massivbau I*), concrete bridges (*Massive Brücken*), analysis and design of concrete structures (*Massivbau II*), and selected topics in advanced analysis and concrete analysis (*Baustatik und Massivbau, Ausgewählte Kapitel*), each a one-semester course. This required Lardy to teach two courses in the winter semester and three in the spring. During the late 1940s, he had to rely heavily on Ritter's curricula although he began to introduce new topics such as prestressed concrete.

The most significant new aspect of Lardy's early teaching was his continued reference to the aesthetics of structure, not as an isolated topic but as one integral to the engineer's design. In his lecture notes there is no mention of collaboration with architects but rather, within engineering, the integration of structural performance and visual choices. Moreover, his lectures regularly included sketches of various structural systems and details he would then analyze mathematically, often with comments on their appearance. His teaching, therefore, centered on a strong mathematical treatment of structures with emphases on simplified analytic methods, discussion of practice including

Figure 6
Straight beams and columns from a standard American text of the 1940s.

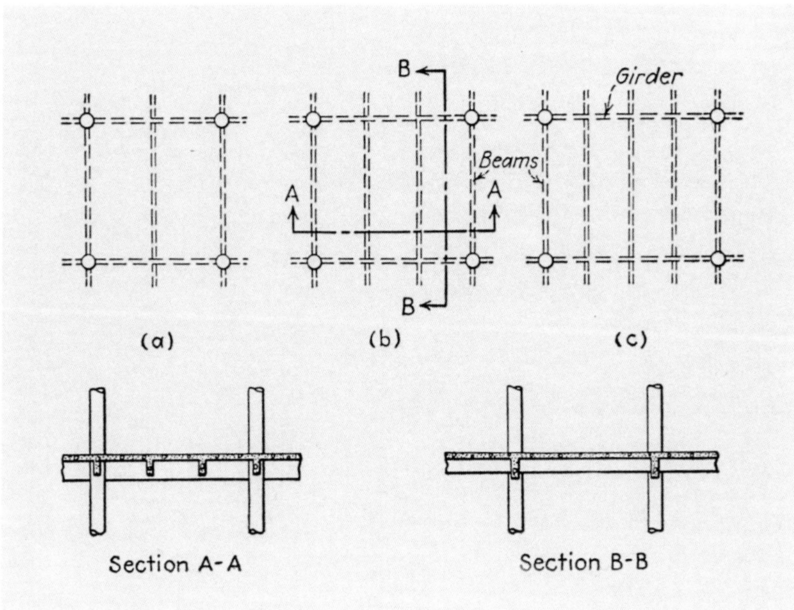

Figure 7
Sketch of beam and column from Isler's notes on Lardy's lectures.

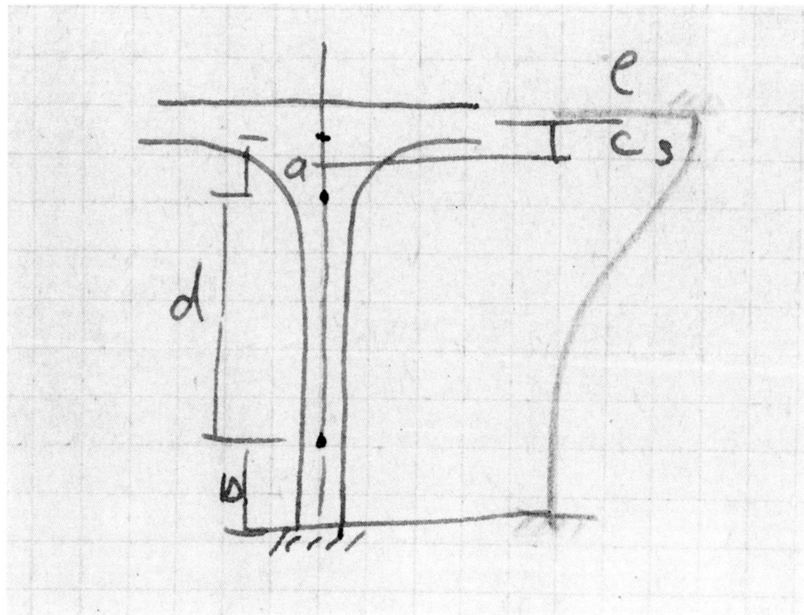

aspects of economy, and continual reference to the aesthetics of form. Lardy was the teacher of concrete structures while another professor, Fritz Stüssi, taught steel structures. Lardy's teaching about concrete structures differed radically from contemporaneous courses given on the same subject in the United States; this accounts, in large measure, for some main differences in concrete design in the two countries. This contrast still exists. Lardy's seminal approach can be illustrated by a few examples taken from the elegant notebooks of Heinz Isler.[12] Three structural forms—frame buildings, flat-slab floors, and arch bridges—exemplify how Lardy's presentation differed from what was being taught to students in the United States.

It is rare to find a complete set of lecture notes on technical courses, and it is almost without precedent to discover notes that so intimately bond the technical and the aesthetic in engineering. Therefore, it is crucial to the argument for the pervasive influence of teaching on practice to present examples from Isler's notebooks that demonstrate Lardy's method (figs. 5, 7-12). This method has nothing to do with the philosophical study of aesthetics but rather with an artist's viewpoint on what most educators would call purely technical issues. Lardy, in these notes, reveals himself to be an artist in teaching and an art critic in practice. There is no evidence that Lardy's students thought he was diluting or perverting his technical teaching by including aesthetics. The evidence from Isler and Menn is strong in claiming a deep influence on gifted designers. Therefore, the examples that follow here must be recognized as unique in engineering education and as the product of an educational artist.

For frame buildings (where beams and columns are integrally connected), Lardy began with beam theory for constant depth beams but quickly moved to beams with variable depths (fig. 5), whereas in the standard American text of the 1940s, there are only straight constant depth beams and columns (fig. 6).[13] Lardy thus stimulated the Swiss engineers to think about form as they learned about formulas, in this case especially about the connection between beam and column (fig. 7).[14] It is true that for most buildings straight beams are economical, but the complete lack of emphasis on form encouraged the student to restrict imagination to standard ideas. Even in the illustrations of figure 5, Lardy presents two slightly different forms for the same problem. Such refinement in no way detracted from the theoretical analyses and practical dimensioning that Lardy taught.

Figure 7 shows the most integral form of connection where the transition between beam and column is achieved by a smooth curve with not even the slight discontinuity of line seen in the examples of figure 5. This distinction has negligible influence both on the structural performance and on the construction cost, but it can change the visual expression noticeably.

Figure 8
Sketch from Isler's notebooks of Maillart's mushroom columns for flat-slab floors.

Figure 9
Lardy's examples of nonrectangular column sections at the bottom of a page from Isler's notebooks.

Figure 10
Lardy's illustrations of the advantages, including aesthetics, of Maillart's system.

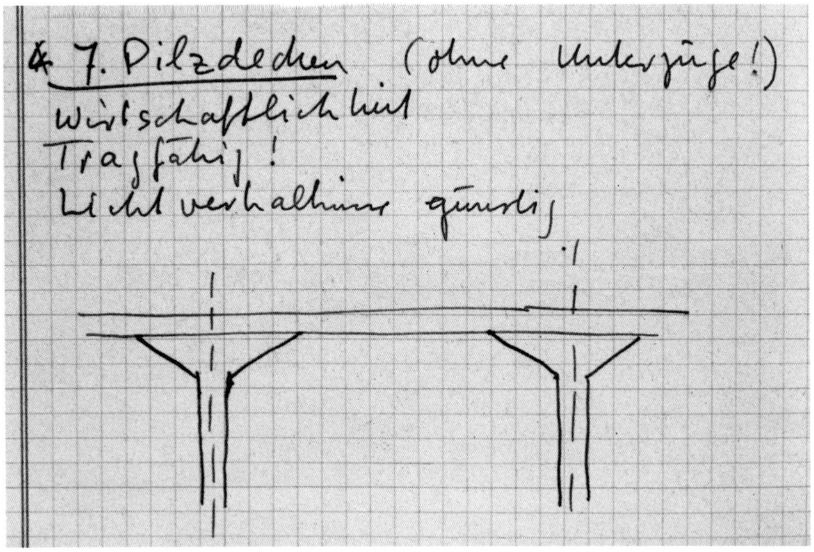

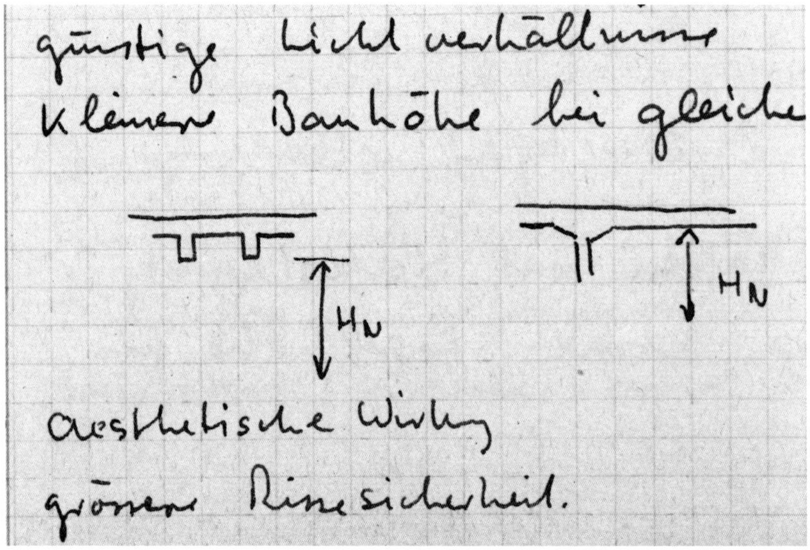

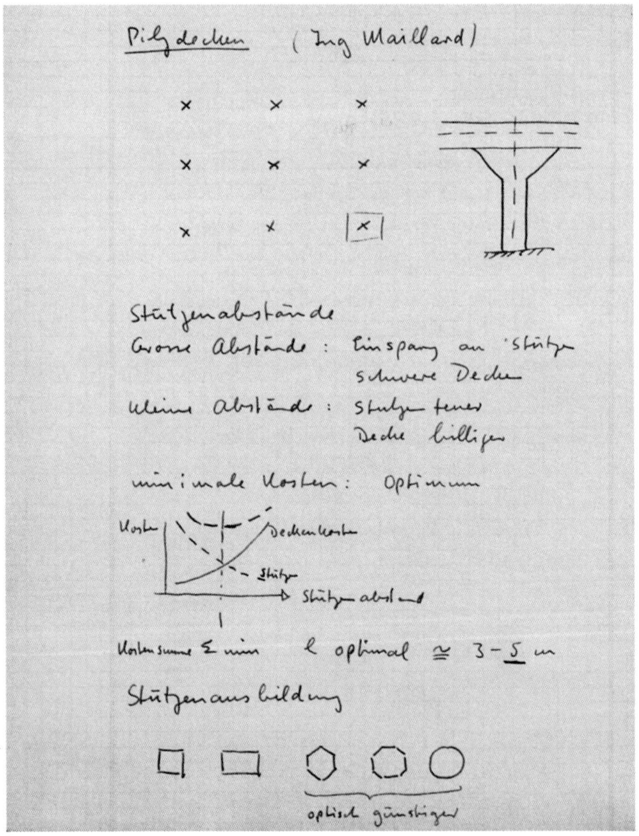

For the second example, flat-slab floors (floors, without beams, supported only by regularly spaced columns), Lardy emphasized the floors supported by mushroom columns (*Pilzdecken*) pioneered by Maillart (fig. 8) and noted that the column sections were more striking if they were not rectangular (fig. 9). He stressed the better proportions and improved aesthetics of Maillart's system (fig. 10). Lardy proceeded to contrast various systems of capitals (fig. 11), observing that the drop panel (Doric column) was cheaper but that the Maillart design was more pleasing aesthetically and more sound structurally.[15] The standard American treatment was to present only one form (with drop panels) and to make no mention of appearance or even economy (fig. 12).[16] American acceptance of the drop panel, originally copied directly from the Classical Doric capital, assumed its economy, as did Lardy. This assumption was untested because most designers—unlike Maillart—were not builders, and thus the cost difference was rarely, if ever, tested in direct competitive bidding.

The Maillart design in figure 11 shows the same improvement in form over that in figure 9 as did the example of figure 7 over those in figure 5 for beams. With the graph in figure 9, Lardy illustrated in principal how to judge the influence of column spacing on the cost of the flat-slab floor system. As the spacing between columns increases–leading to fewer columns– their cost goes down but the slab cost goes up due to the greater spans. The sum of these two effects leads to the top curve, which has a minimum cost identified by the vertical line.

Figure 11

Isler's illustrations from Lardy's lectures of a Doric capital and the more aesthetically pleasing Maillart capital.

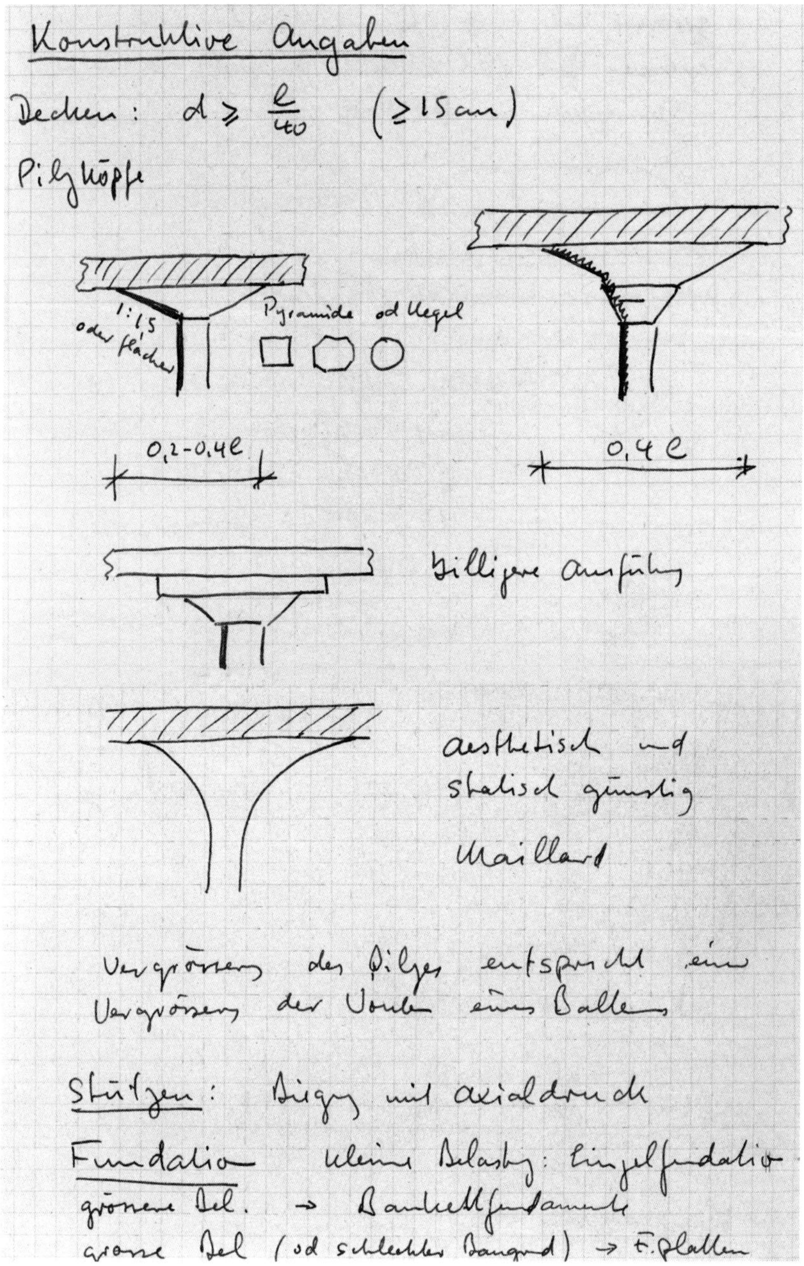

Figure 12

American treatment of capitals, showing only one form and avoiding a discussion of aesthetics or economics, from a standard American text of the 1940s.

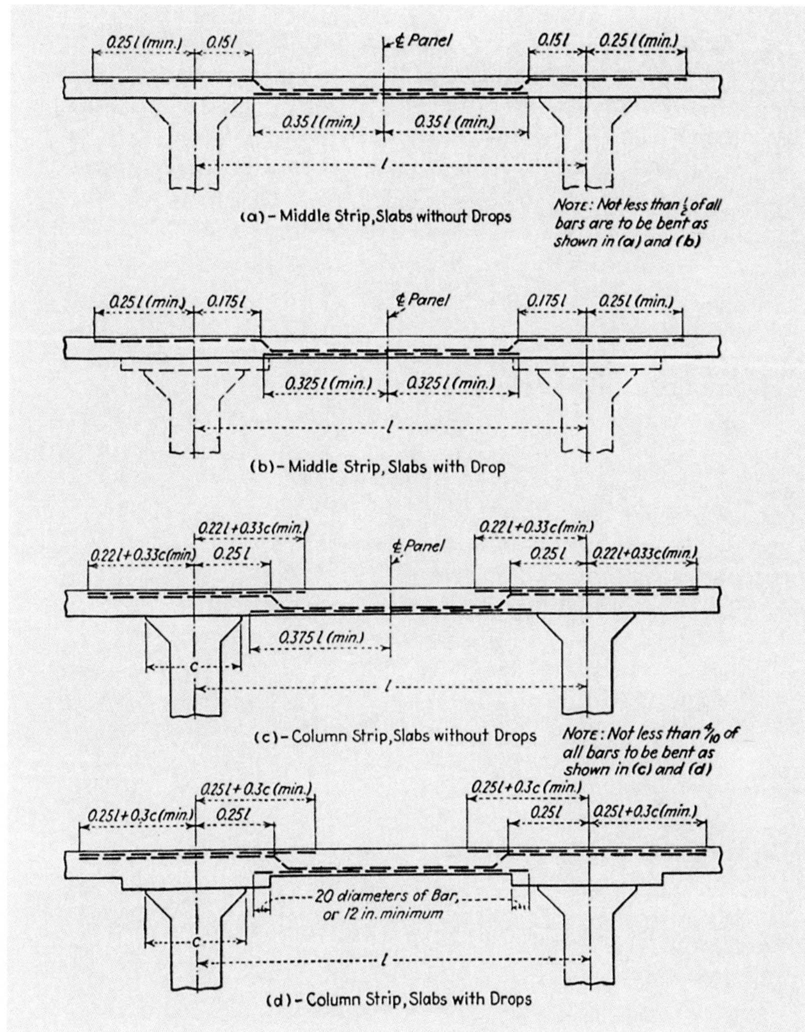

Figure 13

Three major bridge forms Lardy presented to his students, from Isler's notebooks.

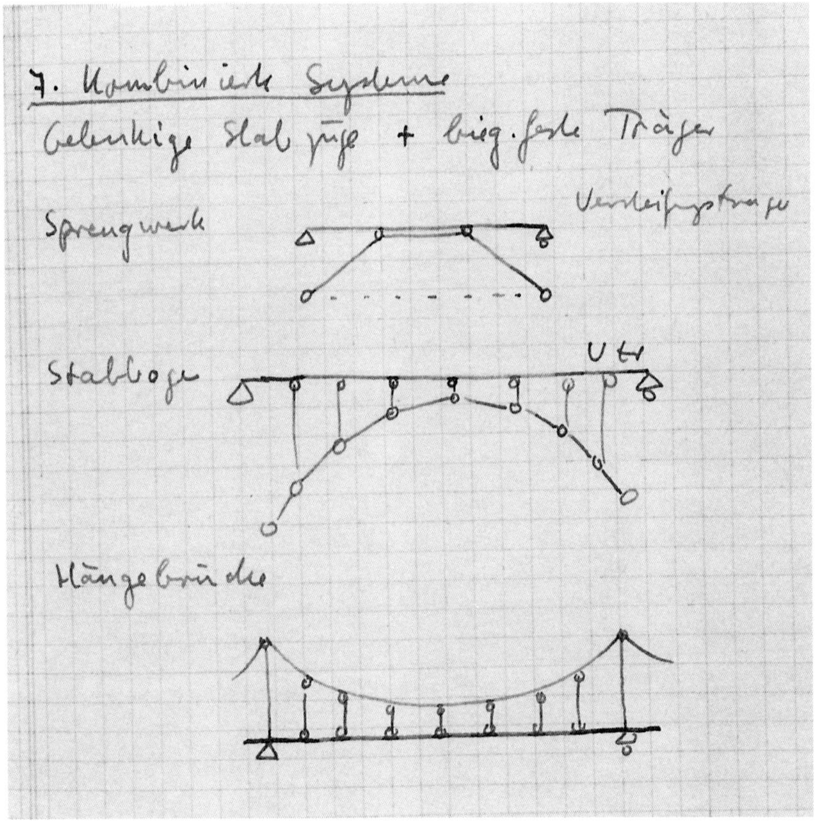

In arch bridges, the differences between Lardy's teaching and that in the United States are even more dramatic. Lardy introduced his students to a variety of bridge forms (figs. 13, 14) and illustrated these with specific designs, including Maillart's Tavanasa form (fig. 15), for which he referred to Max Bill's book on Maillart, which had just appeared.[17] The students from 1949 onward were introduced to Maillart's works in the first book-length exposition not only of photos but also of engineering drawings. By contrast, the American presentation of arch bridges consisted only of analysis with detailed formulas for determining dimensions; there was essentially no discussion of aesthetics. In fact, detailed examples were of standard designs (fig. 16) and of one major design with a strange unbalanced set of arches (fig. 17).[18] A comparison of this design to those shown by Lardy (fig. 18), given in a section entitled "Form and Aesthetics of Arch Bridges," demonstrates that Swiss students were introduced to the proportions in structure, an approach that was largely missing in America.

Figure 14

Many different bridge forms Lardy presented to his students, from Isler's notebooks.

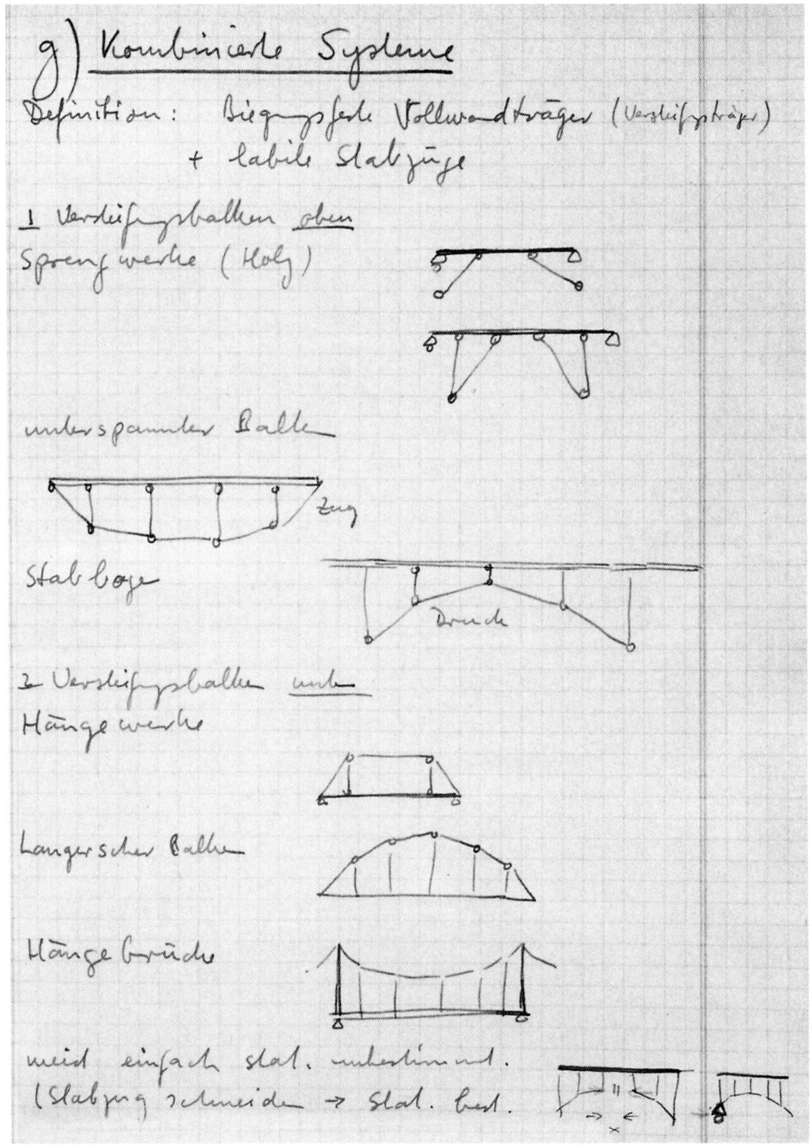

In this last figure, Lardy illustrated poor integration in an arch-bridge profile with three separate structural systems compared to one with a single system. Below that he contrasted a trellis support with a much simpler and more expressive shape, designated *früher* (earlier) versus *heute* (present day) designs. He next showed a polygonal arch with a light profile (*Frontalansicht leicht*), below which are a few visual details.

Figure 15

Isler's sketches from Lardy's lectures on Maillart's hollow-box, three-hinged arch system.

Figure 16

Standard design from an American textbook.

Figure 17

Bridge design with a series of not well-integrated arches from an American textbook.

Figure 18

Lardy's "Form in General," from Isler's notebooks.

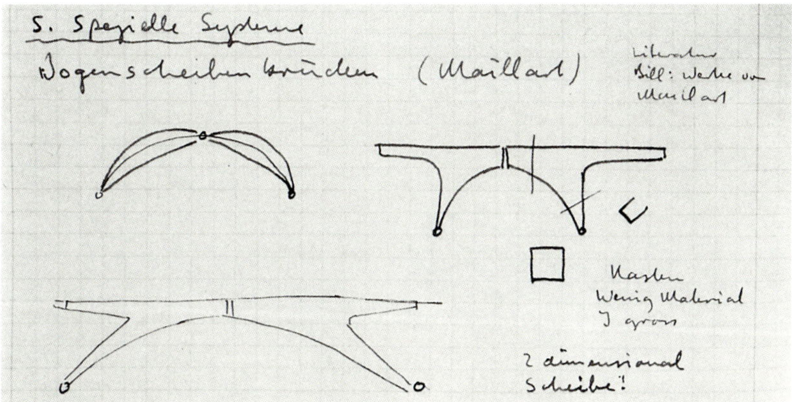

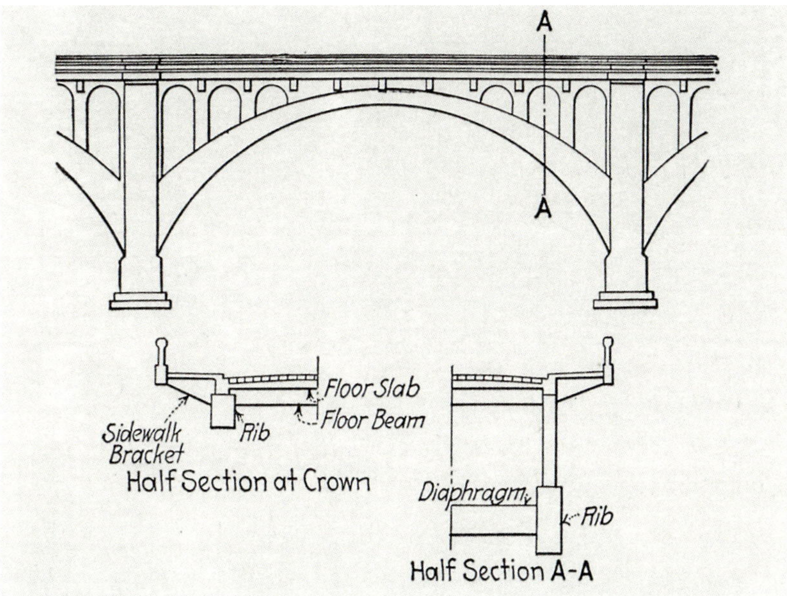

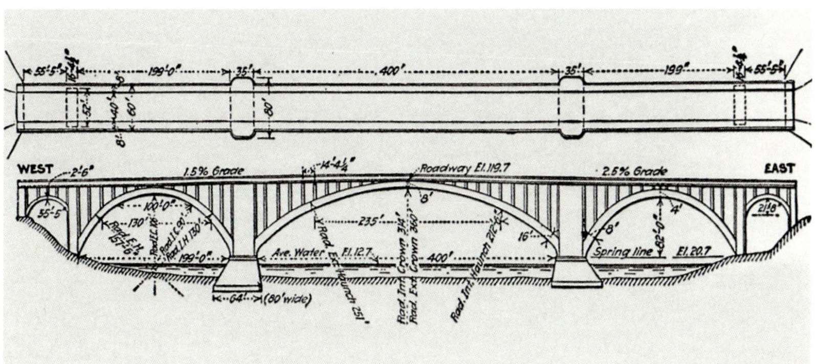

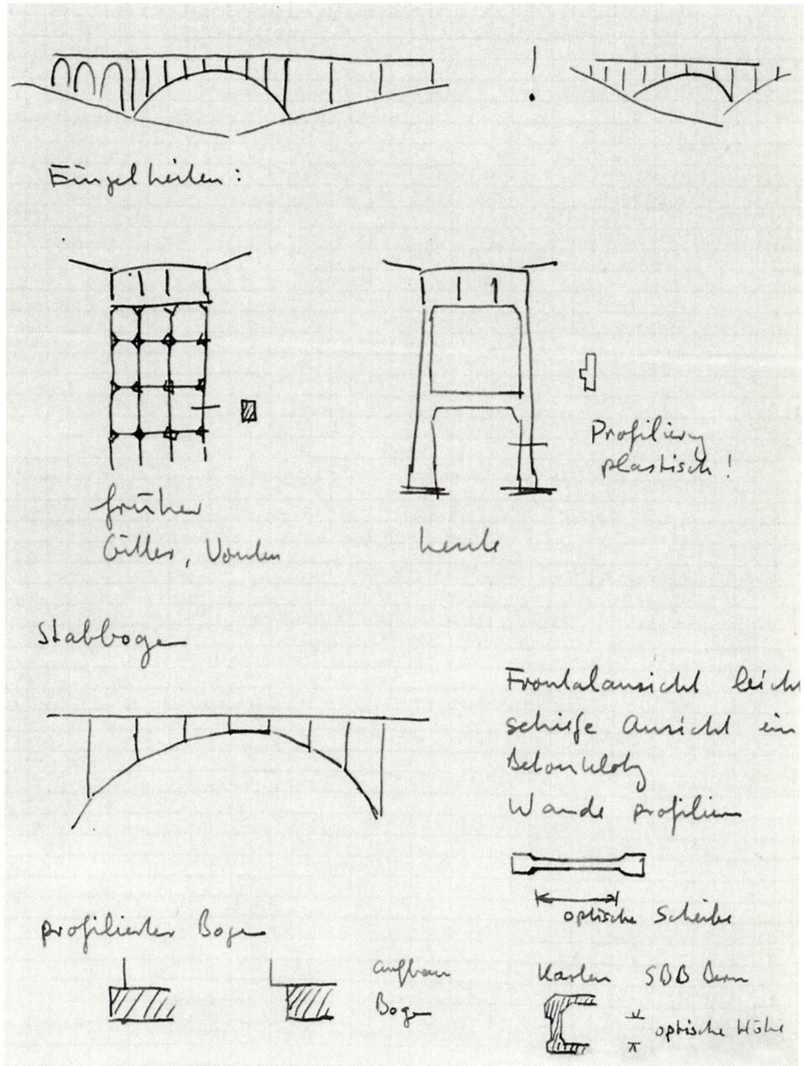

Figure 19
Drawings from Isler's notebooks, showing relative spacing of the verticals, and "terrible" practice of thick double columns over the abutments.

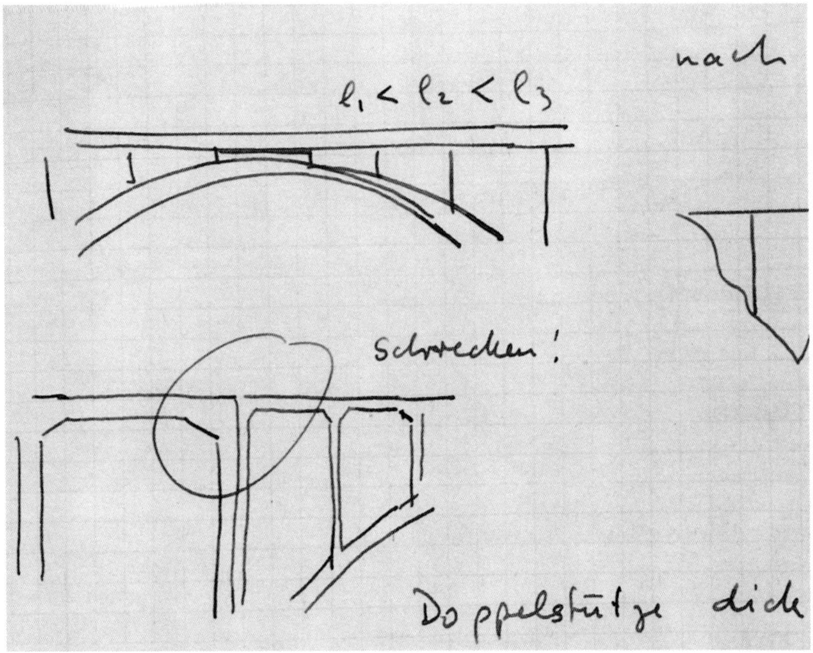

Figure 20
Drawing of an open spandrel arch from Isler's notebooks, showing the relationship between the height of verticals and their horizontal spacing.

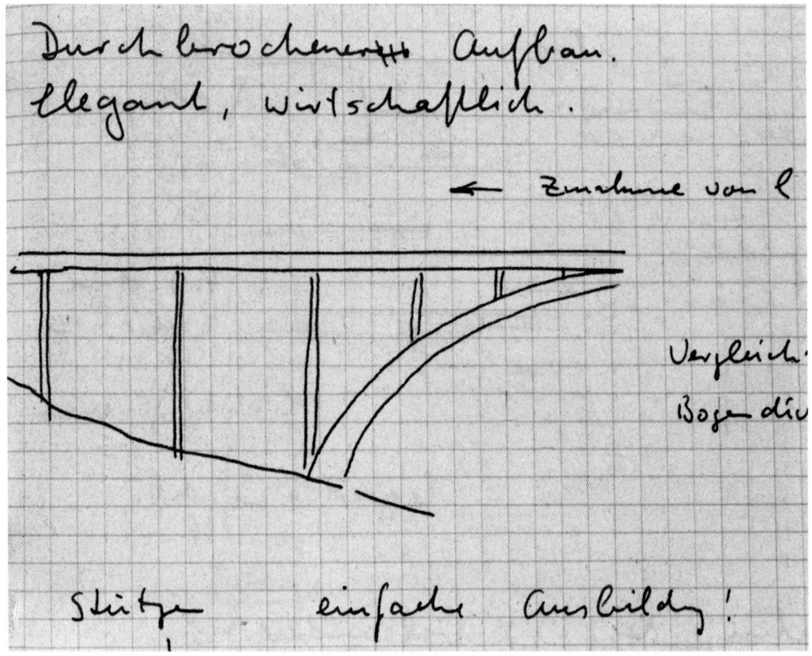

A set of drawings follows (figs. 19, 20) that illustrate careful relative proportioning, first of thicknesses for deck, verticals, and arch, and second of the desirability for spacing these verticals more widely apart as their heights increase. He noted emphatically that the double thick supports at arch abutments were terrible (*Schrecken!*) (fig. 19), and he praised the lightness of the open spandrel arch as elegant and inexpensive, especially when combined with an increase of the horizontal spacing of verticals as the height increases (fig. 20). He continually connected elegance with both lightness and integration, a negative example of the latter being the enlarged or double column over the abutments (fig. 21).[19]

These aesthetic examples do not obscure the mathematical rigor Lardy brought to his lectures, for example, with a full analytic theory both of thin cylindrical shells and of thin plates. He devoted the last parts of his senior spring course (eighth semester) to an advanced study of bridges, including deck-stiffened arches, an introduction of prestressed concrete, and a more theoretical discussion of reinforced concrete.[20]

The hallmark of Lardy's teaching, however, was his influence on a small but highly distinguished group of students whom he singled out to be his assistants after they had graduated. At the time, formal graduate education was not offered in European engineering schools; instead, professors chose certain gifted students to be teaching assistants, research assistants, and aides to the professor's consulting practice. The assistants were paid by the state and held positions as full-time engineers rather than low-paid graduate students. Some, like Christian Menn, did research and completed a doctoral dissertation while acting as an assistant, but many, like Heinz Isler, who had no desire for an advanced degree, entered engineering practice with considerably more background than the undergraduates (or diploma engineers, as they were called). Pierre Lardy's teaching fell, therefore, into two distinct parts: undergraduate lectures and the mentoring of his assistants. It is clear from the complete and elegantly recorded notes of Heinz Isler that Lardy's undergraduate teaching was well organized and presented in a lively and aesthetically sensitive manner. The record of mentoring his assistants has not been preserved, but his influence is clear. In only twelve years as professor, he attracted outstanding young men as assistants, and the recollections of the most distinguished all tell the same story. One of these, Hans von Gunten, later professor and subsequently rector of the ETH, described Lardy as one who "could inspire and sweep students along with his enthusiasm… he was a genius in analysis but he never stopped there, rather he sought synthesis and indeed found it."[21]

Figure 21
A double column, designated ugly (*unschön*), from Isler's notebooks.

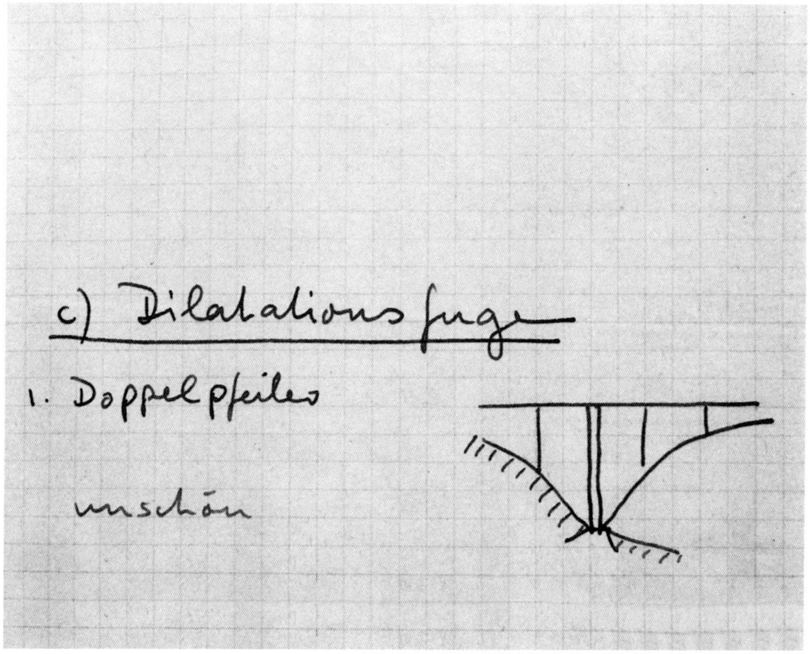

Hans Hauri, Lardy's chief assistant in the early 1950s and later professor and then president of the ETH, described his mentor: "For me, his merit is based on his outstanding personality. He was intelligent, quick in understanding, and open to new ideas." Hauri went on to illustrate Lardy's openness: "As an example: whilst he was plunged in his solutions of differential equations, I got permission to try out my ideas for measurements on models, a method I could develop first for plates, then for shells, and finally for three-dimensional concrete hydraulic dams. In a similar way, Heinz Isler could develop his ideas on shell-structures."[22]

In March 1948, the well-known Spanish engineer Eduardo Torroja (1899–1961) lectured at the ETH on his concrete designs, primarily three major thin-shell concrete buildings built before World War II. He emphasized the role of small-scale models in their design and described his laboratory where finely built models were carefully tested. Torroja was an academic and a designer in reinforced concrete. Like Lardy, he was strong mathematically and even wrote a book on the theory of elasticity, but for design he relied on simple ideas and clearly thought out patterns of how forces would flow in a structure. His most famous structure is the Zarzuela Hippodrome Roof near Madrid, a thin-shell cantilever completed in 1935 (fig. 22). Its scalloped edges and swooping form were the product of the engineer's imagination, and that excited Lardy. Torroja described how he had arrived at the form by emphasizing that its design was neither purely rational nor purely imaginative, "but rather both together. The imagination alone could not have reached such a decision unaided by reason, nor could a process of deduction, advancing by successive cycles of refinement, have been so logical and determinate as to lead inevitably to it."[23] This design process led Torroja to choose finally a doubly curved surface with a hyperbolic form. It is a structure without ribs. This roof cantilevers out 42 feet from the main support and goes 23 feet back in the other direction. At about 15 feet back, a vertical tie keeps the cantilever from falling over. The shell is only 2 inches thick at its free edge, and thus it has an extraordinarily light appearance.

Lardy responded enthusiastically to Torroja's lecture and the new ideas presented. As Lardy later wrote, Torroja's "conception, powerful and original, allied with a profound sense of structural aesthetic, opens a way for new and fertile possibilities." Lardy noted Torroja's Fronton Recoletos, his Hippodrome roof, and "we cite again the elegance of line in the covered market hall of Algéceris, where the periphery is prestressed to improve the state of stress" (fig. 23).[24] Torroja's focus on model analysis impressed Lardy and also stimulated his newly appointed assistant, Hans Hauri, who suggested that they follow the Spaniard's methods in Zurich. Encouraged by Lardy, Hauri set up such a lab at the ETH in the Institute for Structural Analysis and Concrete Structures,[25] and immediately embarked on tests of plates, using Plexiglas models, followed by experimental studies of the interaction of plates and beams for bridge decks. The following year he began model tests for studies of dams especially related to plans for the Grande Dixence (fig. 24). Lardy had been asked to study the static behavior of this immense project high in the Alps above the Rhone Valley. Hauri began with a planar model (a thin vertical slice from a model of the full dam) and explored the influence of loading on the uncertain ground and rock upon which the massive structure would rest. Construction on the actual dam started in 1951, ended in 1962, and resulted in the world's highest concrete dam (285 meters compared to 221 meters for the Hoover Dam).[26]

Still under Lardy's general direction, Hauri extended his studies of models to a thin cylindrical shell roof. In late 1951 Hauri formed a partnership and began his own private practice. Lardy wrote a letter to the canton engineer of Zurich in which he strongly recommended the two young engineers for future design contracts with the canton.[27] But Hauri still kept close to his mentor.

Figure 22
Eduardo Torroja's Zarzuela Hippodrome Roof, near Madrid, completed in 1935.

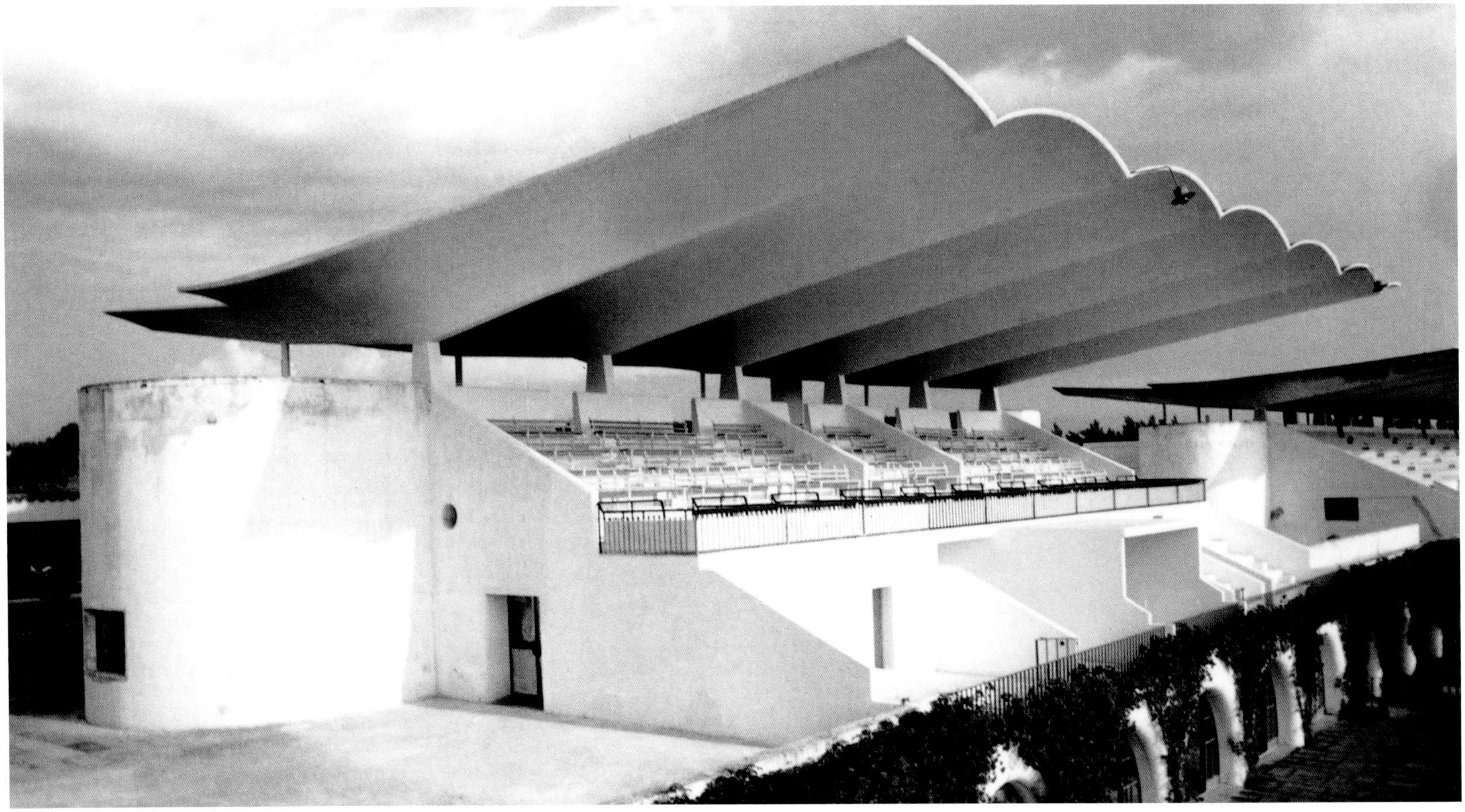

Over the next several years he carried out a series of model studies under Lardy's supporting direction for projects that included arch dams and the foundation for the CERN cyclotron in Geneva.[28]

Pierre Lardy's research reflected his mathematics background but focused on simplifying analysis for use by engineers. He published papers in three areas of structural engineering: prestressed concrete, plates and shells, and concrete dams. Although his contributions were not especially influential in academic circles, they did impress his students because he always wrote about new developments and used them in his teaching.[29] Lardy became involved with concrete dams because many large works were being planned, designed, and built in Switzerland after 1946. He worked as a consultant more often than as a researcher even though Hauri's model work was carried out under his direction.[30]

Two other activities were also significant: his leadership in developing a new Swiss code for concrete structures[31] and his major role in the direction of the International Association for Bridge and Structural Engineering. The development of a standard code of practice for such an important building material as concrete demands considerable time and a talent for working with a variety of engineers, many of whom compete with each other for design or construction projects. There may often be a tension between academics and designers, reflecting their differing perspectives: theory or practice, research or design. Lardy was an ideal person to mediate such conflicting views thanks to his engaging personality, his strong theoretical background, and his ability to communicate. He did not try to overwhelm the designers with needless complexity, but he was not overshadowed by academic theoreticians. His goal was to keep the code short and clear.[32]

Figure 23
Torroja's covered Market Hall in Algéceris, Spain, completed in 1933.

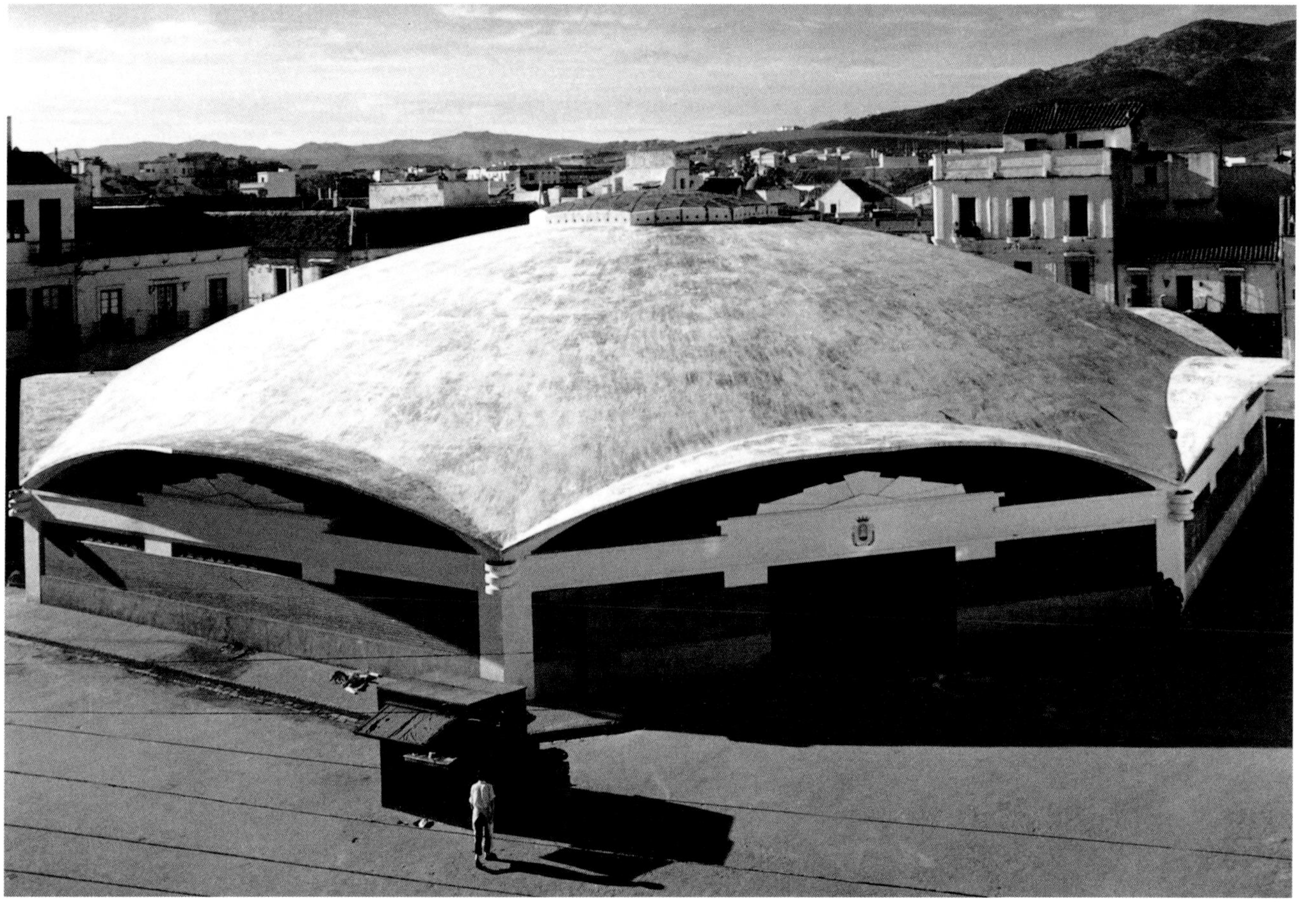

His gift for mediation was also crucial to his popularity and success as general secretary of the Swiss-founded International Association for Bridge and Structural Engineering (a mistranslation of the German *Brückenbau und Hochbau*, which means bridges and buildings). Here he had to deal with the international community—primarily the French and the Germans. He was a diplomat for structural engineering throughout Europe during the revival of international cooperation. But the essence of Pierre Lardy's significance for the generation to be educated after the centennial of his birth lies in the legacy of design that he did so much to nurture during his brief tenure in Zurich.

Thanks to Robert Maillart's handwritten notes, we have a sense of Wilhelm Ritter as teacher, and thanks to Heinz Isler's beautiful and complete notebooks, we can grasp Lardy's educational ideas. Here are the interactions of teacher and student at their most unselfconscious moments. The carefully prepared manuscripts of Lardy's Massivbau autograph is dry and dull—even as it is systematic and clear. But Isler's notes tell a different and far more human story; they are observations from the front lines; they are the undoctored expressions of teaching at its finest. When Lardy's colleague Fritz Stüssi wrote the obituary for the *Schweizerische Bauzeitung*, he stressed that Lardy's contributions were "in the best sense a continued furthering of [our] tradition founded by Karl Culmann and Wilhelm Ritter."[33]

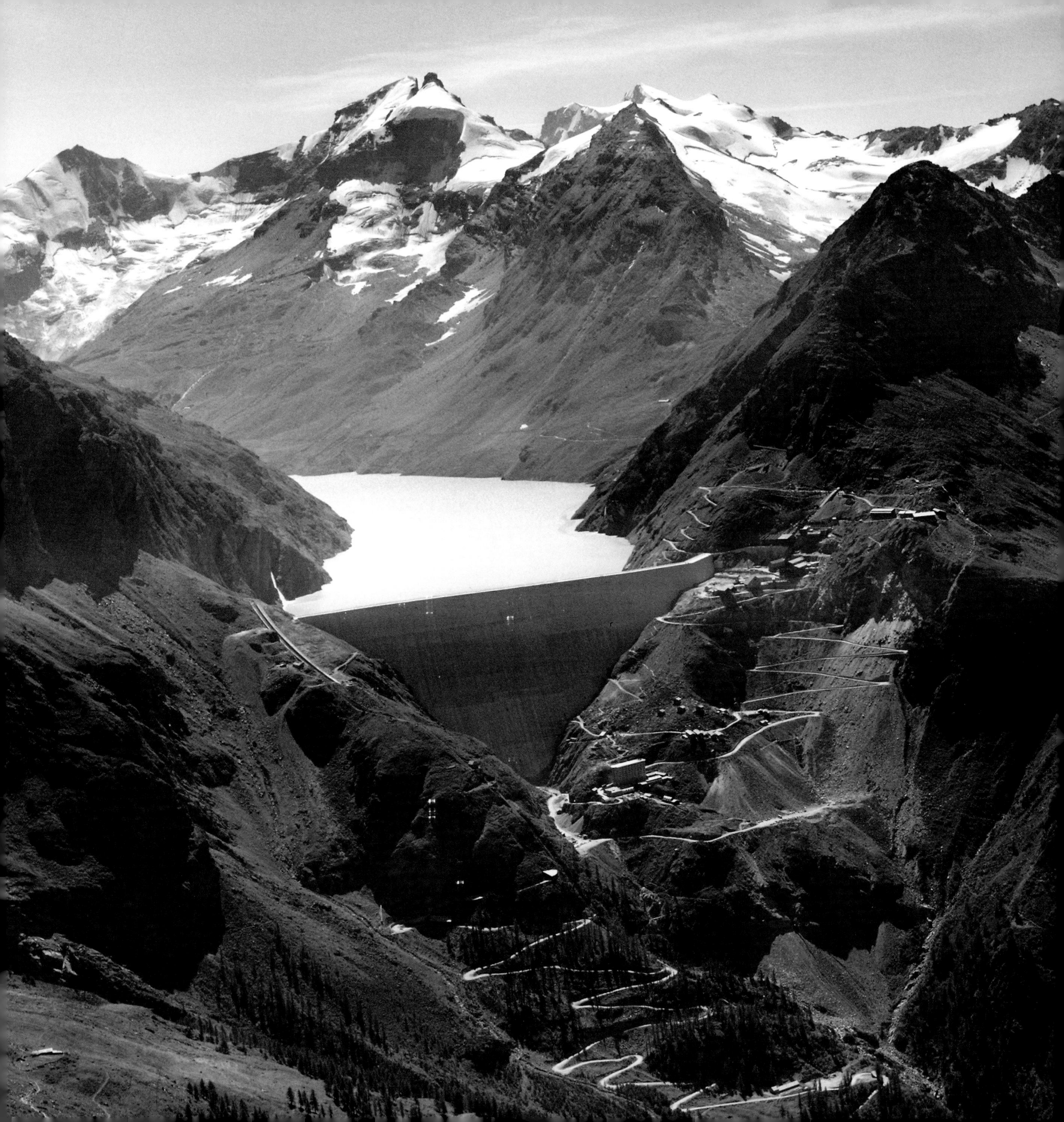

Figure 24

Grande Dixence Dam, Switzerland, 285 meters high.

Notes

1. Letter from Maillart in Geneva to his daughter, Marie-Claire, in Serang-Djaja, Nov. 12, 1930. Once again, on October 29, 1937, Maillart would dine with Ammann, who spoke warmly of Maillart's work. Ammann was in Zurich to give a lecture; see the brief notice in *Schweizerische Bauzeitung* [hereafter *SBZ*] 111, no. 20 (May 14, 1938): 260. Maillart described his meeting with Ammann in a letter from Geneva-Zurich to his eldest son, Edmond, in America, Jan. 15–16, 1938, Princeton Maillart Archive (hereafter PMA).

2. For a detailed discussion of these debates, see D. P. Billington, *Robert Maillart: Builder, Designer, and Artist* (Cambridge, 1997), 102–13. Emil Mörsch, a German, directly followed Ritter at the ETH, but left in 1908.

3. Ibid., 210–14.

4. F. Stüssi, "Nekrologe für Pierre Lardy," *SBZ* 76, no. 50 (Dec. 13, 1958): 762–63.

5. For a contrast between Maillart and Max Ritter, see Billington, *Robert Maillart*, 76–77.

6. M. Ritter and P. Lardy, *Vorgespannter Beton*, 2d ed. (Zurich, 1950), 52–53. (In the first edition of 1946, the committee is described on pages 48–49.)

7. Ibid., 9–10.

8. Ritter and Lardy (ibid., 32) show the losses in steel stress due to creep. The behavior of the beam with fs = 20,000 psi is more complex than noted here because as the steel loses stress, the force P is reduced and so is the creep.

9. P. Lardy, "Eigenspannungen und vorgespannter Beton," *SBZ* 121, no. 5 (Jan. 30, 1943): 51–53; "Allgemeine Dimensionierung und Spannungsberechung vorgespannter Eisenbetontraeger," *SBZ* 121, no. 20 (May 15, 1943): 239–42; "Eisenbetonkonstruktionen mit eingebetteten, vorgespannten Elementen als Armierung," *SBZ* 123, no. 18 (April 29, 1944): 209–12; "Vorspannung durch expansiven Beton," *SBZ* 124, no. 8 (Aug. 19, 1944): 95–96.

10. See note 6 above.

11. H. Hauri, "Recollections of Pierre Lardy," August 5, 2002, p. 1, PMA.

12. H. Isler, Notebooks from courses by Pierre Lardy, 1947–50. There are eight notebooks: 1. Baustatik II, first part, 1947; 2. Baustatik II, second part, 1947; 3. Massivbau I, first part, 1947–1948; 4. Massivbau I, second part, 1947–1948; 5. Massive Brücken, first part, 1949; 6. Massive Brücken, second part, 1949; 7. Massivbau II, 1949–1950; 8. Baustatik und Massivbau Ausgewählte Kapitel, 1950.

13. Isler, Notebook 3; and L. C. Urquhart and C. E. O'Rourke, *Design of Concrete Structures*, 4th ed. (New York, 1940), 278. This edition was the standard text in use from 1940 to 1954 in the United States.

14. Isler, Notebook 4.

15. Isler, Notebook 4, and for figures 9, 10, and 11, see Isler, Notebook 7.

16. Urquhart and O'Rourke, *Design of Concrete Structures*, 339.

17. Isler, Notebook 1; for figure 14, Notebook 2; and for figure 15, Notebook 5; see also M. Bill, *Robert Maillart* (Zurich, 1949).

18. Urquhart and O'Rourke, *Design of Concrete Structures*, 425, 470. For a full treatment of American ideas on concrete arch bridges, see C. B. McCullough and E. S. Thayer, *Elastic Arch Bridges* (New York, 1931); also D. P. Billington, "History and Aesthetics in Concrete Arch Bridges," *Journal of the Structural Division of the American Society of Civil Engineers* 103, ST11 (November 1977): 2129–43.

19. Isler, Notebook 5, where Lardy devoted one section to the aesthetics of arch bridges.

20. Isler, Notebook 8.

21. Letter from Hans von Gunten to D. P. Billington, March 25, 2002, PMA.

22. Letter from Hans Hauri to D. P. Billington, March 25, 2002, PMA. See also Hauri's "Recollections," 1–5.

23. E. Torroja, *The Structures of Eduardo Torroja: An Autobiography of Engineering Accomplishment* (New York, 1958), 7. For his theoretical work, see E. Torroja, *Elasticidad*, 4th ed. (Madrid, 1967).

24. *SBZ* 66, no. 9 (Feb. 28, 1948): 129, and P. Lardy, "Slabs and Various Curved Structures in Reinforced Concrete," in *Preliminary Publication for the Third Congress of the International Association for Bridge and Structural Engineering* (Zurich, 1948), 494–95.

25. Letter from Hauri, March 25, 2002, PMA. See also his "Recollections," 2–3.

26. P. Lardy with H. Hauri, "Modellstatik, optische Untersuchung von Platten, 1948–1950" and "Modellstatik Scheibenprobleme mit Dehnungsmesstreifen, 1949–1950," published by the Zurich Institut für Baustatik und Massivbau (ETH, Zurich), PMA.

27. Letter from Lardy to the canton engineer of Zurich, September 13, 1952, PMA.

28. P. Lardy with H. Hauri, "Modellstatik, Dehnungsmessungen anw3 dreidimensionalen Modellen, 1951–1956" and "Modellstatik Modellmessungen für Staumauern, 1951–1956," published by the Institut für Baustatik und Massivbau, (ETH, Zurich), PMA. For a summary of Hauri's works see C. Menn, "Hans H. Hauri zum 60. Geburtstag," *Schweizer Ingenieur und Architekt* 25 (1984): 501–503.

29. P. Lardy, "Die strenge Loesung des Problems der schiefen Platte," *SBZ* 67, no. 15 (April 9, 1949): 207–9, and "Die Elastizitaetstheorie der parallelogrammfoermigen Scheibe," *SBZ* 67, no. 31 (July 30, 1949): 419–22.

30. P. Lardy, "Ueber einige neue Untersuchungsmethoden mit besonderer Anwendung auf Gewichts-Staumauern bei nachgiebigem Baugrund," *SBZ* 71, no. 9 (Feb. 28, 1953): 121–22.

31. P. Lardy, "Die neuen S.I.A.-Normen fuer die Bauten in Beton, Eisenbeton und vorgespanntem Beton," *SBZ* 73, no. 42 (Oct. 15, 1955): 618–19.

32. In a letter of July 5, 2002, Heinz Isler stated that "Lardy's aim was to keep codes small, concise, clear, and general… and Lardy always stressed that the most important paragraph had to be added at the end… that always *exceptions* from the rules were allowed when scientifically sustained. This opened the view for development. It was not a code of prohibitions."

33. Stüssi, "Nekrologe," 763.

CHAPTER 5

HEINZ ISLER: STRUCTURAL ART IN THIN-SHELL CONCRETE

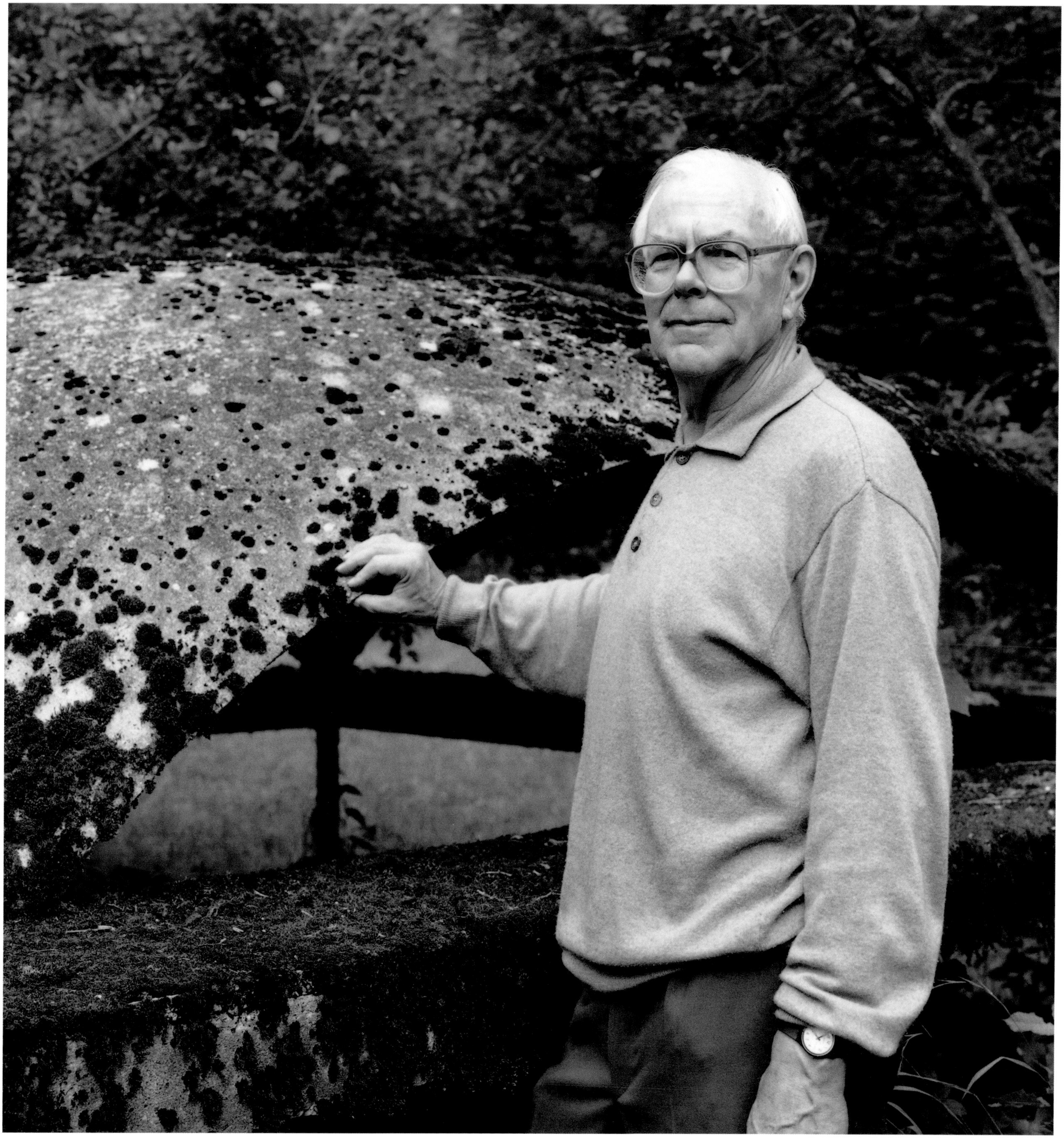

Figure 1 (previous page)
Heinz Isler (born 1926).

Figure 2
Isler standing beside glue-laminated wooden arches, Leuzlinger Sons Company Building, 1979.

Figure 3
Light, tubular scaffolding supporting wooden boards, Leuzlinger Sons Company Building.

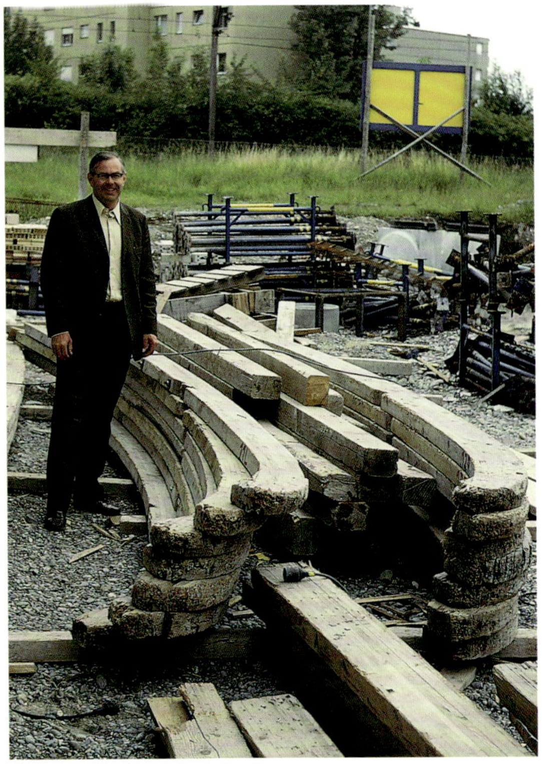

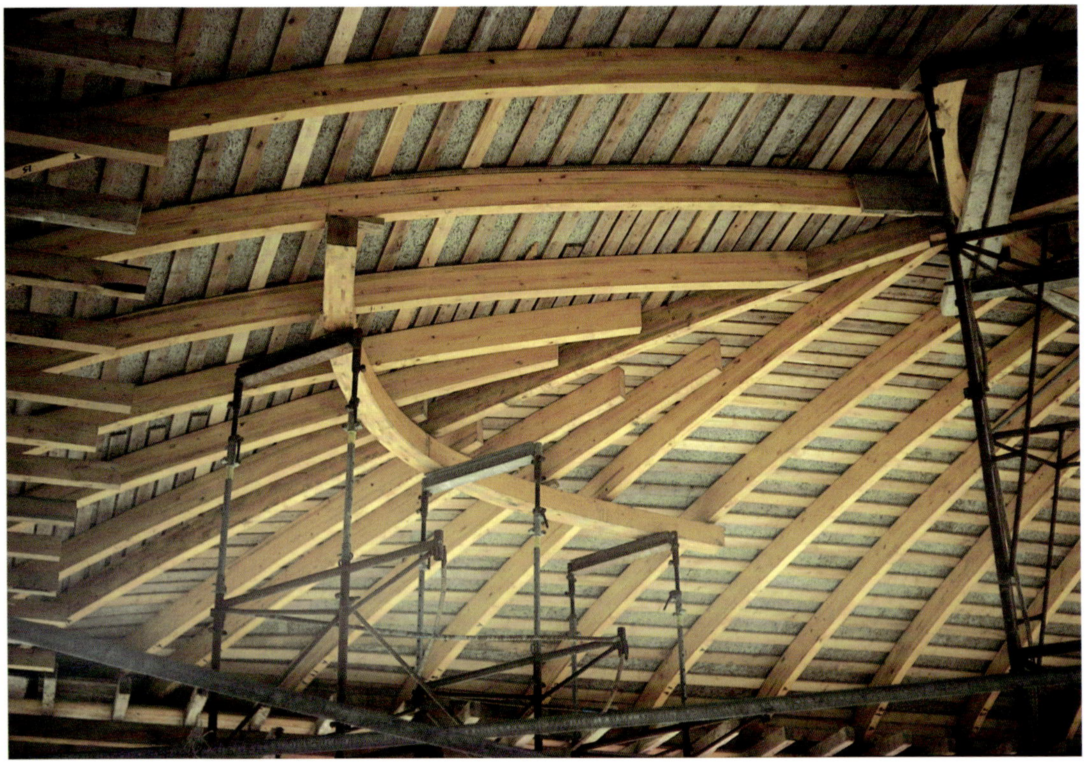

The year after Pierre Lardy's death in 1958, Eduardo Torroja organized and ran the first congress of a new organization, the International Association for Shell Structures. Torroja had stimulated Lardy in 1948 to develop and support a models laboratory in Zurich, and now he was encouraging structural engineers to exploit more fully the potential for thin-shell concrete structures. In retrospect, one event from this congress stands out as surprising and profound—the late arrival of one of Lardy's former students, Heinz Isler (fig. 1). He had come directly from military service and presented the last paper of the congress. A total of twenty-five papers were presented, all of which appeared later in the first eight issues of the association's new *Bulletin*. Torroja's goal for the congress was to present new developments in thin shells rather than to summarize the major works completed since the 1920s.[1]

The first twenty-four papers were not unlike most conference papers, presentations of individual structures or groups of them, which elicited a few questions from the audience. The longest published discussion was two pages for a fourteen-page paper on Polish shell designs. Nothing quite prepared the reader for paper twenty-five. It was entitled simply "New Shapes for Shells," and it contained slightly more than one page of text and nine illustrations.[2] In the text, Isler briefly described three ways to arrive at shell shapes: (1) the freely shaped hill, where, for instance, molded earth is the form; (2) the membrane under pressure, where an inflated rubber membrane gives the shape; and (3) the hanging cloth reversed, where a draped fabric defines surface shape just as a hanging cable defines a funicular line for suspension bridge cables. His illustrations were of some models and a few completed concrete shells, and a full page showed thirty-nine possible shapes with a fortieth replaced by the abbreviation "etc." The implication is of unlimited possibilities.

The paper had an immediate impact and brought forth a rash of discussion, which in print was about five times as long as the actual text. The verbal discussion was substantially longer than the printed version. The quality of the discussers was as remarkable as the relative length of the commentary, which was dominated by the most distinguished designers present: Torroja himself, Nicholas Esquillan of France, and Ove Arup of Great Britain. Isler's

Figure 4

Formwork of fiber board insulation placed on wooden boards, Leuzlinger Sons Company Building.

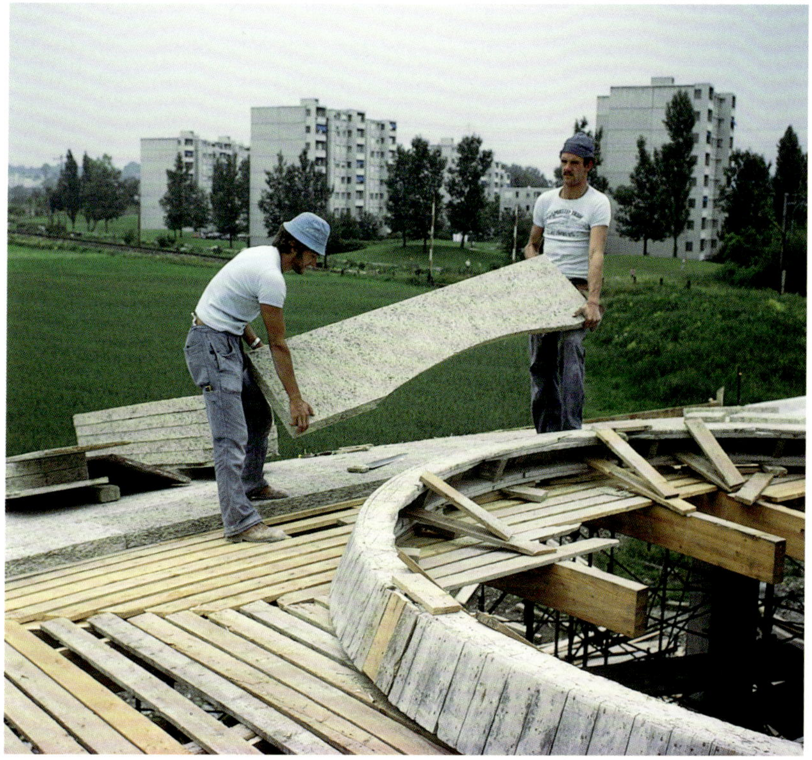

brief presentation struck three themes that the discussers addressed: that new shapes come from simple models; that the resulting nongeometric shapes could be economically built; and that the shapes were designed without reference to architecture.

Torroja concentrated on the question of models, worrying about the dangers of their being used without experience. He emphasized the risks involved in forming ideas based on visually stimulating models that are not related to actual full-scale structures. He also noted the problems of scaling in going from small models to full-size designs. His examples were his own stunning thin shells such as the Market Hall in Algéceris. By contrast, Esquillan, fresh from completing the world's longest spanning (206 meters) thin-shell roof in Paris, concentrated on the economy of Isler's nonmathematically defined shell shapes: "I say that the cost of the scaffolding and formwork is at least fifty percent of the total construction cost of a thin shell. That is how it is!" Isler claimed those costs were only twenty percent in his shells, to which Esquillan replied: "Are the prices in Switzerland so different than in other countries?"

Finally, Ove Arup, head of the largest firm of consulting engineers in Great Britain, spoke about collaboration with architects and about buildings being "functionally right, architecturally right and aesthetically right, and it is not any great consolation that we get something else cheap." He mentioned the problem of shape only in the context of architecture, something Isler had omitted entirely. Undoubtedly, on Arup's mind was the Sidney Opera House, an edifice on which he was to struggle for years as structural engineer.

Isler's responses are at the heart of structural design, revealing the true function of basic research in design. Torroja's concern about models reflected his vast academic experience, which was based on the scientific analysis of structures. Isler's response met that concern directly by emphasizing that shaping (with models) comes first, but it is only "the first link in a whole chain of investigations; the other links are (small-scale) model tests and measuring the first structure, i.e., model tests at full scale as we have it out there; these are of primary importance." Isler noted that technical education in mathematics and geometry did not directly provide the means for studying new shapes. What is important, he stressed, are *physical analogies*, such as the inflated membrane or the draped cloth. But that was only a first part of the problem; a second part was the cost. To answer Esquillan, Isler said: "We use a system which maybe has a few new ideas." These consisted of very carefully made glue-laminated wooden arch pieces (fig. 2) placed on light tubular scaffolding (fig. 3) and supporting wooden boards, over which fiber board insulation rests as formwork (fig. 4). In short, Isler had used both new shapes and new construction methods at the same time.[3]

When asked how he could build shells to accurate shapes when his design used such small models, Isler responded: "We measured the coordinates with very subtle, accurate instruments and then it (the shape) is drawn in a larger scale, and you get a series of curves. If there is a little mistake, it is seen immediately and then you can correct it." This is machine precision controlled by visual observation. Isler made no response to the question of architecture. He merely emphasized that finding good solutions to thin-shell shaping was an engineering problem that could be solved only by imaginative shaping controlled by full-scale and small-scale measurements and by extreme economies resulting from careful construction procedures.

For Isler, the way to Madrid had begun years before in Zurich. Born in Zollikon, now part of Zurich, on July 26, 1926, Heinz Isler attended public primary school in Zurich, and entered the canton school (high school) in 1939. In

1944, at the urging of one of his teachers, the school held an exhibition of Isler's drawings and watercolors, which reflected his love of the natural world in the Swiss landscape. All through school he had the reputation of working alone and doing his work in an unusual way. Following graduation in 1945, he spent nine months in compulsory military service, where he worked on outdoor construction; this he enjoyed immensely. He was attracted to a career in painting, but his father, a geodesic engineer and later director of roads and water for the city of Zurich, insisted that he first study for a professional degree. Therefore, on November 4, 1945, he entered the ETH in Zurich, from which he graduated in 1950 with a degree in civil engineering (his four-year education had been interrupted by a year's service in the army). For his final-year design project, he chose to study thin shells; he was the only one of his more than one hundred classmates to choose that subject.[4]

One of the highlights of Isler's education was Lardy's lectures on structures; he was impressed particularly by Lardy's "chamber of horrors," a series of slides showing ugly structures.[5] Following graduation, he accepted Lardy's offer of a position as assistant at the ETH, where, from January 1951 until May 1953, he helped Lardy with teaching. He also worked on the many cases of structural failure that Lardy had been asked to explain. Lardy's direct influence on Isler was evident through the student notebooks. Deeper even than that mute testimony is the summary that Isler himself gave half a century after leaving Lardy's assistantship:

> I think the most important contribution we students got from our teacher Lardy is this:
>
> He reminded us, the engineering students,
> a) that we have in us a sense for esthetics
> b) that we have the right to use it
> c) that we are allowed to mention our opinion
> d) and that we can find and express it in our projects.
>
> This to my opinion was the invaluable, great and *unique contribution* he gave to us. Not the statics, not the theories, not the investigations were his greatest and lasting influence but encouraging us to find and apply esthetics from within us.
>
> And for this I am very grateful to him.[6]

The four principles noted by Isler are almost never part of engineering education. Engineers and especially academics often argue that aesthetics are not part of their profession. If you want beauty, hire an architect or, more radically, a sculptor. This attitude does not reflect the reality that Lardy knew and Isler expressed, a reality that includes an aesthetic sense within all of us. But it can be suppressed, avoided, even ridiculed, and one major objective of education in engineering should be to encourage students to see, accept, and begin to use that elemental sense. That teaching, however, is only viable when practiced by someone like Lardy, whose disciplined mathematical training (a training that most engineering professors now have) provided the sole context within which the search for engineering aesthetics could flower.

There are many educators who think that beautiful diagrams, elegant models, and free forms will stimulate engineering students to create beautiful works. They are wrong. Beautiful structures only grow out of disciplined engineering. Lardy knew that and Isler's shells express it. Lardy may appear to us in the early twentieth-first century as a unique, inimitable character of another age and more specifically of another background. The legitimacy of his teaching practical aesthetics rested on his serious mathematical background combined with some direct experience in traditional art. It has been overlooked that, fifty years later, surprisingly large numbers of engineering teachers have a similar background. Like Lardy, they are grounded in mathematics, and they have not come from sustained experience in practice. Some have deep connections with painting, music, and literature. But it does not occur to them that they could infuse their teaching with the sense of aesthetics that Isler got from his professor.

Why does this incomplete education persist? Evidence from Isler and his colleague Christian Menn support the imperative for such education in our time. This imperative requires that the case for engineering structure as art be made as clearly and as convincingly as possible. It can be made only through a study of structural artists. Therefore, it is essential in explaining Isler's work to recognize how it avoids needless complexity or mere novelty and rises to art. His major works follow the theme of finding and expressing aesthetics and doing so in a way that avoids form-making pitfalls of computation and of complication.

Thin-shell concrete structures first arose in Germany during the 1920s, when a talented group of engineers, working for the designer-builder firm of Dyckerhoff and Widmann, developed the mathematical basis for domes and barrel shells.[7] They built many roof shells before World War II and wrote extensively about this work. Lardy referred to their work during his advanced concrete course in

Figure 5
Wood frame holding a rubber membrane inflated to form a pillow-like shape.

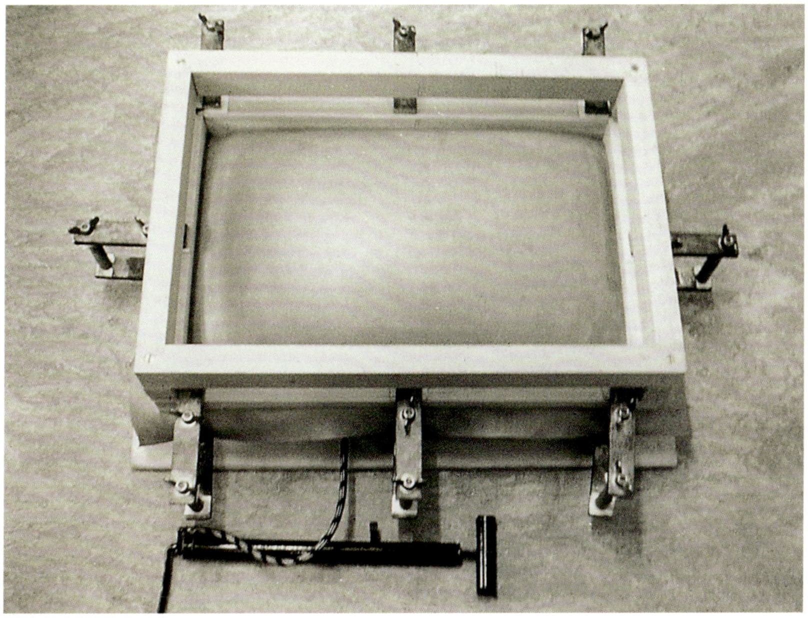

Figure 6
A student in Isler's models laboratory, ca. 1965.

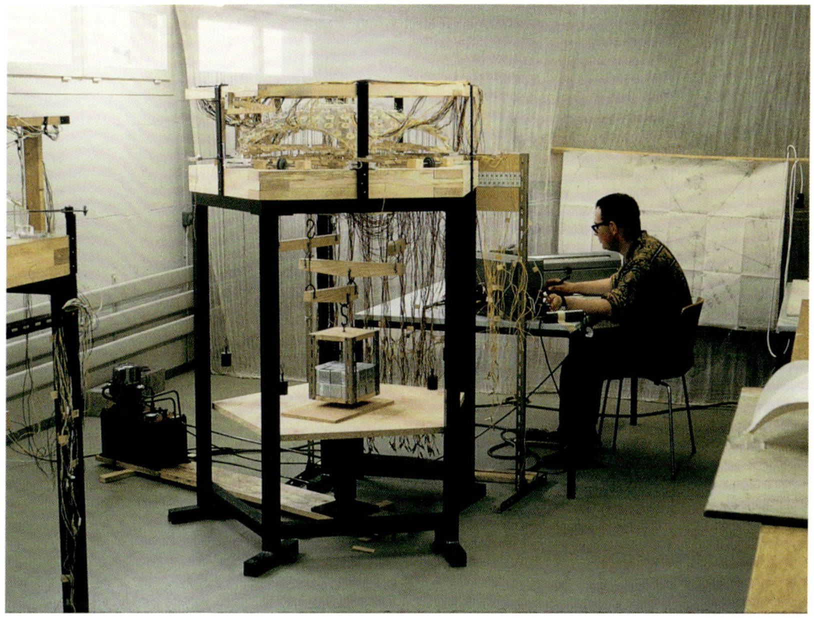

which he presented the mathematical theory for cylindrical shell walls. He drew on the pre-1950 standard published works that reflected the mathematical complexity the German engineers believed essential to justify their new, thin structures that were still restricted to elementary geometric forms. After carefully studying and designing in these traditional forms, Isler found a way in the 1950s to free himself from both geometric constraints and mathematical complexity.

The German ideas came to the United States through an Austrian engineer, Anton Tedesko (1903–1994), who became America's leading shell designer and who moreover sought to simplify the mathematically constructed shell forms.[8] Other European works, including designs by Pier Luigi Nervi, Torroja, and Maillart, seemed to overshadow Tedesko's proven designs so that many architects and engineers thought their spectacular new forms brought a freedom to make any shapes out of fluid-cast concrete. The resulting undisciplined designs led to complexity both in analysis and in construction, and also produced a series of defective structures, designed without recognition of the austere discipline that had guided the European engineers and that became the hallmark of Isler's work.

Before Madrid, Heinz Isler had studied geometric forms and, finding them unsatisfactory, arrived at two revolutionary ideas, both of which have an aesthetic origin that connects them to Lardy's influence. But both are fully the product of Isler's own imagination. At twenty-eight, the same age at which Maillart conceived the hollow box in concrete, Isler pioneered his first idea, the pneumatic form. In 1954 he designed a complex shell roof that grew out of a barrel form; although it was successful in performance, it did not satisfy the designer. One evening on returning to his garret room, he noticed the pillow on his bed and suddenly saw a form that stirred his aesthetic sense—a form that looked right and elegant and natural. He immediately devised an experiment by building a simple wood frame that held a rubber membrane, which he inflated to achieve the pillow form (fig. 5). This device gave him a shape that was in pure tension under pressure loading from below; thus, such a loading from above would put that same shape into pure compression. That load, on a flat shell, is so close to gravity loading (purely vertical) that Isler was sure the form would still be in pure compression and thus be ideal for reinforced concrete.[9]

By 1955 he successfully completed a pneumatic structure for the Trösch Company, which led to other commissions for industrial buildings and stimulated Isler to develop a profitable sideline. The pneumatic shells, as continuous surfaces, blanked out light and led Isler to invent a plastic domed skylight, manufactured first by himself and later by the Eschmann Company in Thun. With money from his skylight, Isler created his own models laboratory (fig. 6), where he could test his new forms by careful measurements of deflections and stresses.

Figure 7
Interior of the pneumatic-form factory roof of the Eschmann Company, completed in 1958.

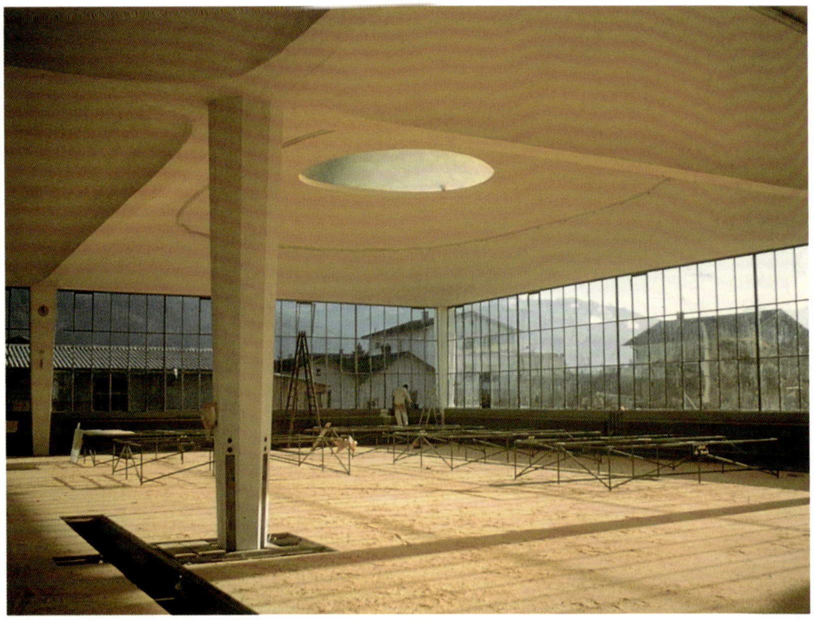

Figure 8
Wet burlap cloth hanging in pure tension over wire mesh.

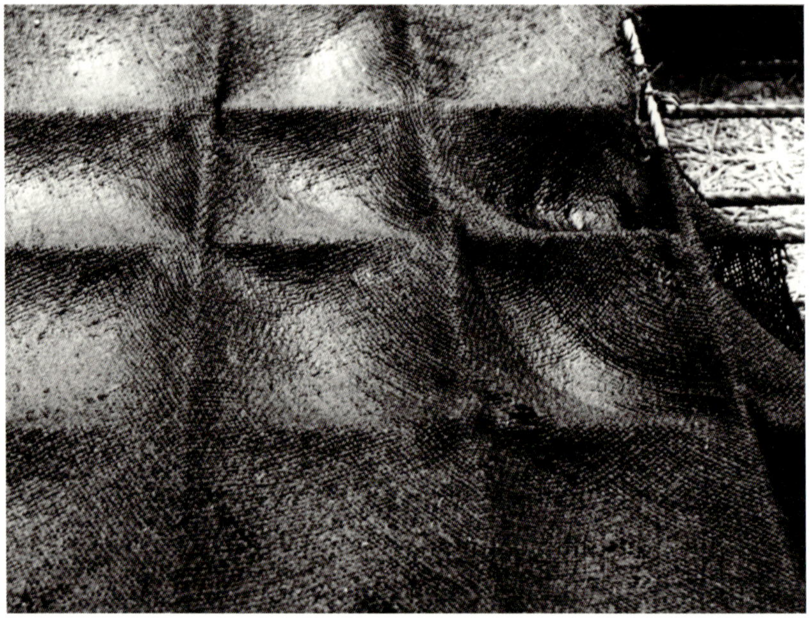

The Eschmann concern proceeded to award him the contract to design a factory building roofed by his shells, which was completed in 1958 (fig. 7). For the first time, Isler was able to design a structure without involving an architect.[10]

Significantly, the pneumatic form required Isler to think about the nature of concrete as a watertight material. His pillow idea had produced a roof in practically pure compression under gravity loading that therefore would not crack the shell and cause leaks. But there were two other immediate issues: edge support and temperature stresses. For the former, Isler designed straight-edge beams whose tension stresses would be counteracted by prestressing—easily done through the straight-edge members (see fig. 2, chap. 4). The prestressing had the added virtue of lifting the shell off its forms, making their removal far easier. The forms also contained the solution to the temperature problem, as he had hinted in Madrid, by using fiberboard as the concrete forming and leaving it in place anchored to the concrete. In this way, the fiberboard acted as insulation so that the concrete exposed to the elements would have no thermal gradient, i.e., the inside concrete surface temperature would be nearly the same as that outdoors. Thermal gradients would cause tension stresses and cracking even if gravity loads caused only compression. Isler's construction method thus avoided any cracking whatsoever and provided an acoustical ceiling as well.

By 1959 Isler had discovered a method of designing and building that would serve him throughout his career. More crucial, his pneumatic shells were so obviously desirable for industry that they became his main line of standard building. Just as Maillart's flat-slab floors sustained his unprofitable bridge business, Isler's pneumatic-form, thin-shell roofs would sustain his adventures into unique custom-made thin shells. But pneumatic shells did not satisfy his urge to exploit the full potential of thin shells.

Isler's second great discovery occurred in the early summer of 1955. On a building site he saw wet burlap draped over a mesh of steel bars in such a way that within one square opening (fig. 8) the cloth hung in a domelike shape solely under its own weight.[11] From this observation, he immediately concluded that the cloth carried itself in pure tension so that when reversed, it could be the form for a concrete shell under pure compression. During the winter of 1957, he sprayed water on such a tension membrane to make it freeze; and when it was inverted, it became his first hanging-membrane reversed thin shell. In principle, this form would be better than the pneumatic shells because the load that formed it was identical to the major load on the concrete shell itself. But to utilize this concept for actual construction required more time, even though Isler was able to describe it at the Madrid conference of 1959.

Figure 9

Felix Candela's Xochimilco Restaurant in Mexico, with its hyperbolic paraboloid roof on eight supports, inspired Isler to think about the expression of thinness.

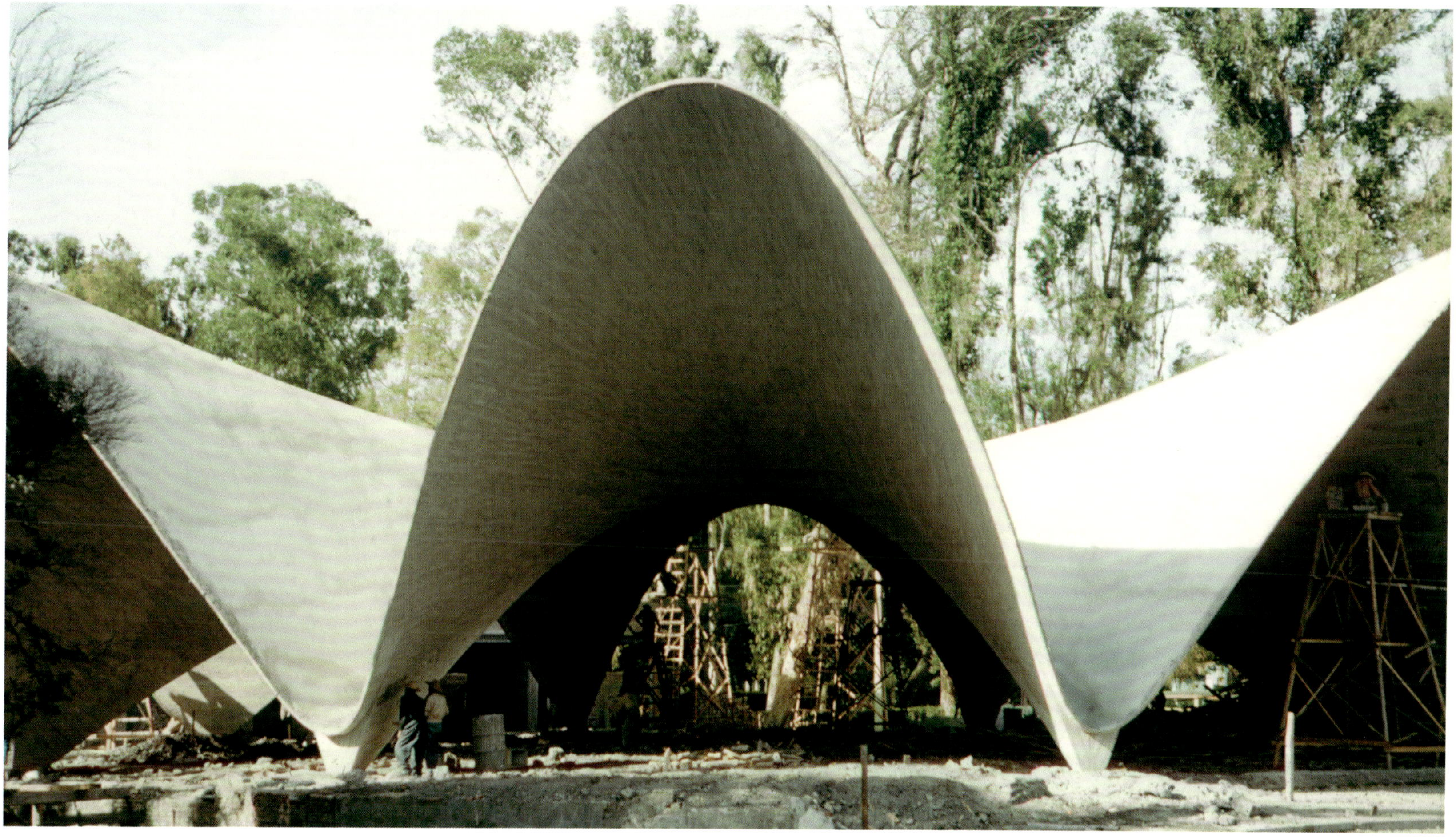

One other key event during the 1950s was Isler's collaboration with the nearby Langenthal building firm of Bösiger, whose personnel he trained in his new building methods. Unlike Maillart, Isler never had a construction business, so he had to design his shells and the procedure for building them himself, then convince a contractor to follow his ideas carefully. This led to both high quality construction and competitive cost structures. But one other incident helped Isler achieve his mature structures of the late 1960s.

In the early 1960s, Isler saw in a Zurich store a book cover with a photograph of the Xochimilco Restaurant (fig. 9) of Felix Candela (1909–1997); the forms that appeared exhibited a completely new shape—almost paper thin—and caused Isler to think about the expression of such thinness. Candela was by then not only a building contractor, but also a recognized structural artist with thin-shell concrete roofs, which he designed using the geometric form of hyperbolic parabolas. He too had devised a roof shell that fit Mexico's economy and provided a steady income from industrial clients. Candela's industrial shells were column-supported like Isler's and, like the Swiss forms, they required prestressed horizontal edge beams. But Candela had found, for custom designs, a way to express thinness with his hyperbolic paraboloids, and that is what so intrigued Isler.[12]

Candela had already become well known by the late 1950s for his stunning shell roofs in Mexico. He had been educated as an architect in Madrid but had always been fascinated by thin shells. He planned to study them in Germany during

Figure 10
Bacardi Rum Factory, completed in 1960. Architect, builder, and engineer: Felix Candela.

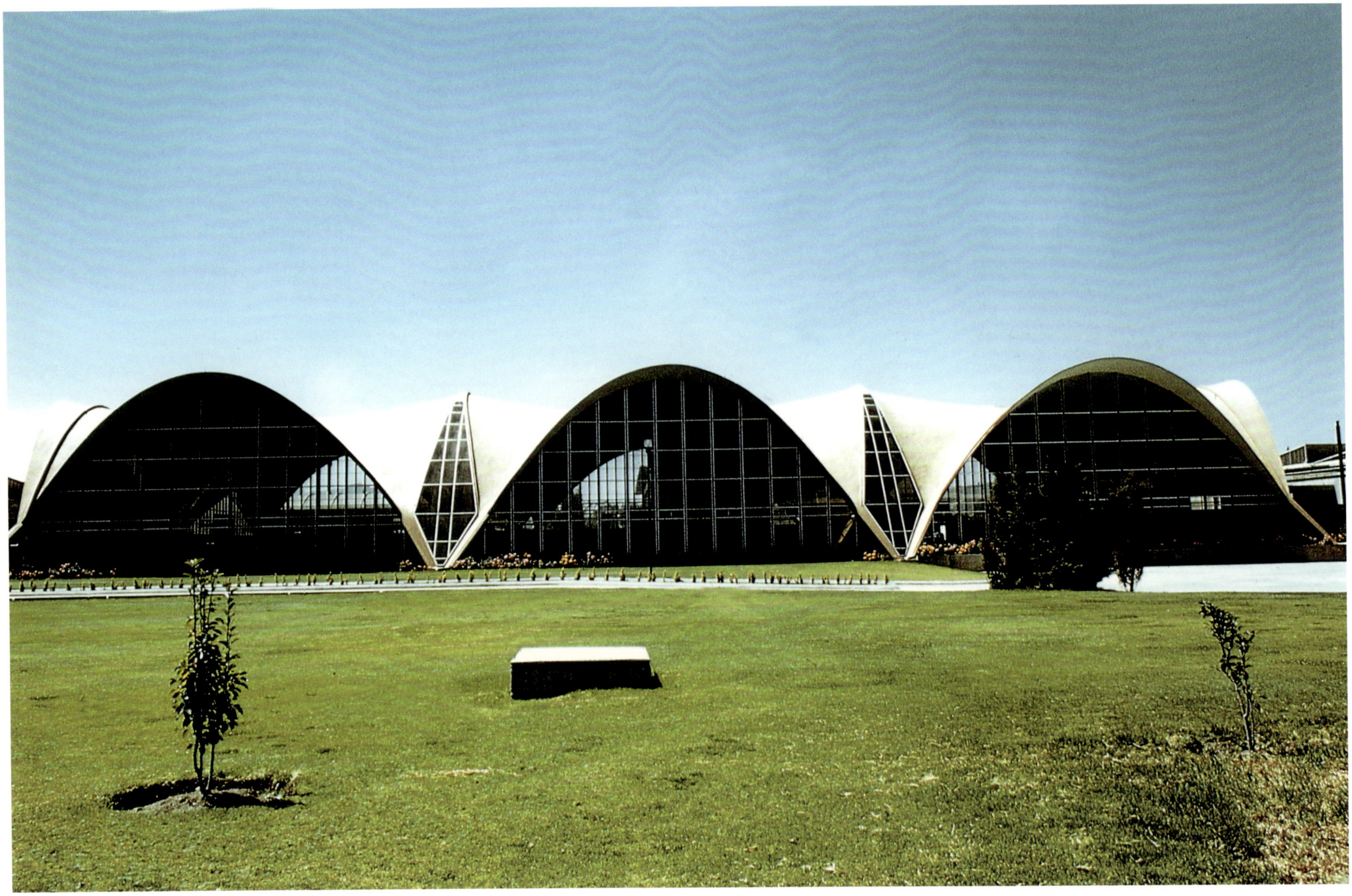

the late 1930s but got caught up in the Spanish Civil War and was exiled to Mexico. There he began a construction business building his own thin shells, and by 1954 his company had become successful. His first major shell was the cosmic-ray laboratory in Mexico City (1951), and seven years later, he completed the Xochimilco Restaurant (1958), which had caught Isler's eye. In 1960, Candela, acting as engineer, architect, and builder, as usual, completed a Bacardi Rum factory building (fig. 10) – again illustrating the thinness and economy made possible by a well-chosen form.[13]

During the 1960s, Isler experimented with highly unusual forms, two of which resulted in roofs that achieved an exceptional lightness while directly expressing extreme thinness. The first of these is the Wyss Garden Center in Solothurn, built in 1962 (fig. 11), where Isler used a geometric form (a translational form) that, being nonstructural, required edge stiffening. He achieved this structural necessity by adding curved elements that connected to the main shell at a sharp angle and therefore did not form a fully integrated surface. Isler saw this imperative as an expression of the flowers that the structure serves.[14]

Figure 11

The Wyss Garden Center, Solothurn, Switzerland, built in 1962.

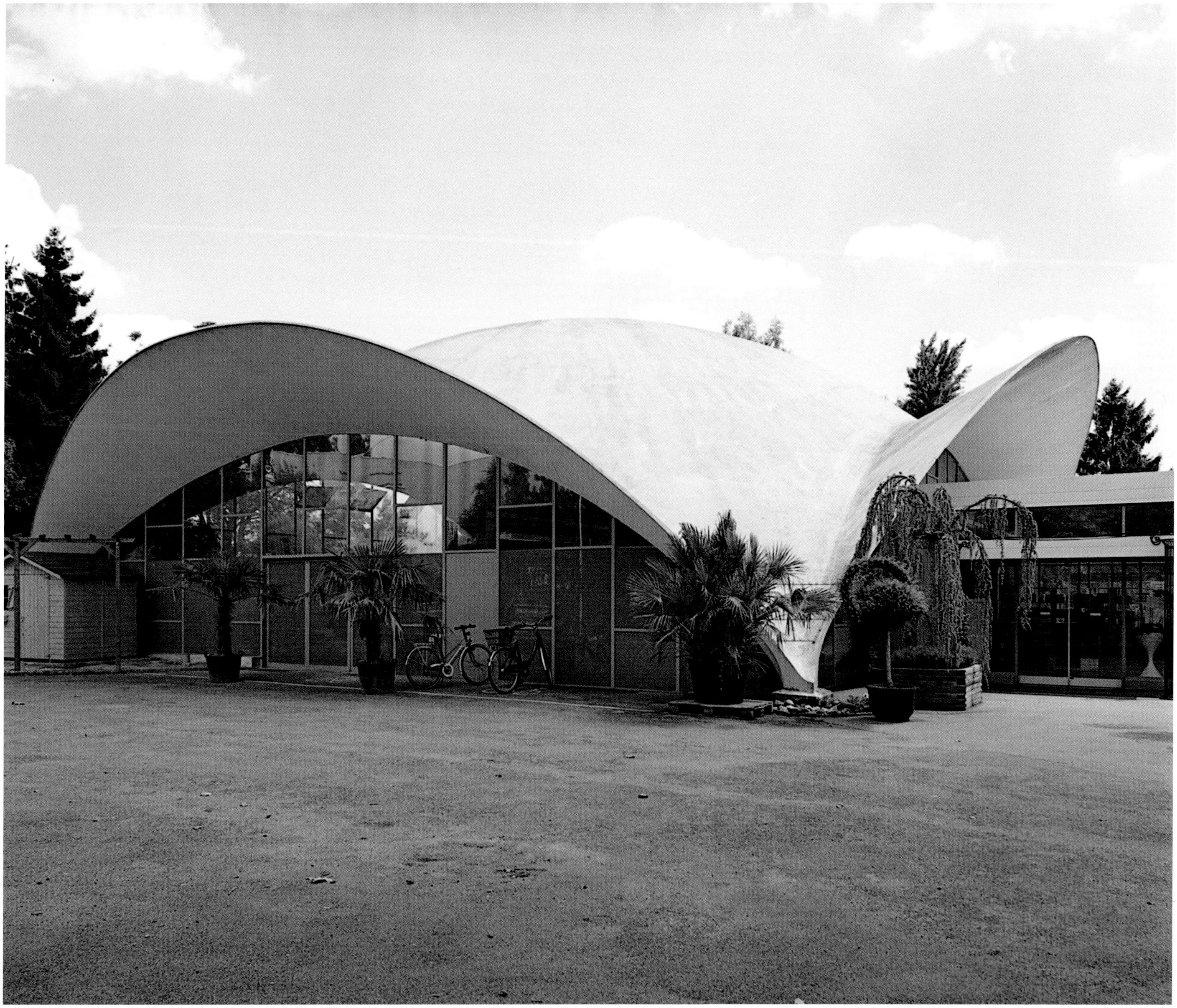

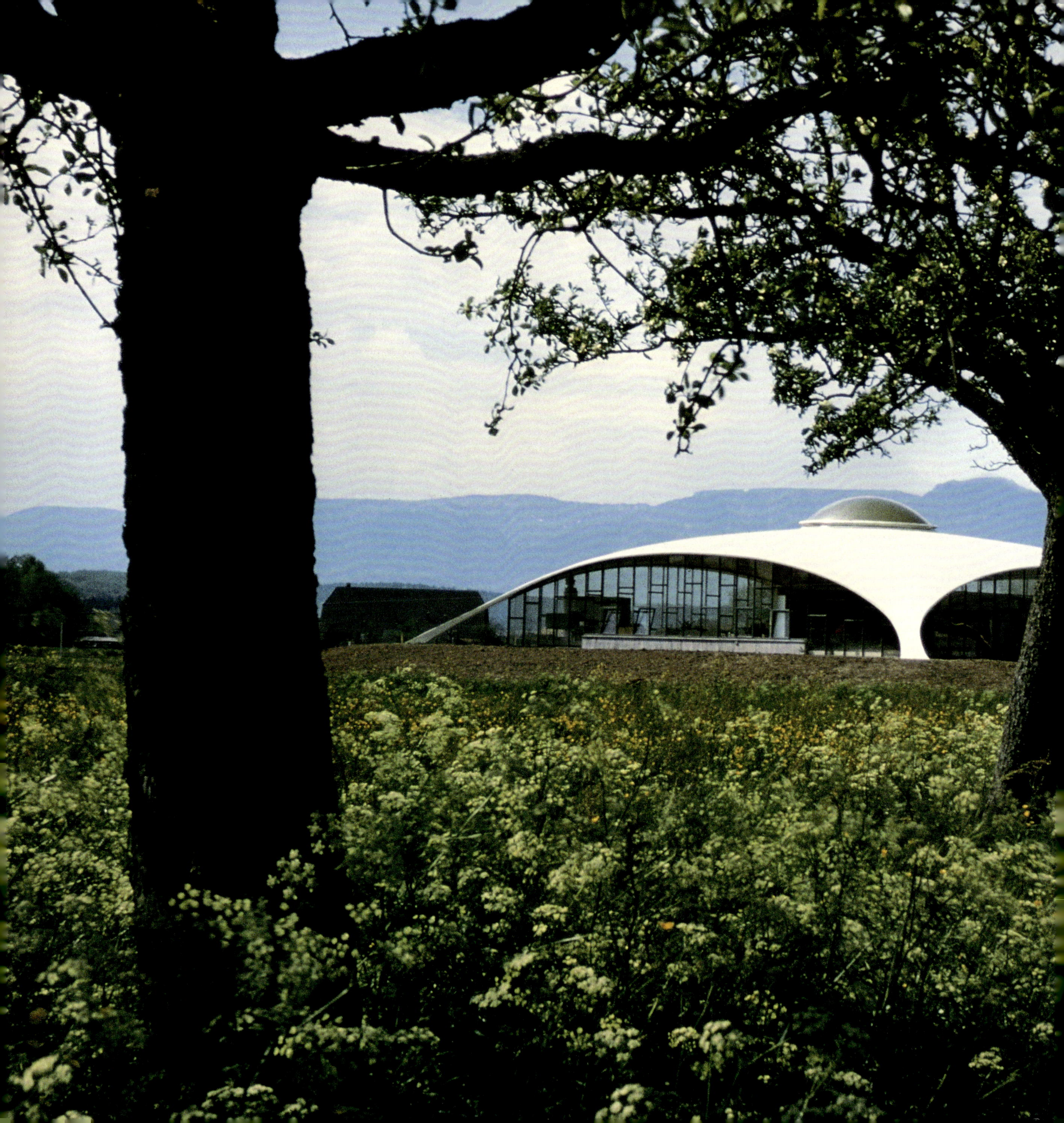

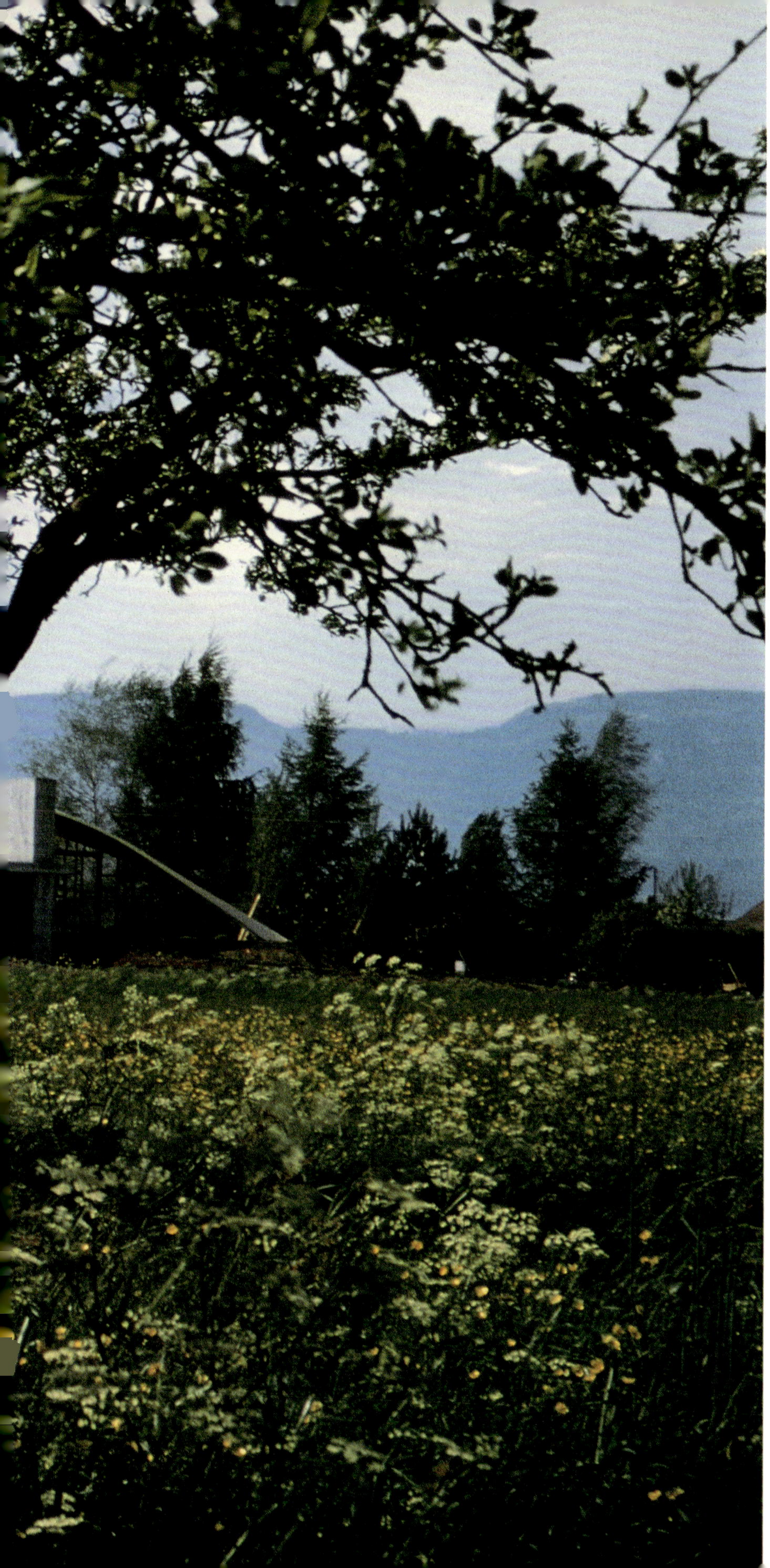

Figure 12
Kilcher Company, Recherswil, designed by Isler using the "flowing method."

He revised this plan in 1965 with a shell roof that was nongeometric in form, which he developed from an entirely new idea. As he wrote in 1980: "Models for shape may also be made by the flow method, by which the advancing velocity of a liquid inside a tube is varied. At the wall, velocity is zero because of friction [and in the corners doubly constricted], whereas in the center there is maximum velocity. A slowly expanding foam leaves a square tube and forms a dome shape. This natural function produces lovely shapes."[15]

This highly unlikely method deviates radically from the pneumatic and the hanging forms because it is not directly connected to design loadings. It is not at all obvious why the velocity of flow should relate to an efficient form, but as Isler made clear, it "produces lovely shapes." The results struck an aesthetic chord within Isler, and he next searched for a way to express it in a project. One soon came: "A factory for the Kilcher Company in Recherswil, near Bern, was designed in 1965 following the flowing method (fig. 12). The shell is 8 to 10 centimeters thick, has no edge beams, and is on a 26-meter square plan. The architect was P. Wirz of Solothurn, but I did the general design and the roof form from model tests."[16]

More such forms followed, including a series of garden centers in France built on a pentagonal plan with sides of 34 meters. For these larger shells, Isler resorted to edge shells for stiffening although they were not so prominent as at Solothurn. As Isler's search for form continued in the 1960s, his business expanded and enabled him to design an office building built in 1965 for his laboratory, and an exhibition and teaching area. As with all his works, he used the structure of his office as a full-scale model to test a new idea for concrete. Having established curved forms that needed no roofing or waterproofing, he decided to design a flat concrete roof that also would be watertight without roofing, just bare concrete. He achieved this unheard-of goal by designing a simply supported slab so that under bending, there would only be compression in the upper part and hence no cracking where rainwater could enter. Finally, he applied a light prestressing to increase the compression and, as with all his works, he insisted on high quality concrete and workmanship.

Figure 13
Shell model on the leak-proof roof of Isler's office building, Burgdorf, designed in 1965.

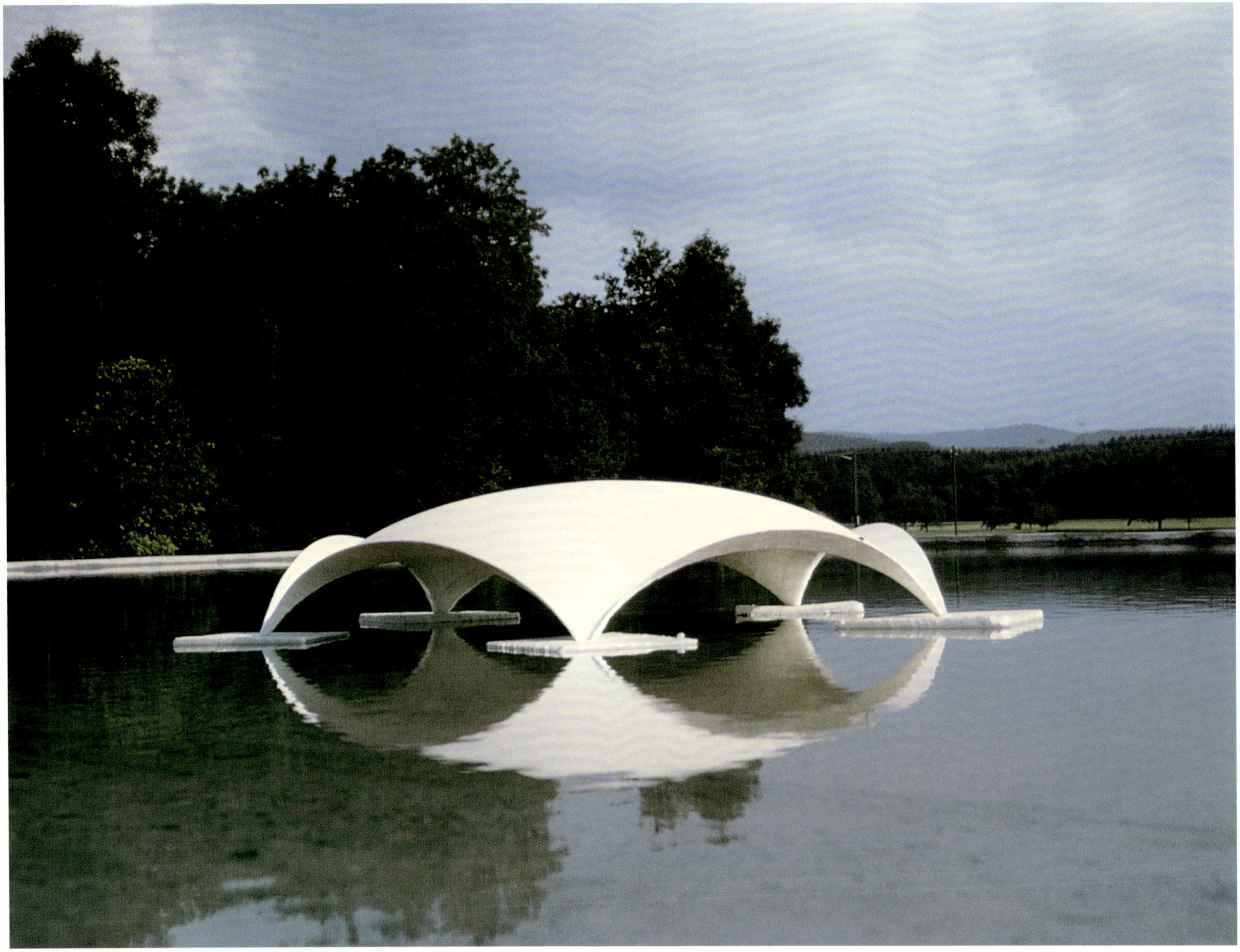

Figure 14
Isler standing in the marsh that gradually developed in the pool on the roof of his office building, Burgdorf, 1983.

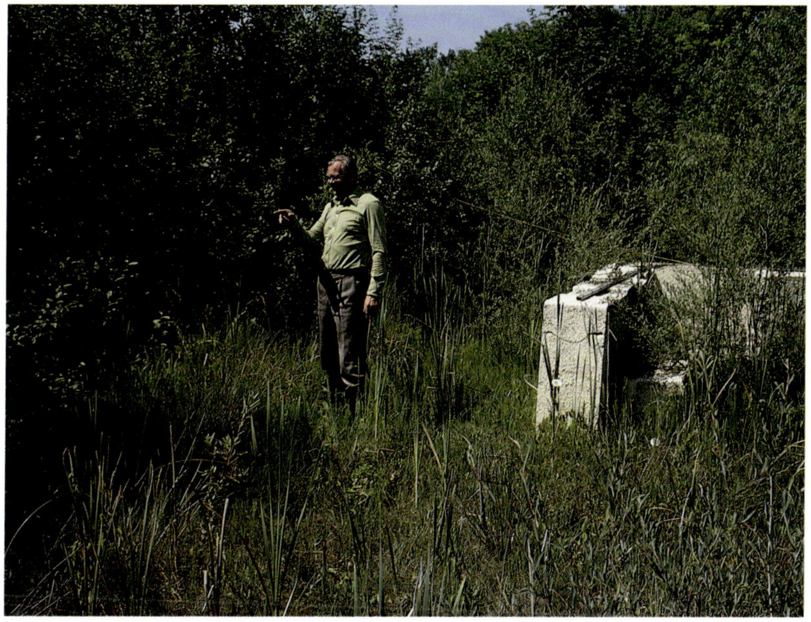

Figure 15
Isler demonstrates his model train set to his wife, Maria, and Marie-Claire Blumer-Maillart, 1983.

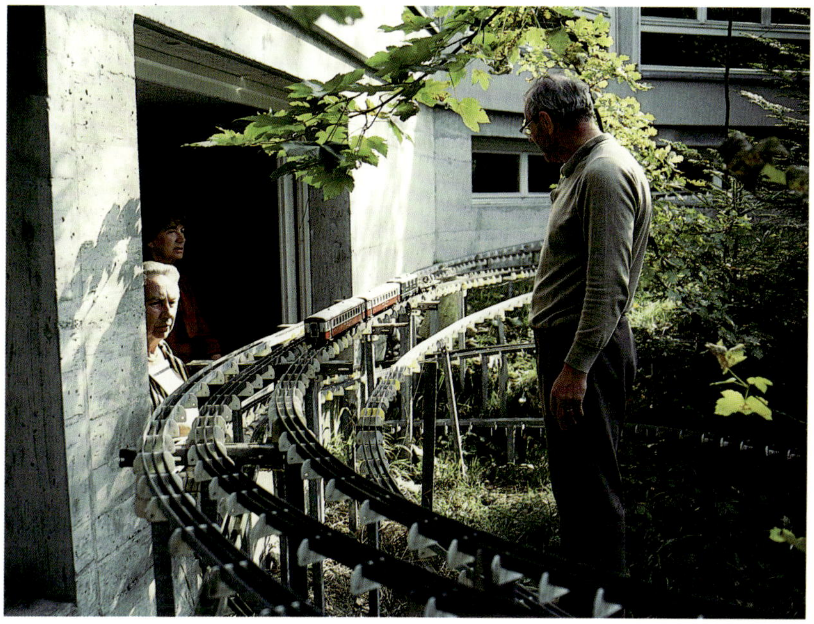

The result was a leak-proof roof, but one that included a surprise. In spite of his usual fiberboard insulation on the underside, the roof absorbed the heat of the sun in summer and caused the office to be overheated inside during the late afternoon. Isler decided on a bold solution: he blocked off the roof drainage and allowed the roof to become a shallow pool (fig. 13). That solved the heat problem, but more surprises followed. After nearly forty years, the roof is a marsh filled with grasses, bushes, and even trees; nature has taken over and Isler exults in showing visitors—who have put on high boots—the nature conservancy above his design studio (fig. 14).[17]

Isler's sense of play extends well beyond the wilderness on his roof and into the wilderness surrounding his office. From one of the lower rooms he built a model train terminal and extended a maze of track from the window into the unkempt woods that has grown up to match the roof garden. The tracks need continued maintenance, but they run smoothly over terrain, trestles, and bridges (fig. 15).

Although Isler began to experiment with his hanging reversed membranes as early as 1957, this idea became a dramatic reality only in 1968. The result was a series of masterpieces, each entirely different in shape, but each expressing the new aesthetic originating in Isler's much earlier observation of wet burlap. These thin shells are comparable in their remarkable efficiency and unprecedented form to Maillart's bridges of the 1930s—the Salginatobel, Schwandbach, and Vessy. As Isler explained: "Now I come to the main category of structural shells: the hanging reversed membrane. This method is, in my opinion, the most efficient one to date for creating new shell forms. If the experiments with models are carried out carefully and with high precision, one gets excellent results."[18]

His first masterpiece was the turnpike service station near Deitingen, built in 1968. This project consisted of two triangular-plan, three-point supported shells 9 centimeters thick, 31.6 meters in span, and 26 meters wide, where two supports rest on a small service building. Because they are designed from the hanging method, they can be very thin and built entirely without ribs. They are watertight and have performed well for more than one-third of a century. Since the shells reach their supports with a relatively small slope, there are substantial horizontal reactions necessary to prevent collapse. Isler provided for these forces by designing prestressed tension ties between supports so that the ground would support only vertical loads. Again, his shells were made possible by the use of straight prestressing elements, a new idea—pioneered in Switzerland by Lardy—that had just come into practice when Isler began his career at the ETH.

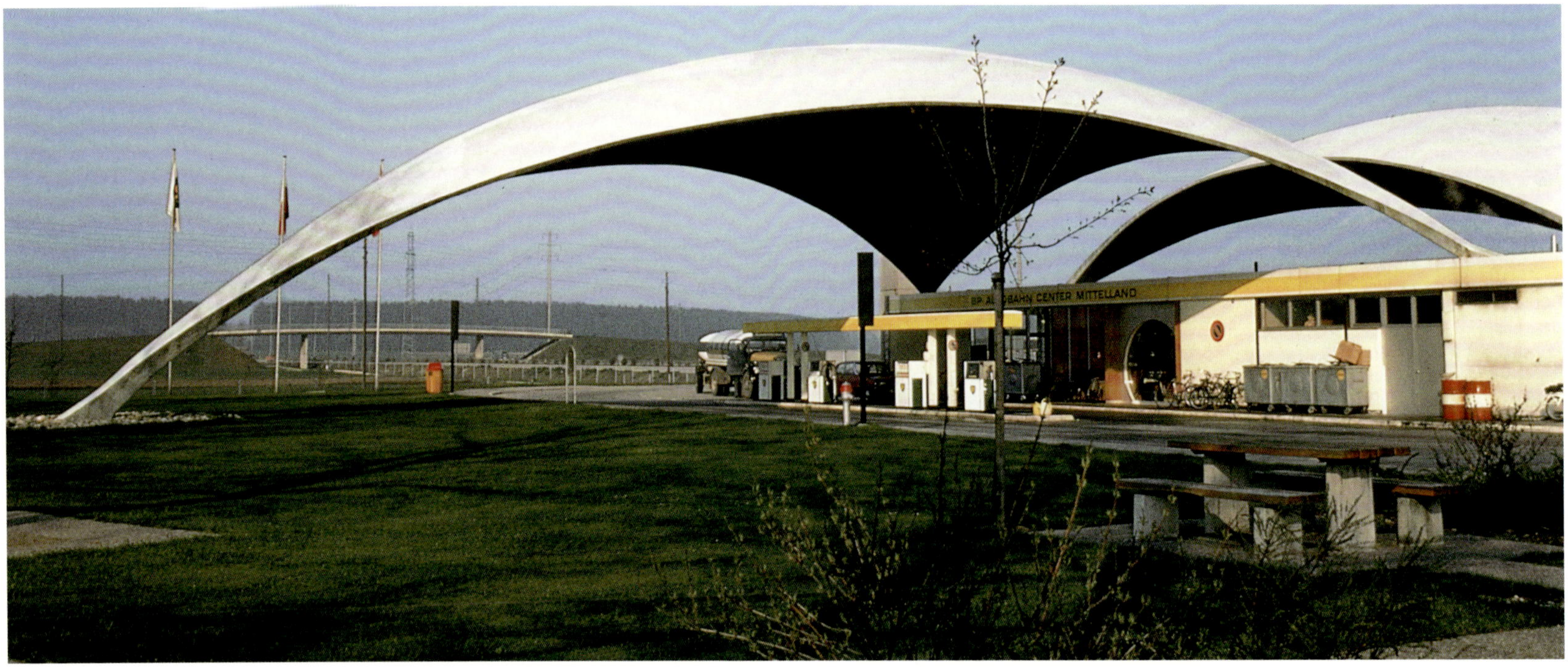

Figure 16
BP service station, Deitingen, built in 1968, with two three-point supported concrete shells.

For the Deitingen shells (fig. 16), Isler followed his well-developed approach of finding a form by models, testing the form by means of other models, and then having it built following careful design instructions. This last was a crucial necessity for these triangular shapes because the final design required that the two shells work together through connecting ties beneath the ground. But it would have been highly uneconomical to build both shells at the same time because, having the same shape, the formwork could not otherwise be reused. Thus, Isler had to design temporary prestressing to allow one shell to stand alone before the second was built.[19] This is a mark of a structural artist who must think deeply about performance and construction while making the design. Overwrought or contorted forms, however visually arresting, do not rise to the level of structural art if their form has been projected with little thought as to construction economy or long-term performance.

There are many examples of overwrought designs of which two stand out. The first is a theater designed in 1961 by the architect Eero Saarinen, who insisted on a three-point shell in the form of a spherical segment. Since this geometric form is quite wrong for a shell surface, the roof has performed poorly: it has required substantial copper roofing, it has leaked badly, and it has had to be shored up around its periphery.[20] A second example is the series of flat spherical segments designed by the architect Roger Taillibert for a complex at Chamonix built between 1970 and 1975. The structure is so badly conceived that Isler found it necessary to add heavy edge beams and huge masses of concrete below ground to equilibrate the forces arising from the designed shape.[21] Here, the supposed simplicity of form led to immense complexity of performance.

The original idea for the Deitingen shells was to provide a thin-shell logo for the oil company British Petroleum (BP), which in 1999 decided to tear down the shells. Fortunately, a prestigious group of Swiss architects protested, and the shells were saved.[22] Sadly, the engineering profession is still behind in recognizing its greatest works and working to ensure their protection.

The year after the BP shells appeared, Isler completed his greatest work and his best example of a design where complexity was transformed into simplicity by the hanging reversed membrane. The Sicli Company of Geneva wanted a factory and office building constructed on a constricted site. Unsatisfied by the architect's proposals, the owner called in Isler to study the design. His solution was both dazzling and original. It did not come easily or quickly.

Figure 17A
Sicli Company Building, Geneva, a continuous thin shell on seven supports.

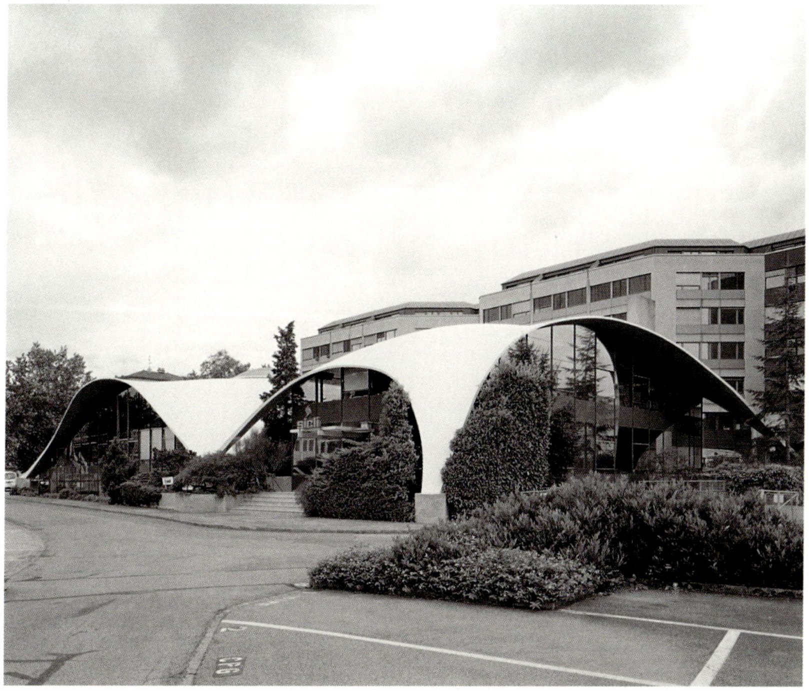

Figure 17B
Oval opening in the roof for the winter garden of the Sicli Company Building.

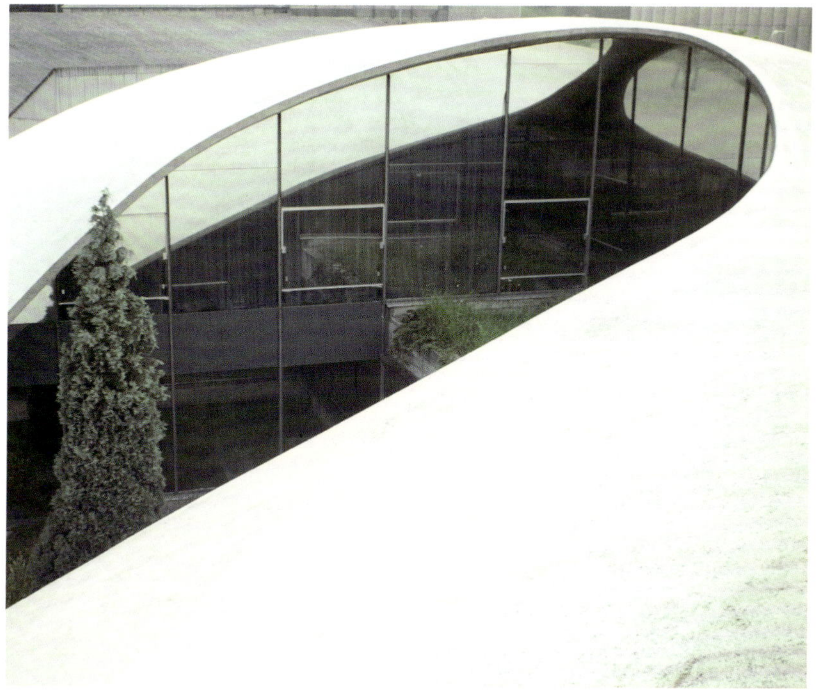

Working intensely and in seclusion over a one-month period, he began to see how the complex program could be satisfied: "A factory and two-story office building for the Sicli Company in Geneva (fig. 17A) has one continuous shell structure on seven supports, with free edges. The irregular plan made maximum use of the available ground. The span of the shell is 33 by 54 meters, with the factory area covering 1,100 square meters. The shell thickness is 9 centimeters. The supports are on mobile elements to allow a displacement during prestressing."[23]

Isler is describing the construction method of "mobile supports" (moving within constraints) along with the design dimensions, especially the radically thin surface, 9 centimeters thick and without any ribs. After the straight horizontal ties were prestressed, the mobility would have been eliminated. But its free sliding was essential during prestressing; otherwise, the prestressing force would have gone directly into the foundation. In that way, the shell could not have been raised to release the formwork. Isler noted the highly complex shape occasioned by the "visually interesting oval opening for a winter garden" (fig. 17B) and then proceeded to describe the model testing: "The statical investigation was made in 1969 by tests on a Plexiglas model because at that time the computers did not yet have the capacity necessary for solving the problem of a completely asymmetric large shell."[24]

Even in the twenty-first century this shell would be difficult to analyze accurately with the computer, especially since the structure's long-term behavior depends upon material properties and environmental influences. Indeed, Isler is far more concerned with full-scale measured deflections than with calculated stresses. For this reason, he carried out a two-decade-long monitoring of deflection that revealed gradual deformation of as much as six centimeters up in some places and four centimeters down in others. These movements have resulted in no observable cracks and thus no leaks.[25] The cracks would be easily visible on the outside (where they would be most detrimental) because the shell has no roofing.

This structure is the most dramatic testimony to date of the inherent potential for thin-shell concrete roofs to be works of structural art. But this can happen only if the architect or owner gives the structural artist full control of making the form. Very few architects would be willing to do this, and, in fact, the Sicli building came about because the owner asked Isler to make the form and the architect accepted (fig. 18).

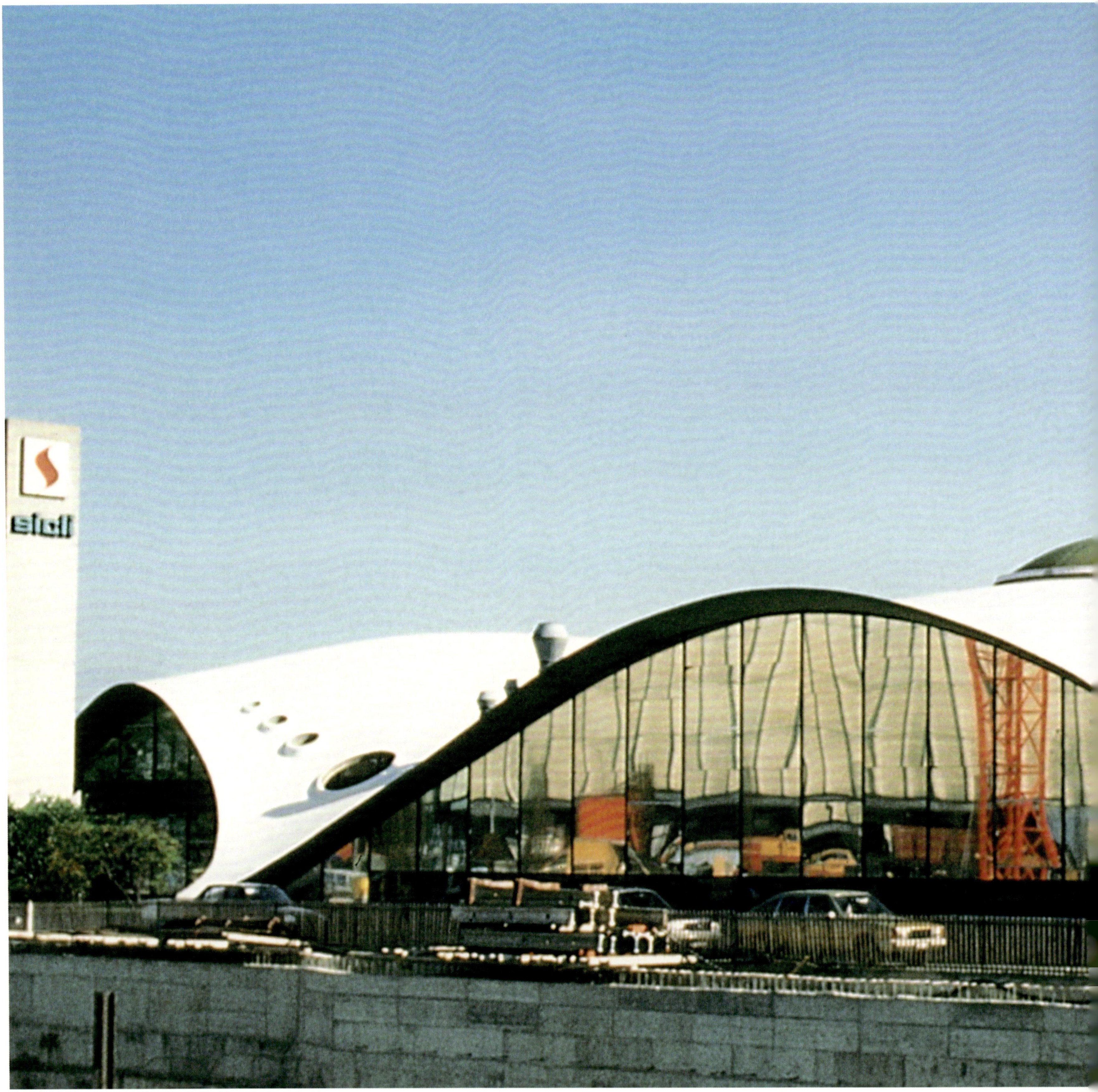
sicli

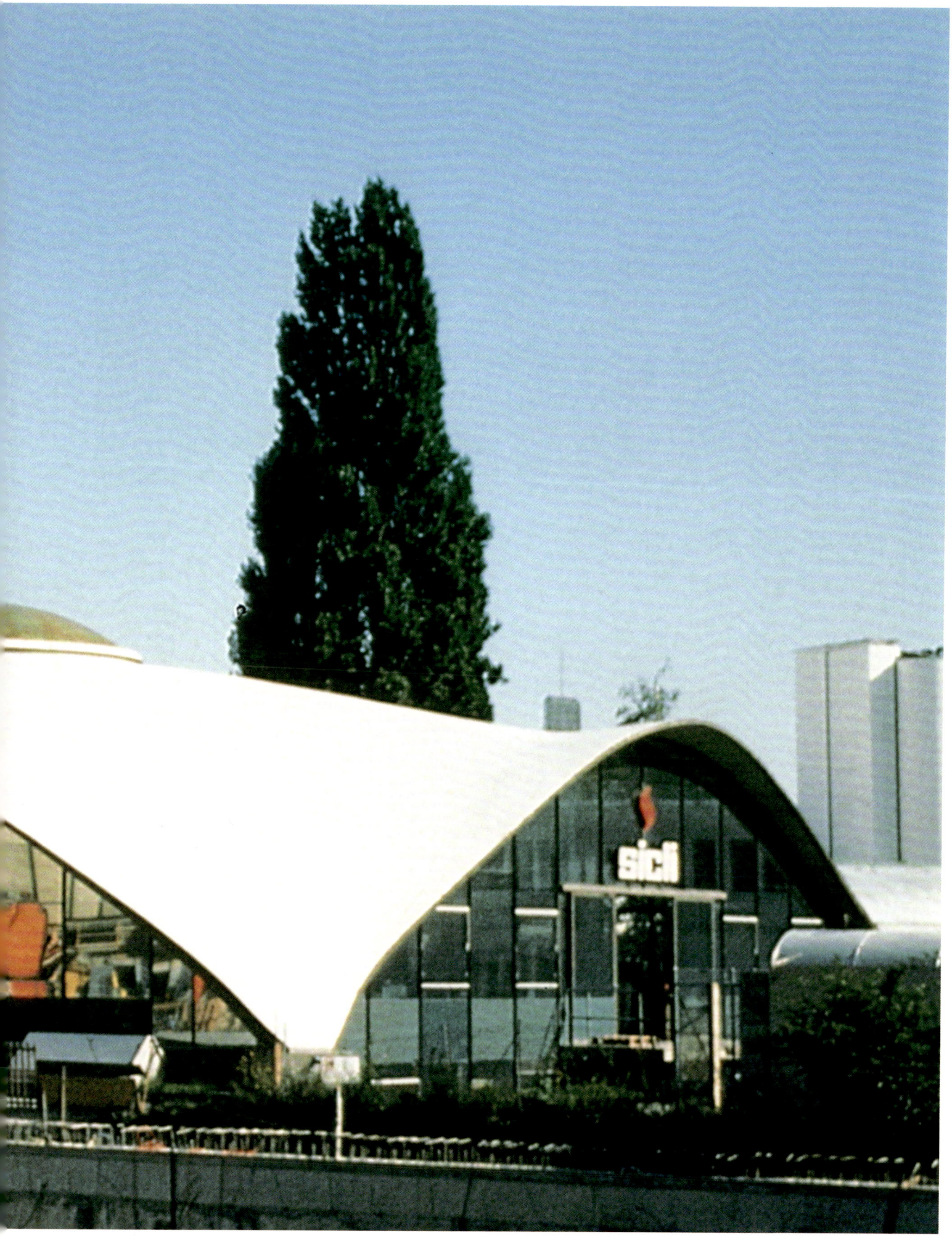

Figure 18
The Sicli Company Building, reflecting adjacent construction.

Figure 19
Creating a square-plan form: pouring a plastic material onto a cloth over a solid surface.

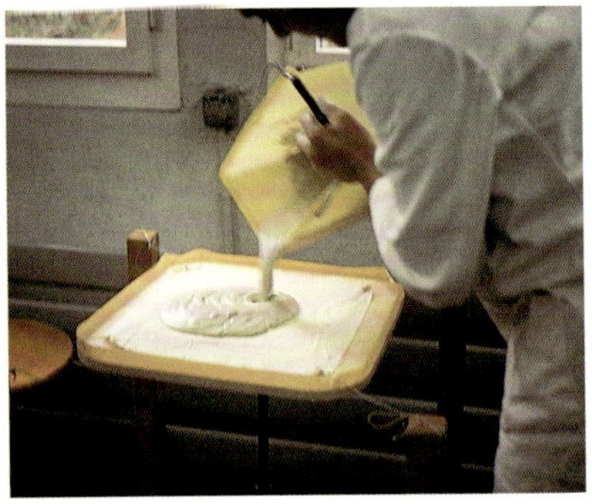

Figure 20
Creating a square-plan form: lowering the solid surface.

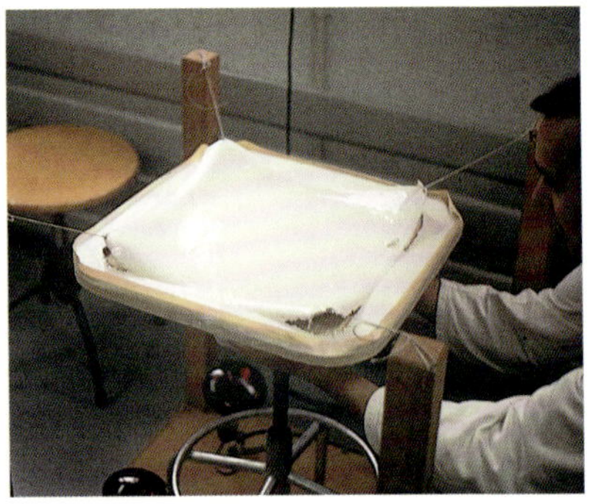

Figure 21
Creating a hanging square-plan form: plastic-covered cloth as a membrane freely hanging from its corners.

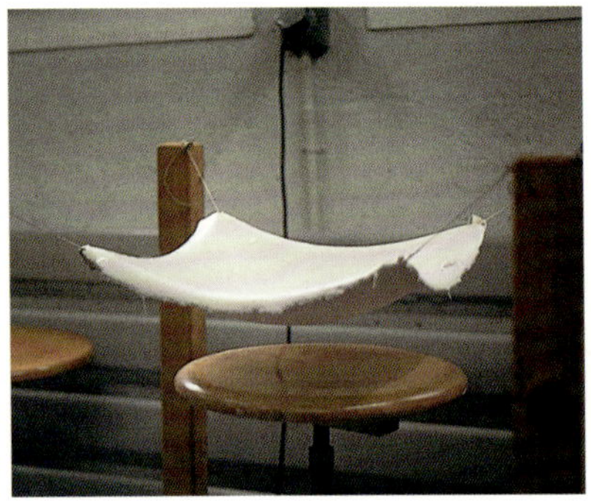

Figure 22
The membrane hardens and is inverted to create a shell form in pure compression.

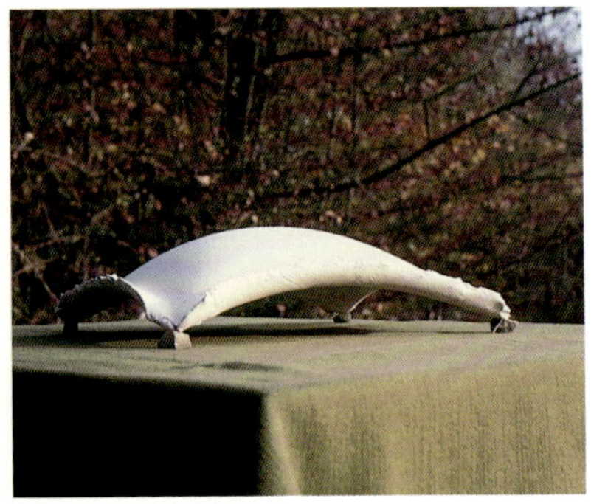

The Sicli design process was vastly different from the Chamonix example, where the architect gave the form to the engineer, who had to make it stand up. Sadly, Isler has rarely been challenged to design roofs with such complex programs as Sicli because fashionable form-making has usually been based only upon nonstructural or antistructural shapes. Many buildings do not have programs that lend themselves to thin concrete shells, but a large number do, and that could well lead in the future to a surprising variety of elegant forms. These will require the care and experience that people like Isler bring to their work. Today a huge effort is underway to create irrational–and hence nonstructural–shapes that are only possible because of highly sophisticated computer programs. If all that time and effort were channeled instead into finding forms like those of Heinz Isler, the results could be dramatic and the quality lasting. The computer has value for engineers, but it can also lead to needless complexity that almost seems to be welcomed as a justification for its use. Moreover, the computer has yet to show any capability for finding new and statically appropriate forms of the same quality as those of Isler.

Some years ago, Isler illustrated how he creates a simple square-plan form, by "an experimental method that is still very modern but that is, to my idea, even better than the computer":

The process (fig. 19) consists of pouring a plastic material onto a cloth resting on a solid surface. Once the material is evenly spread on the cloth, the solid surface is lowered (fig. 20) and the plastic-covered cloth, now in pure tension, is freely suspended from its corners (fig. 21). In that position the plastic hardens and the solid shell model is turned upside down, giving a shell form in pure compression (fig. 22). One does not actually create the form; one lets it become, as it has to according to its own law. It looks so simple because it is simple. These quickly made models can be used both to study shapes visually and to test for structural correctness. If the models are made with high precision, they can be used for the actual form-giving process.[26]

Isler is quick to point out, as he did years earlier in Madrid, that the process does not end there:

Of course, this is only the first step to finding the form. Afterwards one has to do the exact structural analysis, the modeling investigation, the layout of reinforcement and prestressing elements, and the support details. Also, the designer needs to think out the construction problems and finally to observe carefully the structure while it is in use.[27]

This is the program of the structural artist who, above all and after all that study, still seeks a form that produces, as Isler put it, "lovely shapes." There he was speaking of the flow method, an idea he kept exploring while creating the BP and Sicli shells. His masterpiece in this unlikely method was completed at Camorino for the Bürgi Garden Center in 1971 (fig. 23). Here Isler created a thin shell with all edge stiffening gone so that the discontinuity visible at Solothurn is not a factor and the entire shell is one slender form that includes even the light supporting points. Less material was needed to make this integrated structure, and the stresses still remain well below allowable limits; hence, the roof is still structurally correct without requiring edge ribs. The Camorino shell was a bold early attempt to express the ideas of thinness and integration.[28] As Isler recognized, the deflections, while allowable, are much greater than those in the earlier Solothurn shell. In later shapes, Isler arrived at the same expression of thinness with much lower deflections.

Figure 23
Bürgi Garden Center in Camorino, designed using the "flowing method" and demonstrating that edge beams and arches can be eliminated.

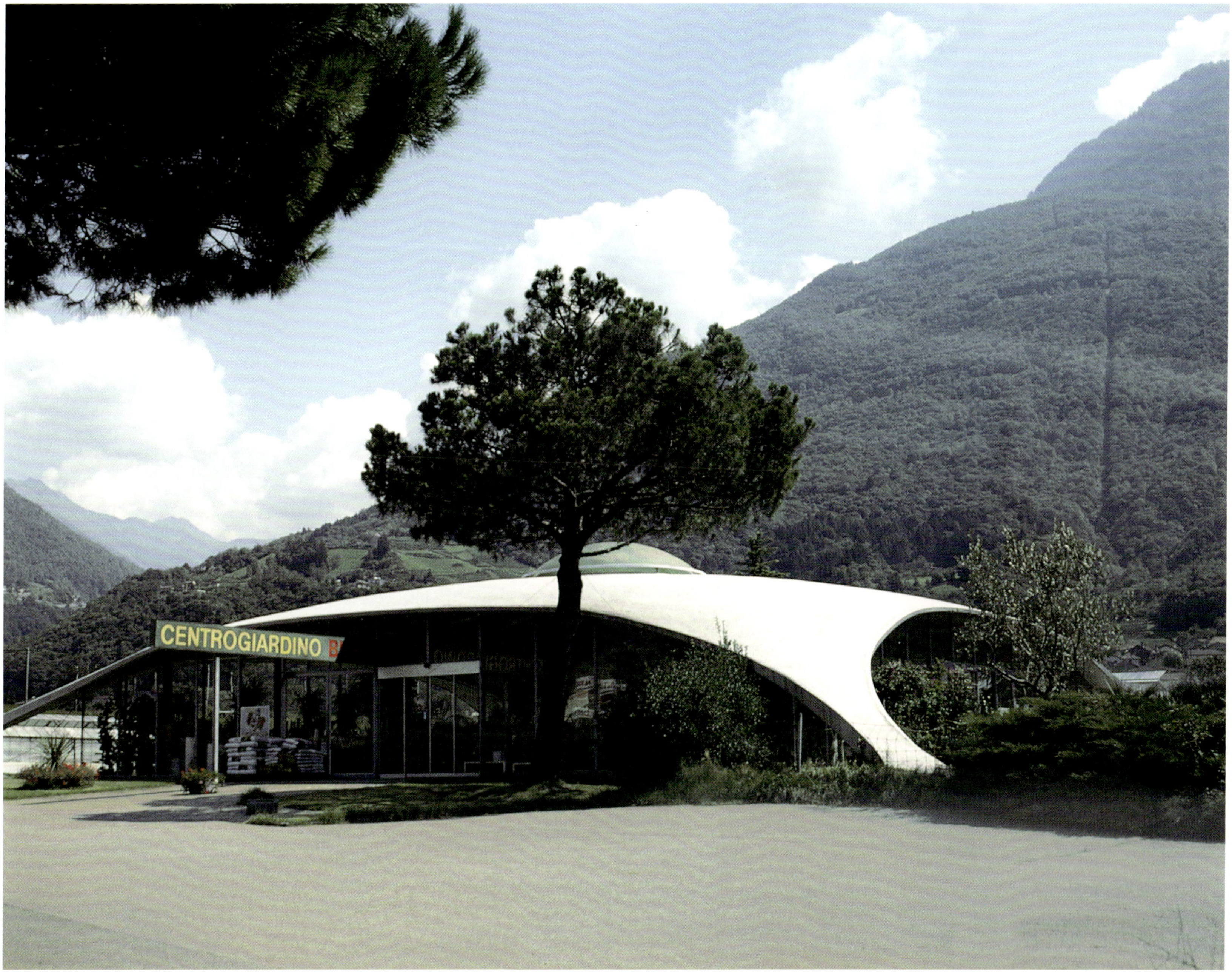

Figure 24
Hanging reversed-membrane model on five supports, constructed in 1977 for the Grötzingen outdoor theater.

Figure 25
Grötzingen outdoor theater, inclined shell designed by the hanging membrane reversed method and resting on five points without any edge-stiffening members, near Stuttgart.

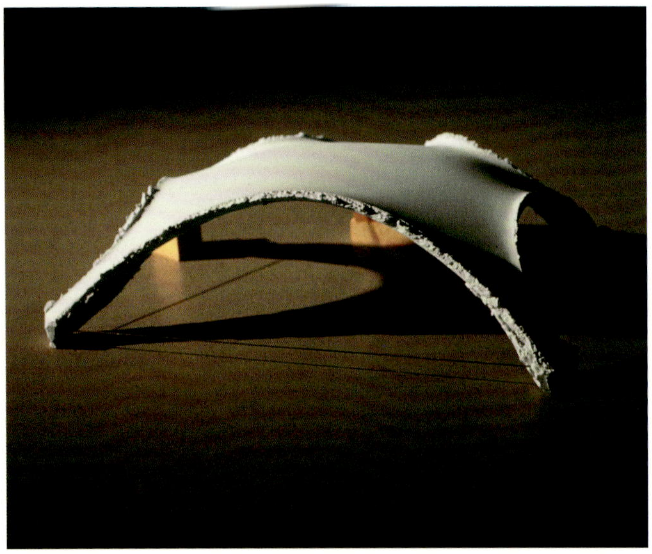

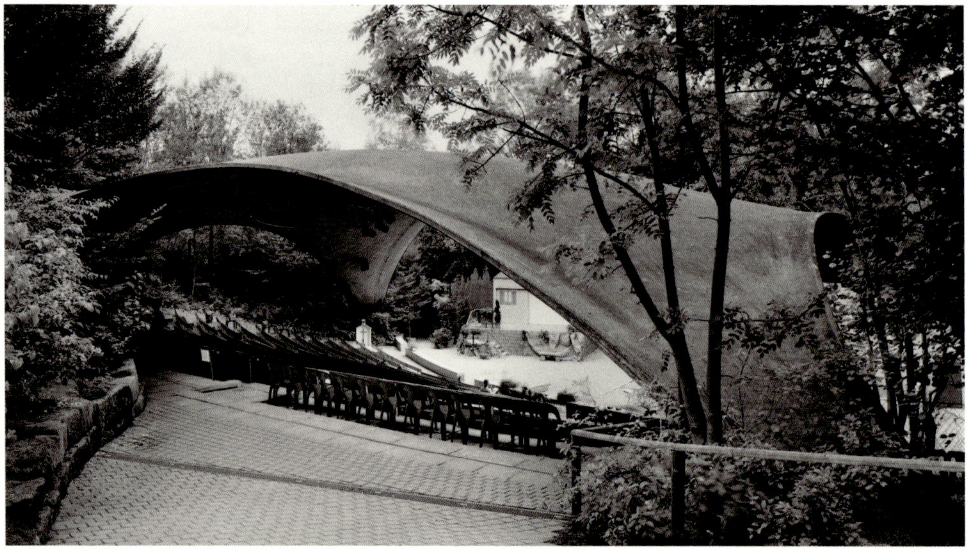

The Camorino shell was a stunning experiment that demonstrated the superfluousness of edge beams and arches. This edge thinness is all the more remarkable because the form is, strictly speaking, not structural, and the free edge is in the shape of a flat arch. Such a shape with so little thickness would normally be in grave danger of buckling, but Isler found that there was no danger because that part of the shell was in tension. Since the tension region is outside the building wall, there appears to be no danger of internal leaking.[29]

Later in the 1970s, Isler found an architect who was eager to cooperate on designs and give a program that the engineer could turn into a structural form. The masterpiece was an outdoor theater in Grötzingen near Stuttgart and the architect was Michael Balz. Isler made a hanging reversed-membrane model on five supports (fig. 24), which was constructed in 1977. It has a main span of 42 meters and a thickness of 9 centimeters. There are no edge-stiffening members; rather, the hanging membrane automatically takes a countercurvature at the edges to provide stiffness without any increase in shell thickness. The structure, as one fully integrated form that includes the supports, sits on sloping ground with the roof height greatest on the downhill—ideal for the audience at an outdoor theater (fig. 25).[30]

Following this trio of masterpieces—the BP station at Deitingen, the Sicli building in Geneva, and the outdoor theater at Grötzingen—Isler entered another stage in his creative career, this time carrying his hanging reversed membranes into a function that would permit repeated use of new forms for swimming pools and tennis courts covered for all-weather activities. In this last stage of development, Isler combined his favorite form with his standard construction methods to produce thin shells on square plans and on elongated rectangular plans. For the square-plan shells, he improved upon his pneumatic and flow forms by creating the hanging shape that fits well for a covered swimming pool. The expensive forming by laminated wooden arches (fig. 26) is rendered competitive by frequent reuse, as with the pneumatic forms. The resulting shell, with the slight countercurved edges, is the purest of structural forms (fig. 27); we can even see the prestressed tension ties connecting the four corners (fig. 28). Here the prestressing is at ground level rather than at the column tops of his pneumatic-formed roofs. In effect, the prestressed edges have been placed below at Badi Brugg or Heimberg so that the shell thickness can be seen, whereas in the pneumatic form the thick-edge members are above and completely hide the thinness of the shell (fig. 29).

The most dramatic of the newer shells are the tennis courts on rectangular plan. Isler's standard designs are about 48 meters long and usually 18.6 meters wide. Between 1978 and 2000 he constructed more than fifty such shells in Switzerland, but he developed his best form in 1979 by close study of the four shells at Heimberg (fig. 30). There, as a parameter study, he varied the shape slightly for each of the shells and carefully measured the deflection of each as the tension ties below were prestressed. The shell that deformed upward the

Figure 26
Laminated, reusable wooden arches for giving the form to the concrete thin shell of the Heimberg pool roof, Heimberg, Switzerland.

Figure 27
Thin shell on a square plan for the swimming pool at Brugg.

Figure 28
The horizontal concrete beams are prestressed tension ties connecting the corners of the swimming pool at Brugg.

Figure 29
Pneumatic roof with prestressed tension ties, which rest on comer columns and conceal the thinness of the shell over a Ford dealership.

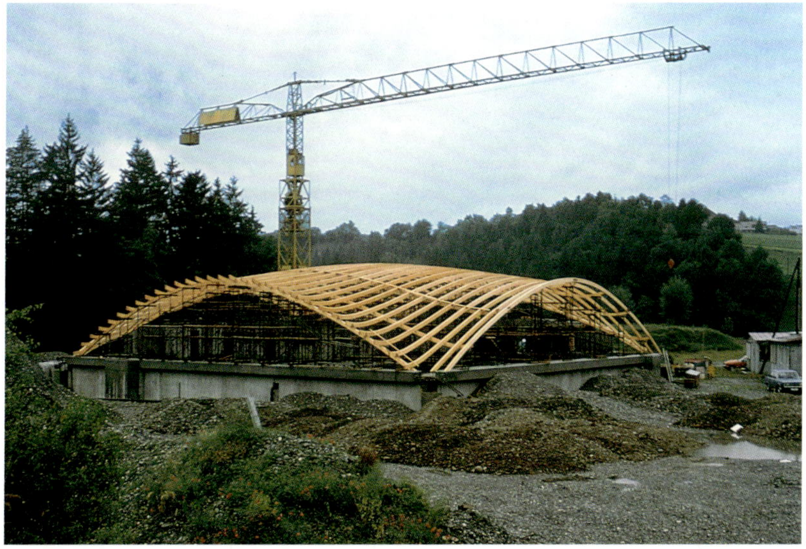

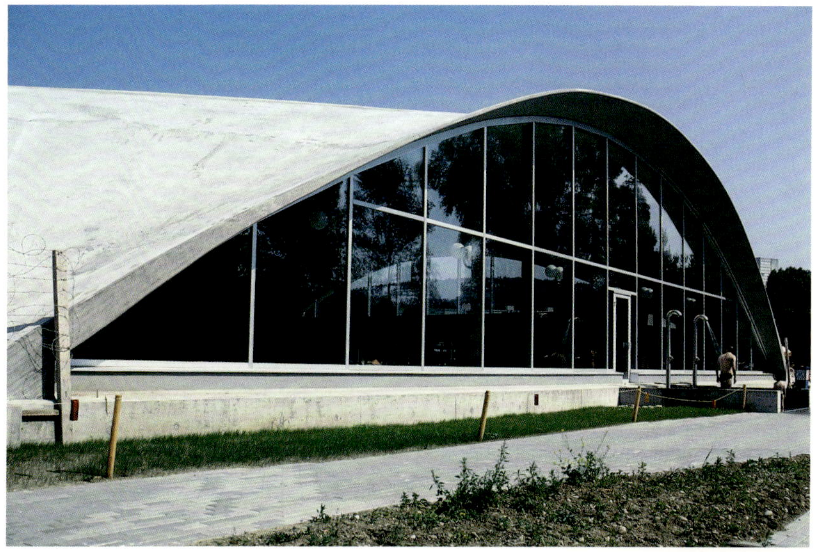

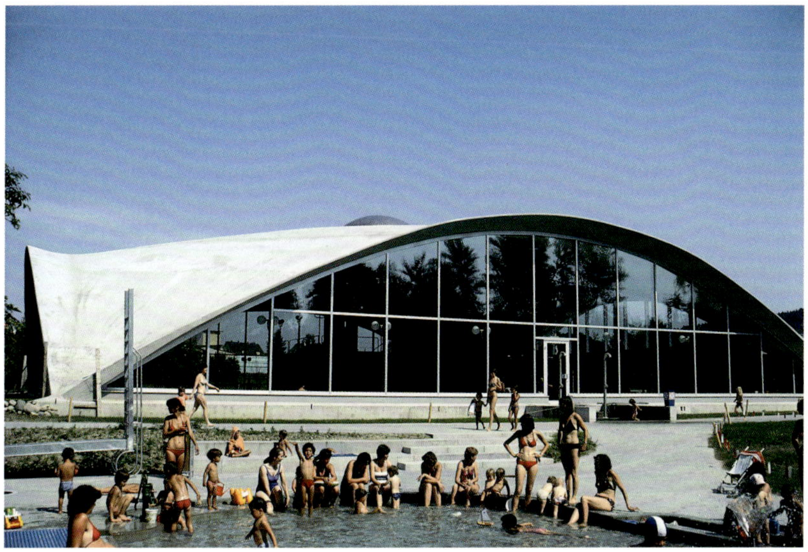

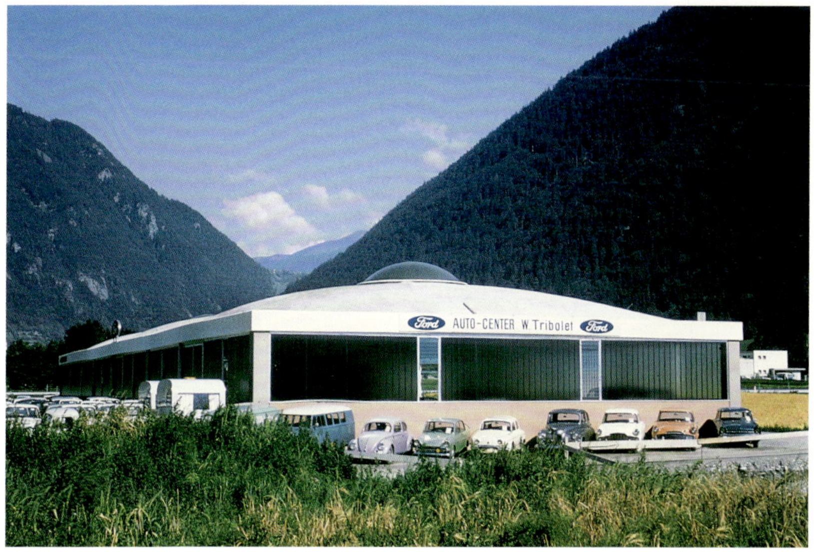

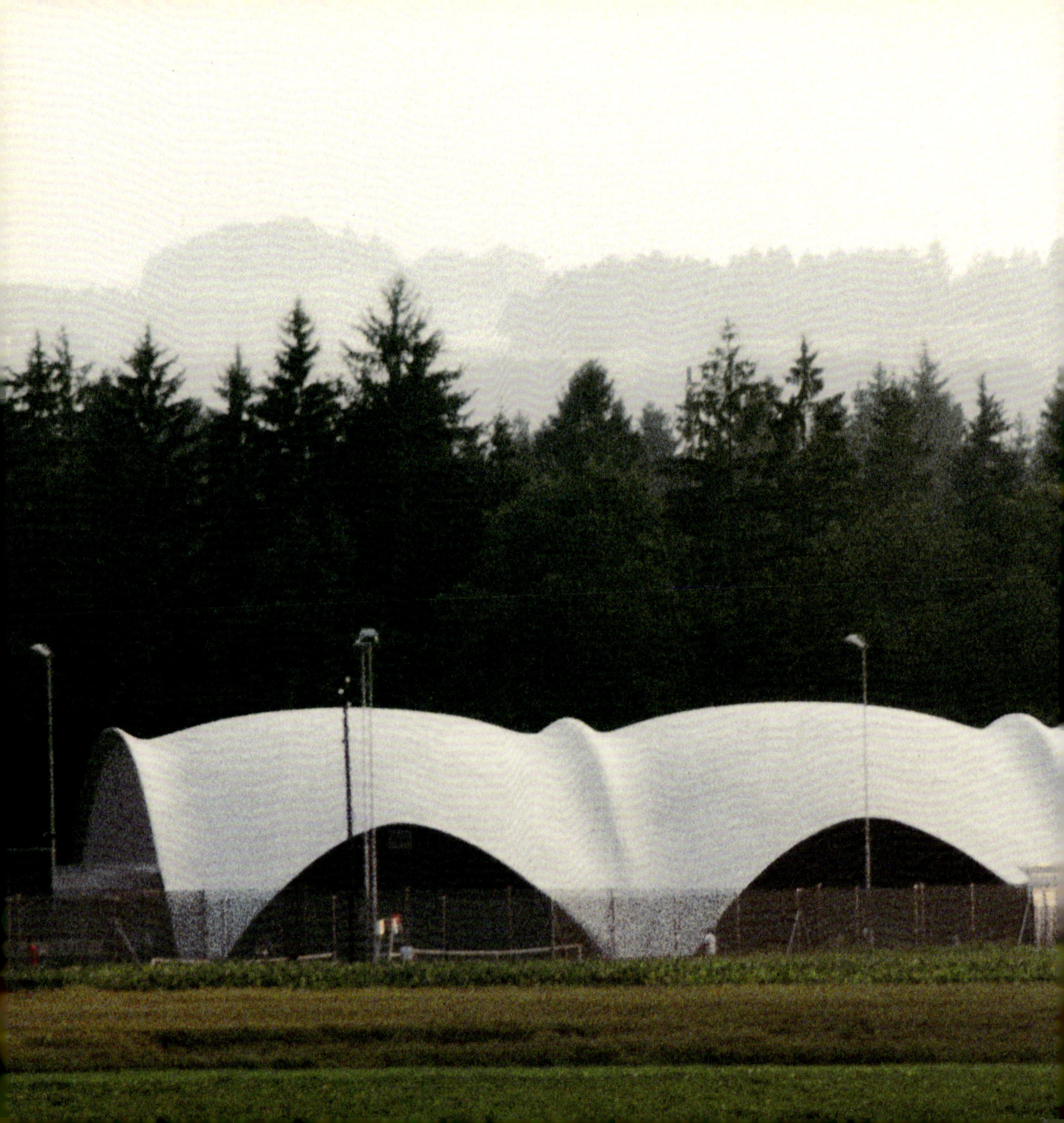

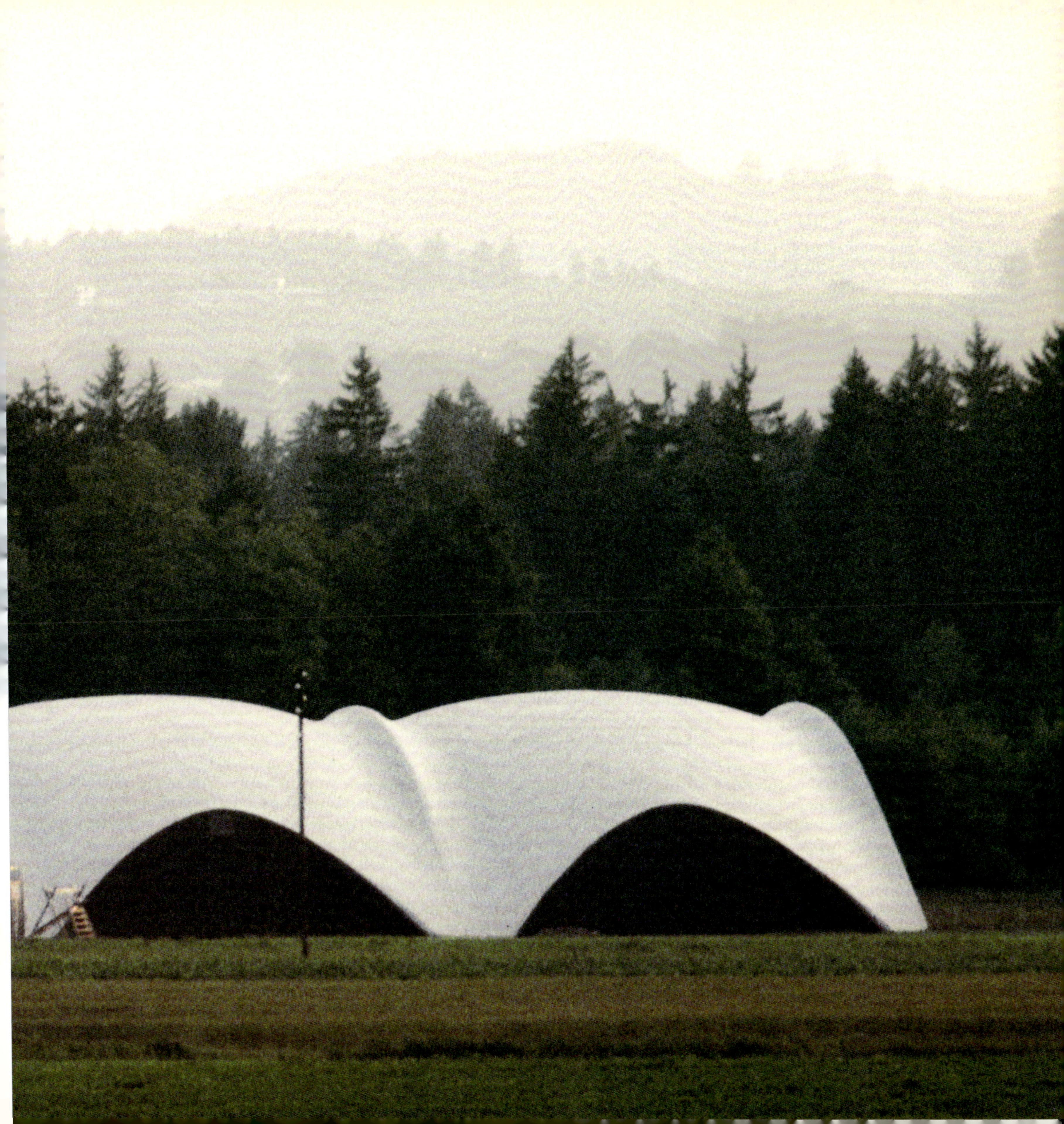

Figure 34

Ice chapel in Isler's yard, Zuzwil.

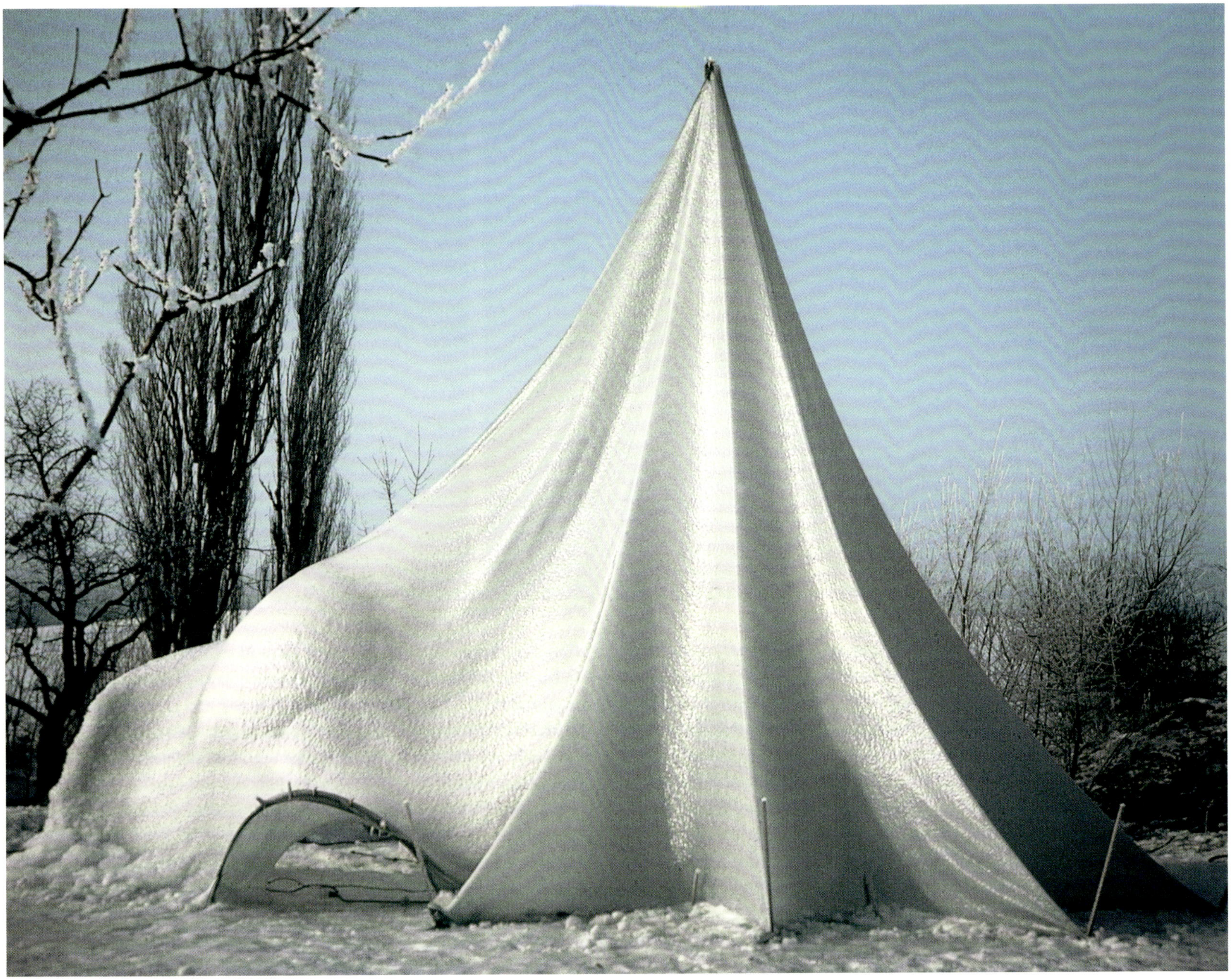

Figure 35

Plan and elevation of a sample tennis hall of 48 x 18.6 meters, the most technically difficult form for the hanging-membrane reversed shell.

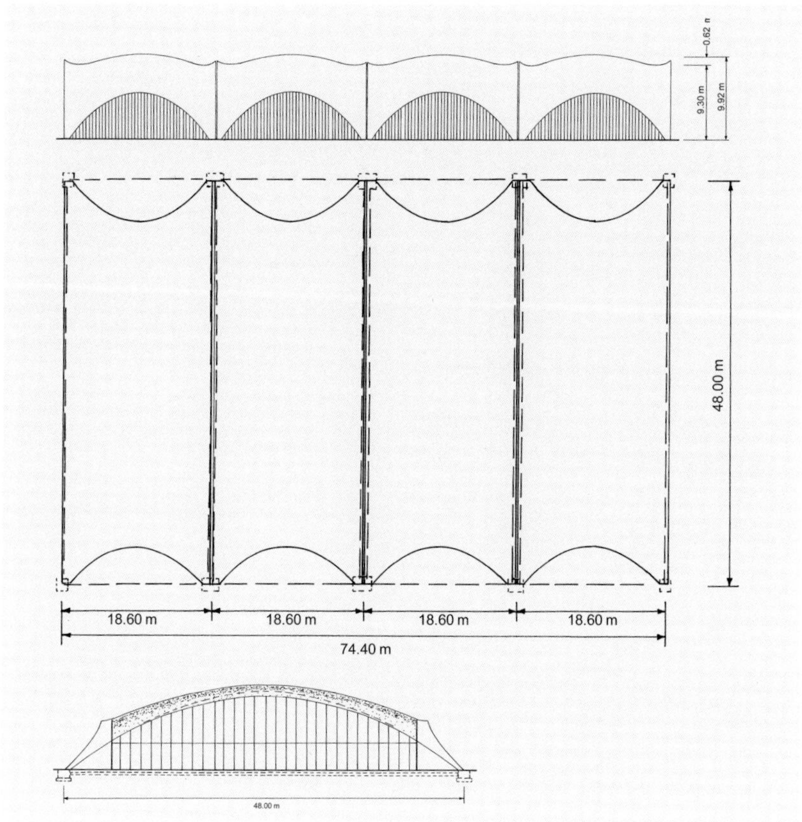

does not at all resemble the superficial notion of making things look like natural objects. It comes from deliberate design.

Isler hung netting from a pole, sprayed it with water, allowed it to freeze, and then removed the pole. The result is his ice chapel, a form that did not come from nature but from an artistic intention (fig. 34). Yet, since they are experiments, the detailed shapes do present the designer with surprises. Some are truly elegant, others less so, but the results call attention to that aesthetic sense Isler described with respect to Lardy's influence. Seeing photographs of the ice sculptures or listening to one of Isler's lectures about them can produce the misleading impression that these works are easy to make. But like all of his works, they require intensive concentration and many trials. The joint products of human actions and nature's reactions, they are entirely different from facile renderings or clever model-making that often passes for creative design. Rather, they are the products of discipline and play, the play that can arise only from within the discipline. Such a collaboration of supposedly incompatible ideals can be nourished by focusing on one at a time: that is, the pure discipline of numbers or materials without any intention of creating new forms, or conversely, the pure play of gardening or model trains with no thought to the disciplined use of those joys.

It is in this separated sense that Isler derives intense joy from natural forms, whether the gigantic mountains that surround his hiking or the delicate flowers and other flora that spring up naturally on his office roof or around his chalet in Zuzwil. Just as the piano did for his teacher Lardy, the recreation of mountains and flowers provides an amateur aesthetic taste that helps satisfy the hunger for professional aesthetic creation. In a lecture, Heinz Isler shows all kinds of natural forms—leaves, flowers, and shrubs—without ever claiming that they were a direct stimulus to his concrete designs. He reacts to the active intelligence of nature in the same way that nature reacts to the intelligence of his designs.

Isler's unique professional constructions, the most prototypical of his mature designs, are the hanging-membrane reversed shells at Heimberg for covering tennis courts. Isler identified the tennis-court form, on a ground plan of 48 x 18.6 meters, as the most technically difficult one for the hanging-membrane reversed shell (fig. 35).[37] This is because the great length to width ratio (about 8/3) makes the shell act more like a single, wide, flat, and very thin arch than a series of intersecting arches acting in different directions as in the square-plan swimming-pool shell. Isler conquered that difficulty by his highly disciplined shaping that produced a form of great stiffness, with a ratio of 1/5000. His long period of study for this new form predicted such rigidity, and his full-scale measurements confirmed it. This result verified the strict technical discipline of efficiency for the tennis-court roofs, whereas the discipline of economy was assured by the construction procedure Isler developed in the 1950s. What remains to explore is the symbolic significance of this form and, more precisely, the form as it appears at Heimberg. It is this mature structure that best serves to explain why Isler is one of the greatest structural artists in the two-centuries-old history of this art.

Symbolic significance means those aspects of the structure that make it a work of art within the disciplines of efficiency (minimum materials) and economy (competitive costs). Within these disciplines, Isler seeks to make an elegant form, one that carries the personal expression of the artist, which at Heimberg we can approach through a visual analysis. The starting point is the overall form,

Figure 36
Heimberg Indoor Tennis Center, showing nearly identical shells, each made from a sheet of concrete.

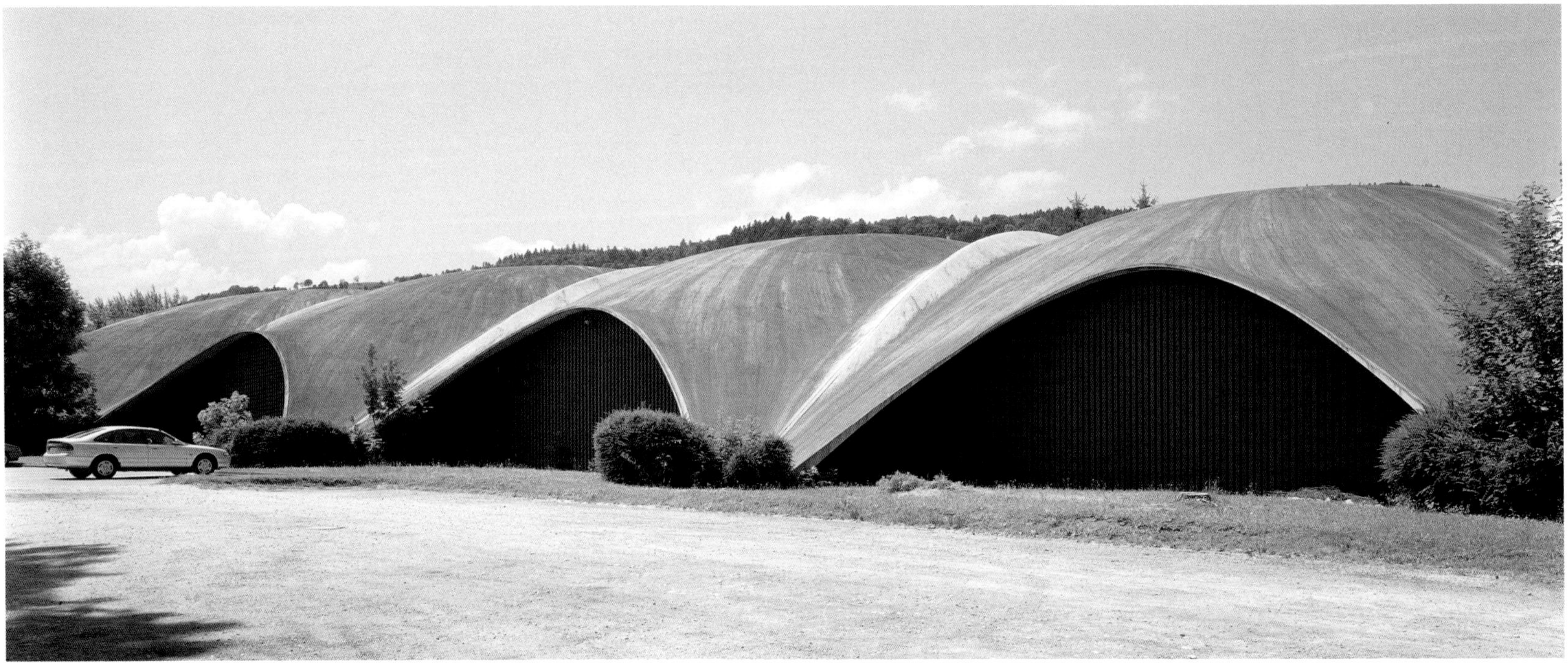

in this case four identical shells each made from a doubly curved sheet of concrete that meets the ground at four points (fig. 36). Between these points are 18.6-meter-span arched openings at each end and 48-meter-span arched openings along each side. We recognize, therefore, each of the four shells as one integral form with no discontinuities. The separate slabs, beams, and columns typical of most buildings do not exist; it is in this sense that we identify the structure as fully integrated visually.

Looking more closely at the arched openings, one is struck by the extraordinary thinness of the structure. The full visual effect is of an almost paper-thin shell, in spite of the nonstructural wall, unseen from a distance but visible up close. It is fortunately clearly differentiated from the structure by its metallic sheen. Closer up, the countercurve along the edges can be seen; without adding thickness, this greatly increases the stiffness to prevent buckling.

In addition to these two primary characteristics of structural art, integration of form and expression of thinness, there is a third visual element that is obvious at Heimberg. In front of the shells is a greenish brown field, close behind is the green forest, and far behind are the high black and white mountains of the Bernese Oberland (figs. 30, 37). The thin-shell structure makes no concession whatsoever to this strong visual environment and in fact is set totally apart by its form and material. Fabricated concrete contrasts strongly with natural surroundings, expressing the intervention of human society into the organic landscape. The hard concrete surfaces stand out against the soft horizontal agriculture of the field and the vertical foliage of the forest.

In shape, texture, and color, the tennis center structure is not natural, leading to the second aspect of symbolism, which is approached through an analysis of function. Function in structural art is always simple, never of the complexity found in architectural art.[38] The bridge is a clear example of one overriding function, the passageway along a line for vehicles and people. Isler's roofs represent another example, where the engineer makes a form to enclose a simply defined space, usually for one carefully organized purpose. The tennis centers cover tennis courts, which have specific dimensions and clearances; at Heimberg and similar places they serve local communities, allowing inhabitants to engage in the sport year-round. One unusual feature is the roof's longitudinal curve, which follows the flight of tennis balls more naturally even than the code requirements for a covered court.[39]

Figure 37
Mountains of the Bernese Oberland loom behind the Heimberg Indoor Tennis Center.

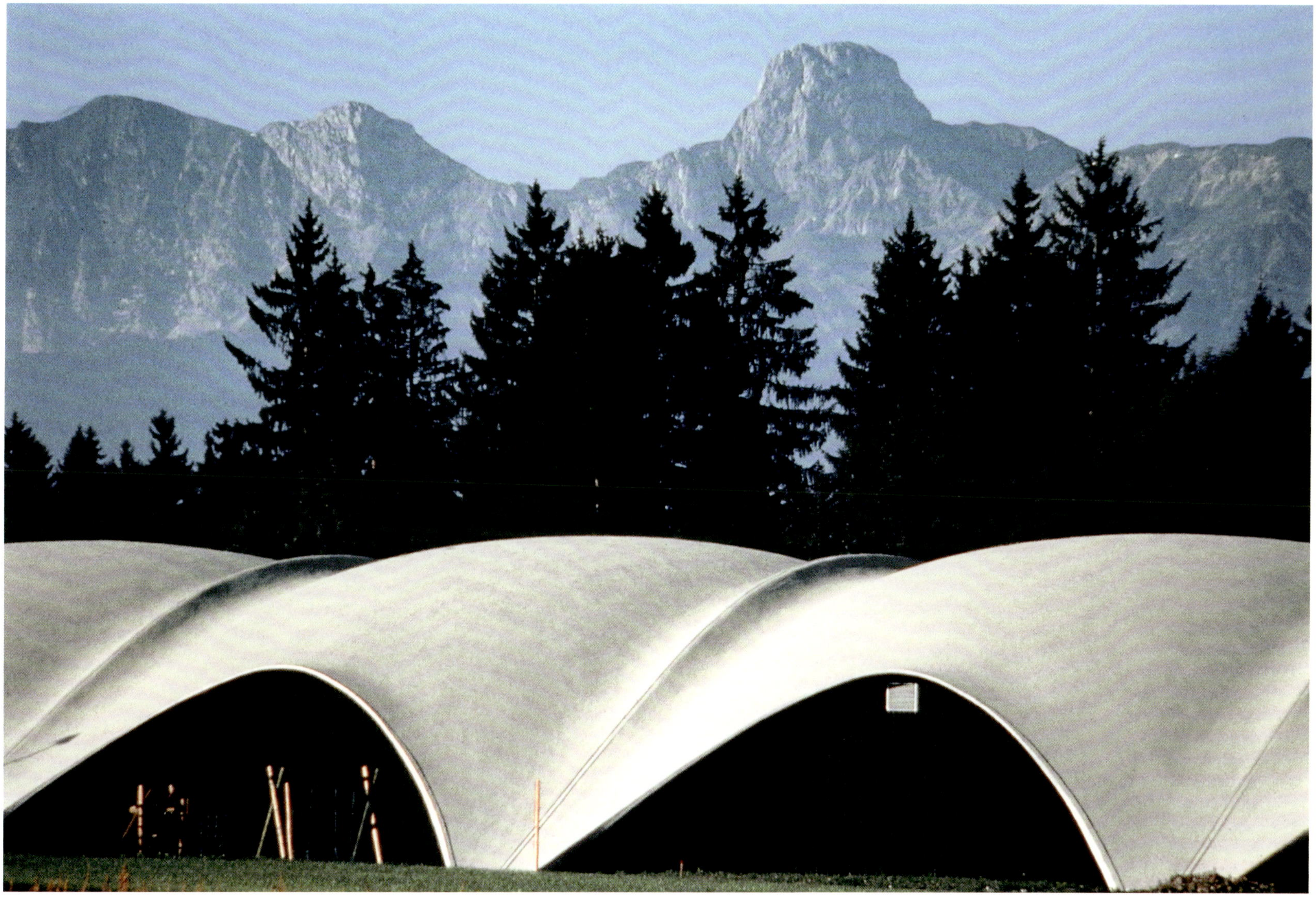

Figure 38

Isler in front of the vertical walls of the Heimberg Indoor Tennis Center, 1980.

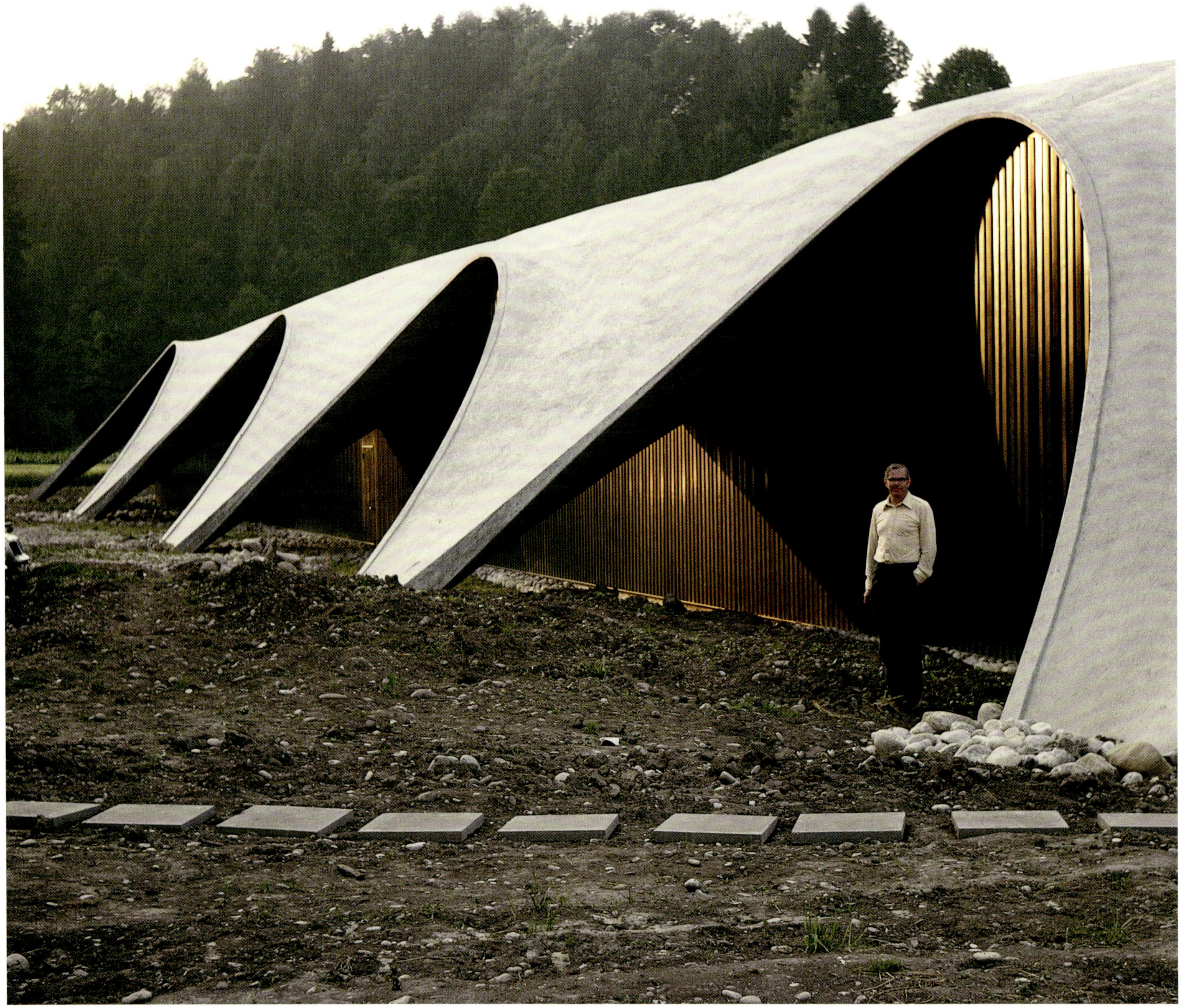

Figure 39
Interior view of the Heimberg Indoor Tennis Center, where the complete shape of the shells is hidden.

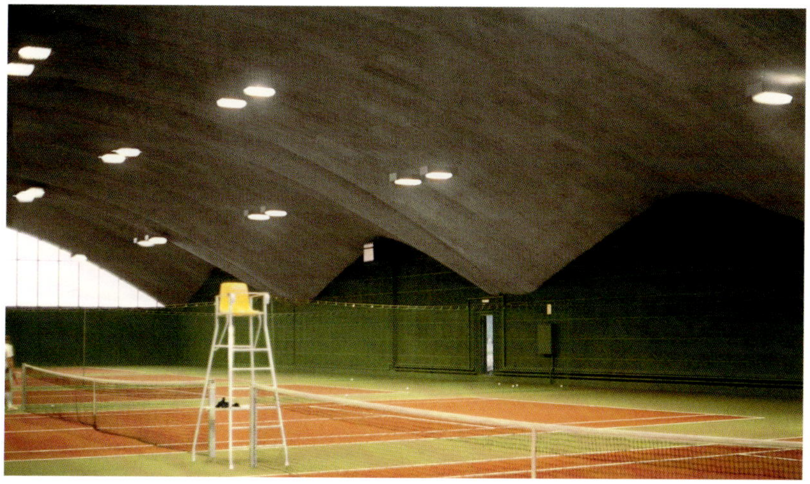

We are led back to Huizinga's theory that "the profound affinity between play and order is perhaps the reason why play… seems to be to such a large extent in the field of aesthetics." This gives a good insight into the playing of tennis and the making of form: "Play has a tendency to be beautiful." Here again Huizinga is talking about creations and performances by people, not by nature. These are conscious activities. His strongest insight comes with the supposition that "it may be that this aesthetic factor [play] is identical with the impulse to create orderly form, which animates play in all its aspects." Isler credits Lardy's teaching to this impulse. The same impulse is symbolized almost perfectly in the out-of-the-way thin-shell concrete roofs where, as Huizinga would say, "the least deviation from it [order] 'spoils the game,'" the precision essential for the game of tennis and the shapes of shells.[40]

The function protected by the shells does, however, necessitate the vertical walls that hinder a full visual appreciation of the integrated form from the exterior (fig. 38); this Isler dislikes. The interior view is disturbed in a different way because the shell undulations end abruptly at the walls, hiding the completion of the form where it reaches the ground (fig. 39). Only with the BP shells at Deitingen could Isler express a pure shell form; in many cases, he could use glass walls, as at the garden centers, swimming pools, and Sicli, so that the purity of form would be clearer visually. Nevertheless, the Heimberg tennis center represents the high point of meaning for Isler's work when we take a third symbolic perspective with more general, cultural analysis. Such an analysis builds upon the visual and functional insights while considering much broader meanings.

These meanings can include the relationship between society and nature, the changes in society characterized by this center, and the role of structural art in the modern industrialized world of a new century. The visual contrast between the shell and its setting stands for Western civilization, where all aspects of urban industrial life spring from the making of things that protect and reorder naturally occurring events and landscapes. These shells provide protection in the harsh Swiss climate, and they reorder the local environment into a center for recreation. Nature itself is not changed except that its influence is diminished locally. More than that, the shells themselves are a product of civilization. Reinforced concrete is a late-nineteenth-century innovation, prestressing entered practice in mid-twentieth century, and the idea of concrete roof structures—in German called *Massivbau*—in thin, integrated forms is primarily a result of ideas stimulated by Maillart and Ammann and made more widespread in the late twentieth century. Isler's shells summarize that progress in engineering, which takes place only in developed regions.[41]

But even in these regions, the advent of such works is new. Modern tennis began in the late nineteenth century, primarily in Britain, and quickly spread to continental Europe, Australia, and the United States. For a long while it was a game for the elite and was originally known as lawn tennis.[42] It was thus an outdoor game, usually played by people of some means. It was only after World War II that the game became more accessible and moved indoors. Late-twentieth-century indoor tennis centers are public recreation for middle-class local communities. Village public tennis has entered where there used to be only suburban private clubs.

Structural art, therefore, having grown up with the Industrial Revolution, is symbolic of free democratic societies.[43] Its goals of efficiency, economy, and elegance correspond to those of societies based upon conservation of limited resources, accountability of public funds, and responsibility for encouraging the creation and preservation of art. The Swiss shells of Isler, like those at Heimberg, exemplify these goals. In addition, they exemplify the influence of a teacher, Pierre Lardy, who instilled in his students that universal goal of "encouraging us to find and apply aesthetics from within us." This goal will always meet resistance from those who are not concerned about minimizing the use of natural resources, being accountable for construction funds, and finding a new expression in art. For those who are concerned, Isler's shell points to a new future just as Isler himself does on his Heimberg roof (fig. 40).

Figure 40

Isler on the roof of the Heimberg Indoor Tennis Center, 1980.

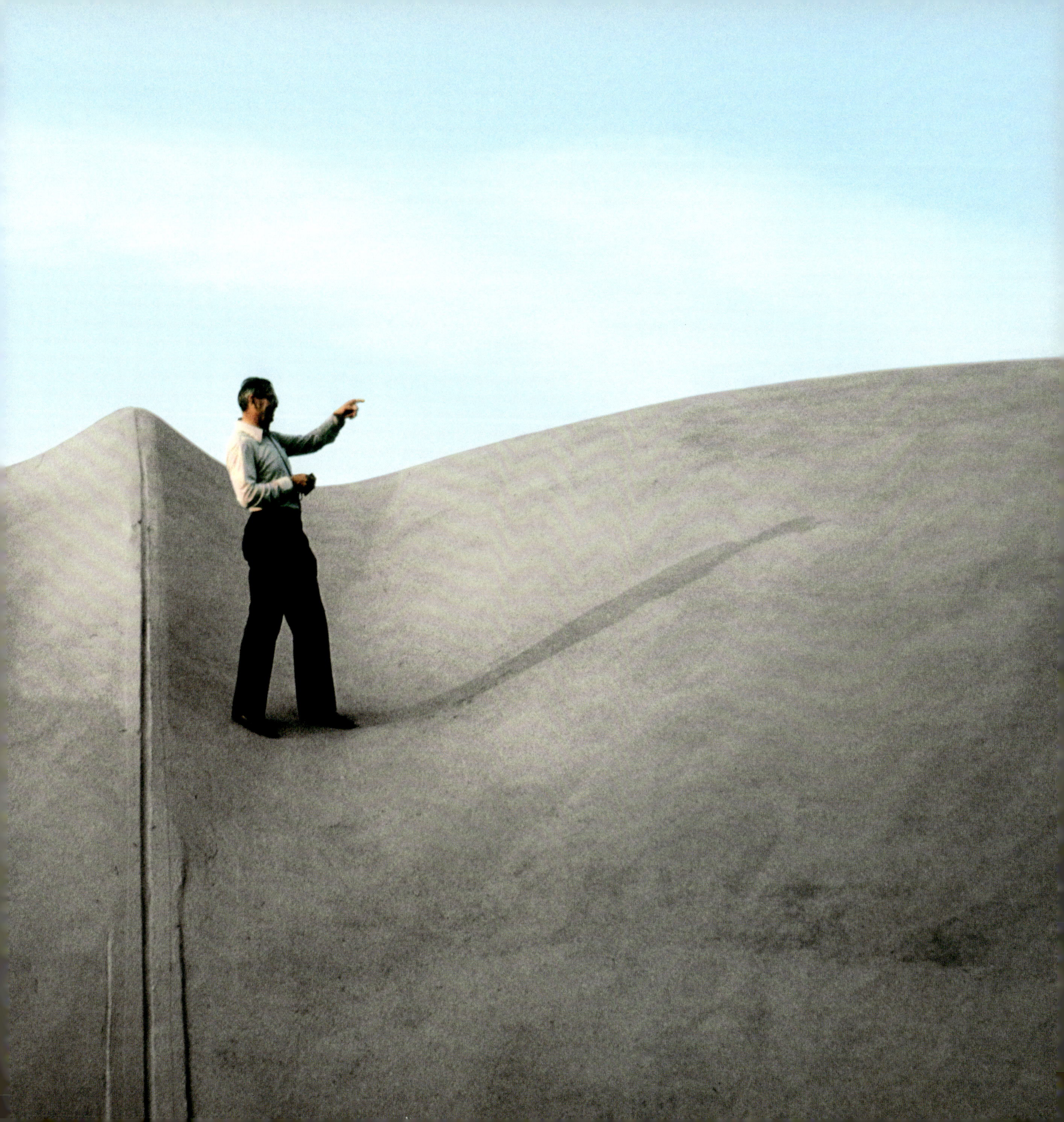

Notes

1. The title of the congress, "Colloquium on Nontraditional Construction Processes of Shell Structures," seemed to express the viewpoint that the future of shells lay in new ideas rather than in looking backward. None of the papers came from West Germany nor did any refer directly to works by the builder-designer firm of Dyckerhoff and Widmann; no work by Felix Candela was presented even though he had gained substantial fame by 1959.

2. H. Isler, "New Shapes for Shells," *Bulletin of the International Association for Shell Structures* 8 (1960), paper C-3.

3. D. P. Billington, "Heinz Isler as Structural Artist," in *Heinz Isler as Structural Artist*, exh. cat. (Princeton, 1980), 16–17.

4. H. Isler, "Textilfabrik," Diplomarbeit (thesis), Zurich, Dec. 1950, 11 pages, Princeton Maillart Archive.

5. Billington, *Isler as Structural Artist*, 9.

6. Letter from Maria and Heinz Isler to Phyllis and David Billington, April 2, 2002.

7. F. Dischinger, "Eisenbetonschalendächer Zeiss-Dywidag zur Ueberdachung weitgespannter Räume," in *First International Congress for Concrete and Reinforced Concrete* (Liège, 1930), vol. 1, 262–91; and U. Finsterwalder, "Die Querversteiften Zylindrischen Schalengewölbe mit kreissegmentformigem Querschnitt," *Ingenieur-Archiv*, 1933, 43–65. The work was submitted for a prize competition to the Prussian Academy of Construction on January 15, 1930, where it won second prize.

8. A. Tedesko, "Z-D Shell Roof at Hershey," *Architectural Concrete* 3, no. 1 (1936): 7–11; A. Tedesko, "Large Concrete Shell Roof Covers Ice Arena," *Engineering News-Record*, Apr. 8, 1937; and A. Tedesko, "Multiple Ribless Shells," *Journal of the Structural Division of the American Society of Civil Engineers* (ASCE) 87 (October 1961).

9. J. Chilton, *Heinz Isler* (London, 2000), 33–35. These pneumatic shells created imperfect corners and had to be adjusted by Isler to avoid difficulties.

10. H. Isler, "New Shapes for Shells—Twenty Years Later," in Billington, *Isler as Structural Artist*, 30.

11. Chilton, *Heinz Isler*, 35.

12. Billington, *Isler as Structural Artist*, 18. The cover was on a book by C. Faber, *Candela: The Shell Builder* (New York, 1963).

13. D. P. Billington, "Felix Candela and Structural Art," *Bulletin of the International Association of Shell Structures* 88 (January 1986): 5–10. See also D. P. Billington, *The Tower and the Bridge: The New Art of Structural Engineering* (Princeton, 1985), 189–93, 217–19.

14. Ibid., 21.

15. Isler, "New Shapes for Shells—Twenty Years Later," 32. The following quote is from the same page.

16. Chilton, *Heinz Isler*, 25–28.

17. The author first visited the office in July 1979 and climbed onto the roof with Heinz Isler.

18. Isler, "New Shapes for Shells—Twenty Years Later," 34.

19. Chilton, *Heinz Isler*, 94.

20. Ibid., 78; see also J. Joedicke, *Shell Architecture* (New York, 1963), 126–30.

21. Billington, *The Tower and the Bridge*, 225–26.

22. Chilton, *Heinz Isler*, 95

23. Isler, "New Shapes for Shells—Twenty Years Later," 35.

24. Ibid.

25. H. Isler, "Long-term Behaviour of Shells," in *Proceedings of the Seiken-IASS Symposium on Non-linear Analysis and Design for Shell and Spatial Structures, Tokyo*, October 19–22, 1993 (Tokyo, 1993), 9–16.

26. Isler, "New Shapes for Shells—Twenty Years Later," 38.

27. Ibid.

28. Billington, *Isler as Structural Artist*, 21.

29. Isler, "New Shapes for Shells—Twenty Years Later," 34.

30. Ibid., 37.

31. Chilton, *Heinz Isler*, 107.

32. Isler, "New Shapes for Shells—Twenty Years Later," 40.

33. Ibid., 41.

34. Ibid., 42.

35. "Eisversuche" (Research with Ice), *Heinz Isler Schalen*, ed. E. Ramm and E. Schunck (Stuttgart, 1986), 87–91.

36. D. Drew Egbert, *Social Radicalism and the Arts: Western Europe. A Cultural History from the French Revolution to 1968* (New York, 1970), 366–68.

37. Chilton, *Heinz Isler*, 106.

38. The Sicli roof is the product of a complex architectural program that Isler solved brilliantly. An anomaly within his work, it demonstrates the creative potential for some complex programs to be solved elegantly by the engineer who is a structural artist. Isler could never have satisfied that program without first having developed his methods of form-finding and construction process along with his aesthetic intentions of integration, thinness, and contrast.

39. Letter from Heinz Isler, June 25, 2002.

40. J. Huizinga, *Homo Ludens: A Study of the Play Element in Culture* (Boston, 1955; completed in 1938, published in Switzerland in 1944), 10.

41. H. Isler, "The Stability of Thin Concrete Shells," in *Buckling of Shells*, ed. E. Ramm (New York, 1982), 652–55; Isler describes the failures and poor workmanship of shells in South America.

42. *The New Encyclopedia Britannica*, 15th ed. (1997), vol. 11, 634, s.v. "Tennis."

43. Billington, *The Tower and the Bridge*, 20–24.

CHAPTER 6

CHRISTIAN MENN: STRUCTURAL ART IN BRIDGES

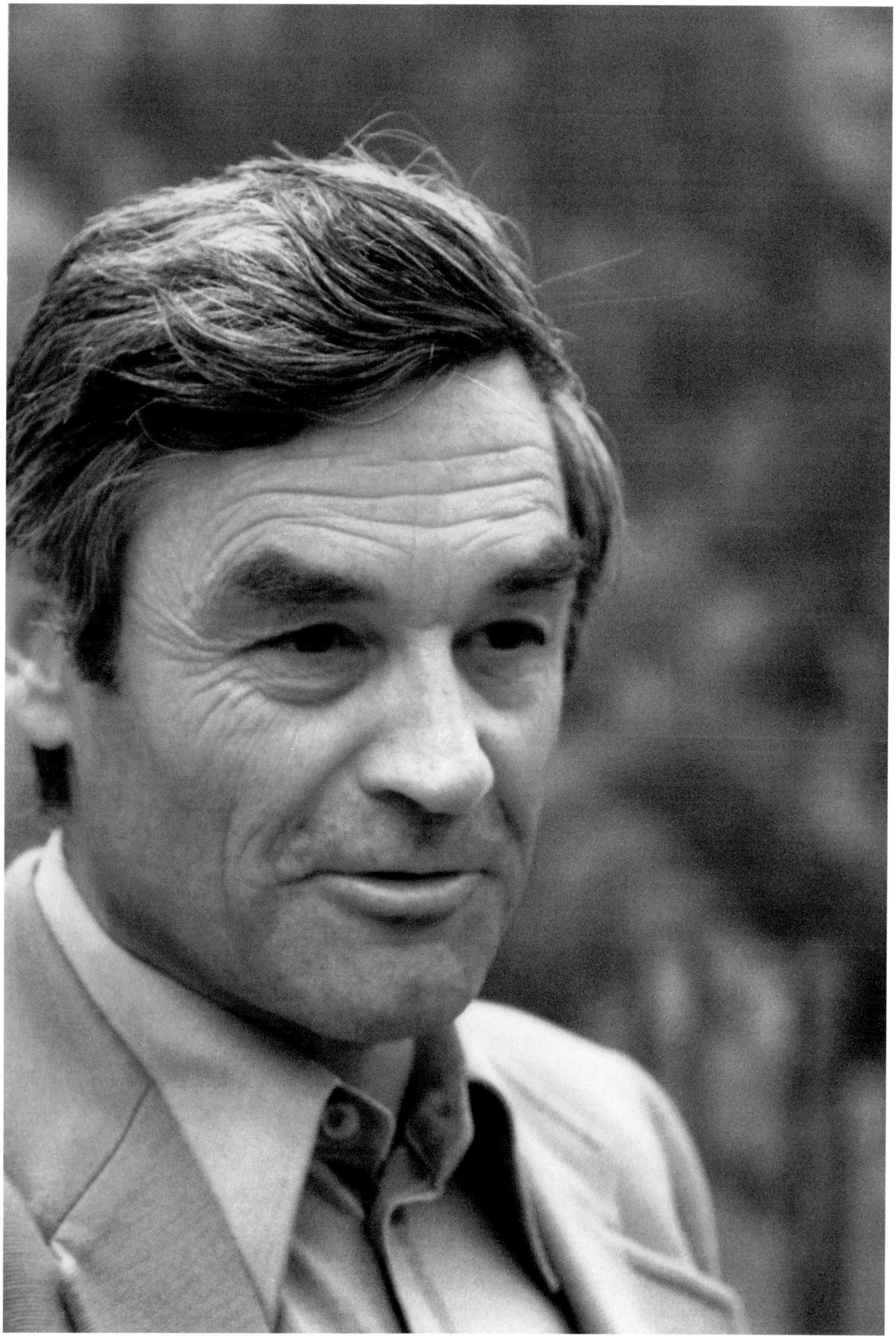

Figure 1 (previous page)
Christian Menn (born 1927).

Figure 2
Target illustrating Menn's design goals, with aesthetics at the center.

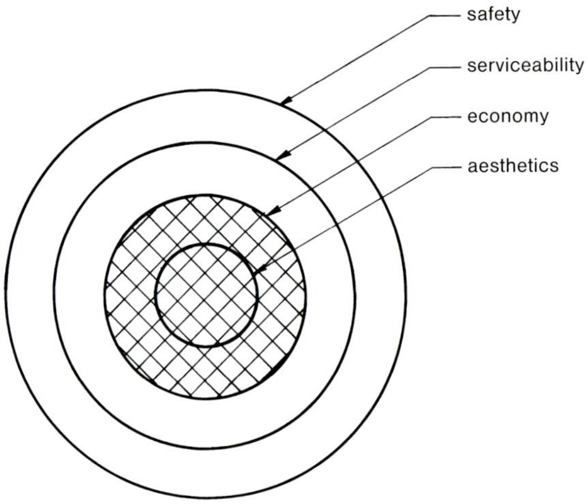

At a special meeting in 1982 to honor Felix Candela, Christian Menn (fig. 1) presented a paper, "Aesthetics in Bridge Design," which, in its simplicity and elegance, reflects the bridge forms he has created over the past forty-five years. He began with a "very roughly summarized philosophical background" in one brief paragraph and quickly proceeded to a discussion of design goals, followed by a section on "Fundamentals of Aesthetics in Bridge Design" and ending with brief remarks on the "Correlation between Aesthetics and Economy."

Published in the same journal as Isler's brief Madrid speech, but twenty-five years later, this paper is less a set of rules for creating elegant bridges than it is a personal explanation of Menn's own designs. He is quick to emphasize that "successful design of a perfect structure can never be performed only on the basis of general rules concerning structural system, dimensions and proportions alone, as long as the design lacks in originality and individuality."[1]

The originality and individuality of his bridges mark Menn as a structural artist and as a student of an exemplary teacher. Menn's paper of 1985 is the clearest late-twentieth-century definition of structural art, which is characterized at the beginning by a target (fig. 2) that defines his design goals—safety, serviceability, economy, and aesthetics.[2] This prototypical Swiss symbol, like Isler's, has a bull's eye, but here the center is aesthetics, whereas for the shell designer the ideal center is minimum deflection. Isler's target indicates the approach to highest technical quality, whereas Menn's expresses the more general ideal of art within discipline (an ideal certainly shared by Isler). The two targets together illustrate how a variety of hierarchies can be put into a disciplinary picture. Both designers give visual form to their abstract ideals just as they try to visualize in their structures the abstractions of forces carrying loads.

One year after the journal publication, Menn expanded upon his ideas and incorporated them into his text *Stahlbetonbrücken* (Reinforced Concrete Bridges), which Paul Gauvreau edited and translated into English for the second edition of 1990 entitled *Prestressed Concrete Bridges.* In chapter 2 of the 1990 edition, "Economy and Aesthetics," Menn explains that the engineer achieves safety and serviceability "through an understanding of the underlying scientific principles," to which all of the remaining chapters are devoted. He then defines the centrality of the two nonscientific ideals:

Economy and elegance, on the other hand, are achieved through nonscientific means. They depend almost entirely on the creativity of the engineer. Economic and aesthetic criteria have not been codified and are largely subjective. Useful guidelines are available to help in improving the cost-effectiveness and visual form of bridges. Proficiency in designing economic and aesthetically pleasing bridges can nevertheless be acquired only through direct design experience, critical observation of completed structures, and full utilization of the engineer's creative talents.[3]

This paragraph, quite original to Menn, emphasizes surprisingly that both elegance and economy are nonscientific, largely subjective, and depend upon the engineer's creativity. In an age that tries to elevate economics to a science with mathematical models, it is crucial to hear this realistic assessment of bridge design coming from its greatest living practitioner. His discussion of economy had two goals: to give the inexperienced designer some means of estimating costs and to place the different elements of a bridge into relative percentage costs. The second goal, summarized in Table 1 (taken from his book), illustrates how Menn uses costs to direct both analysis and design. The table represents average costs of nineteen Swiss bridges built between 1958 and 1985; it allows Menn to discuss analysis by expanding, for example, on the 24.5 percent construction cost for reinforcing and prestressing steel:

Between 60 and 65 percent of the reinforcing steel is normally required for stability of the assembled cage of bars during concreting and for crack control in the finished structure. Only the remaining 35 to 40 percent of reinforcing steel and the prestressing steel can therefore be reduced through calculation. Reducing this steel by 10 percent would decrease the total construction cost by less than 2 percent. Substantial savings, therefore, cannot be achieved by refined calculations alone. The time and effort required for refined calculations is preferably invested in the detailing and layout of the reinforcing and prestressing steel.[4]

Table 1: Breakdown of Average Construction Costs

Item	Cost (% of Total Construction Cost)		
Mobilization			8.0
Structure			
Substructure			
Foundations	18.0		
Piers and abutments	5.5		
Total substructure	23.5	23.5	
Superstructure			
Falsework, formwork	20.0		
Concrete	10.0		
Reinforcing steel	13.3		
Prestressing steel	11.2		
Total superstructure	54.5	54.5	
Total sructure		78.0	78.0
Accessoires			14.0
Total construction cost			100.0

Thus, according to Menn, detailed computer analyses with supposedly more rigorous assumptions will not produce a less expensive structure because other factors are of far greater importance. He stresses the connection between economy and elegance, where "aesthetically pleasing bridges are distinguished by transparency, slenderness, and the lack of unnecessary ornamentation, all of which result in an efficient use of materials and hence low construction cost."[5] He follows that statement with a key warning: "It is incorrect to infer, however, that the most economical design is necessarily the most elegant."[6] Thus, central to structural art is realizing the fundamental error in the simple-minded theory that claims that if a structure is economical (or efficient), it will be elegant. Menn's approach to aesthetics concluded the chapter, and we shall return to that subject after reviewing his major works because his ideals come from his experience rather than the reverse; that is, he did not first formulate a theory and then apply it to create his designs. And yet the seeds for his type of flowering were planted much earlier by his great mentor, Pierre Lardy.

Christian Menn (born 1927) prepared for entrance to the Federal Institute of Technology from 1939 to 1946 by attending the canton school in Chur, capital of his home canton, the Graubünden. He had been born in Meiringen, in the canton of Bern, because his father, Simon Menn, was working in that region as a civil engineer. The elder Menn had worked with both Florian Prader and Eugen Losinger on projects designed by Robert Maillart. Simon Menn knew Maillart professionally.[7] In 1946 Christian Menn entered the Zurich engineering school and in 1948 began taking courses from Lardy. Menn described studies there in the late 1940s: "Structural analysis found itself at that time in the transition from descriptive graphical analysis to abstract analytical statics. The simple, graphic analysis enabled a good and fast qualitative evaluation of the stresses and the distribution of forces in the structure and remained an outstanding and important basis for design work. The abstract analytical statics paved the way for schematic calculations and later for computer applications."[8]

This transition was largely the work of Max Ritter, with whom Lardy had worked and who was a strong opponent of Maillart. Lardy certainly learned from the man he replaced in 1946, but he also developed a different view of Maillart by the time Isler and Menn were students. Menn remembers Lardy's growing independence: "Great engineers and designers like, for example, R. Maillart, placed greatest value on the graphic qualitative analysis as the simplest method to find the distribution of forces in the form, i.e., 'the aesthetics of the constructive truth' permits expression and visualization. Prof. Lardy saw these connections very well and although, or directly because he had first studied mathematics, he placed great value on the overview and estimation of dimensions."[9]

Like Maillart and Menn, Lardy's facility with mathematics gave him a firm basis for recognizing its limitations for engineers as he began to appreciate the deep power of visualization as a prelude to the expression of "the aesthetic of the constructive truth." Lardy began in his teaching to cast skepticism over the supposedly more accurate mathematical complexities of structural analysis: "Over and over again he observed with irony those whose designs were based upon a painstaking and exact theory laid on crude foundations, for example, heavily idealized building material performance."[10] Students recognized quickly the value of Lardy's teaching and enjoyed him as "an enthusiastic and inspiring teacher." But they found him far more than an entertaining speaker when they recognized "his ability to make his theoretical presentations understandable to his listeners."[11] Moreover, Lardy's lack of design experience did not interfere with his ability to discuss and even criticize commonly used forms and details. Isler's notebooks tell the story of Lardy's emphasis on bridge aesthetics (see chap. 4 above). An example of a continuous beam bridge with three possible forms (fig. 3) shows

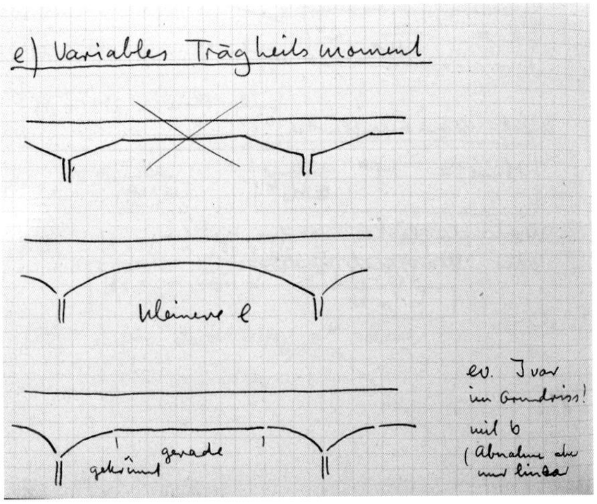

Figure 3
Lardy's three forms for a continuous beam bridge, as recorded in Isler's notes.

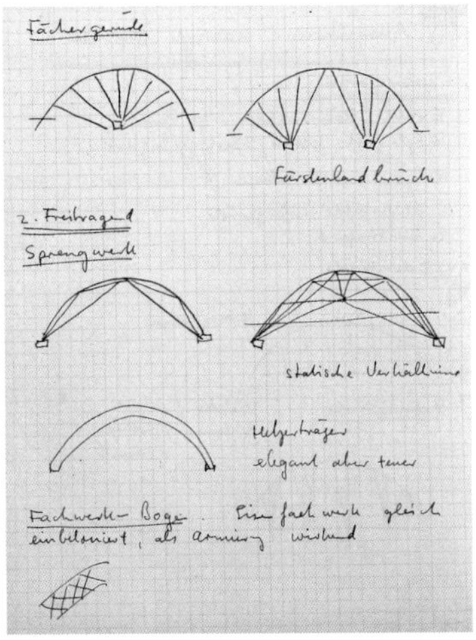

Figure 4
Lardy's examples of false work, or scaffolding, that can be used to support arch forms when casting concrete, as recorded in Isler's notes.

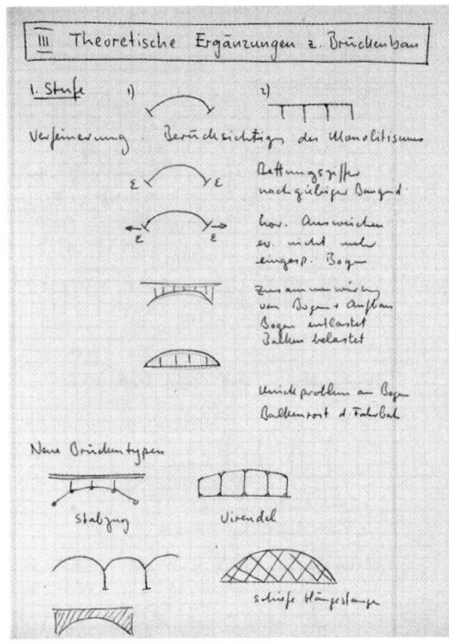

Figure 5
Lardy's general introduction to theoretical extensions of bridge structures, as recorded in Isler's notes.

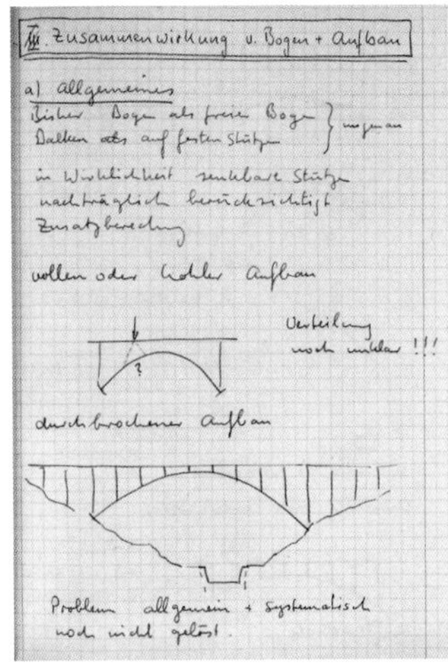

Figure 6
Lardy's analysis for deck-stiffened arches, as recorded in Isler's notes.

that Lardy crossed out the first, noted that the second was only for small bridges, and preferred the third. The choice is primarily aesthetic and demonstrates that there are many technically reasonable forms from which the bridge designer can make a choice based on aesthetics. There is a variety of false work, or scaffolding, that the builder can use to support arch forms for casting the concrete (fig. 4). This figure is followed by details that emphasize the need for the designer to think about how the bridge is to be built.[12] In his most advanced course, the section "III Theoretical Extensions of Bridge Structures," Lardy began by introducing some general ideas (fig. 5) and later presented the analysis for deck-stiffened arches (fig. 6), where the lower diagram fortells Menn's first major deck-stiffened arch at Crestawald (fig. 7). The course concluded with a guest seminar on a simplified method for deck-stiffened arches, which was precisely the method developed and used by Maillart in 1923.[13] Many other examples of Lardy's bridge ideas are discussed in chapter 4, all of which demonstrate the professor's aesthetic sensitivity and the broad viewpoint he had of his subject. Menn recalled: "By far most of the students [in structural engineering] wanted to do their independent written thesis (Diplomarbeit) under his direction… At the end of 1950 I completed my written thesis with him as advisor."[14]

Following graduation, Menn entered the Swiss military service. In 1951 he contracted tuberculosis and spent fourteen months in a sanatorium at Arosa high up in the Graubünden Alps. During this time he began theoretical studies on the structural analyses of spiral staircases, a difficult subject that required a detailed mathematical analysis in structural mechanics. He had yet to focus on bridge design. Upon his recovery, he accepted the offer to assist Professor Lardy, with whom he spent the next three years (1953–56) at the ETH. He, along with two others, acted as the usual teaching assistant for Lardy's courses and helped Lardy with his research and consulting. In addition, he completed his dissertation and in 1956 received the doctoral degree with Lardy as his principal adviser. It was during these years that Menn began a careful study of prestressed concrete as well as of shell-roof design.

The most significant interaction with Lardy, however, came during his consulting on prestressing and on bridge design. The Stahlton Company had asked Lardy to research and determine the validity of a new prestressing system and also to study the design for prestressed concrete bridges using that system. Lardy had little practical experience with bridge design and therefore had to think and study deeply in this new field. He knew the theory of prestressing but not its practice. Here was a great opportunity for a young assistant, and Menn took advantage of it. He participated in Lardy's studies of prestressed bridges and "came for the first time in contact with the practice of bridge structures. That was for me the greatest motivating experience of my assistantship."[15]

In spite of inexperience, "his character and his artistic inclinations led Lardy to reflect on questions of bridge form."[16] From this study he could see thin elements as the most natural components for concrete structures, and with that view he identified himself with "Maillart's structural principles." Lardy often articulated these ideas "in contradiction to his predecessor Max Ritter."[17] Lardy had a profound influence on Menn; this began to take shape shortly after Menn left the ETH and entered practice.

Figure 7
Crestawald Bridge over the Hinterrhein, built in 1959, with deck-stiffened arch that spans 71.5 meters.

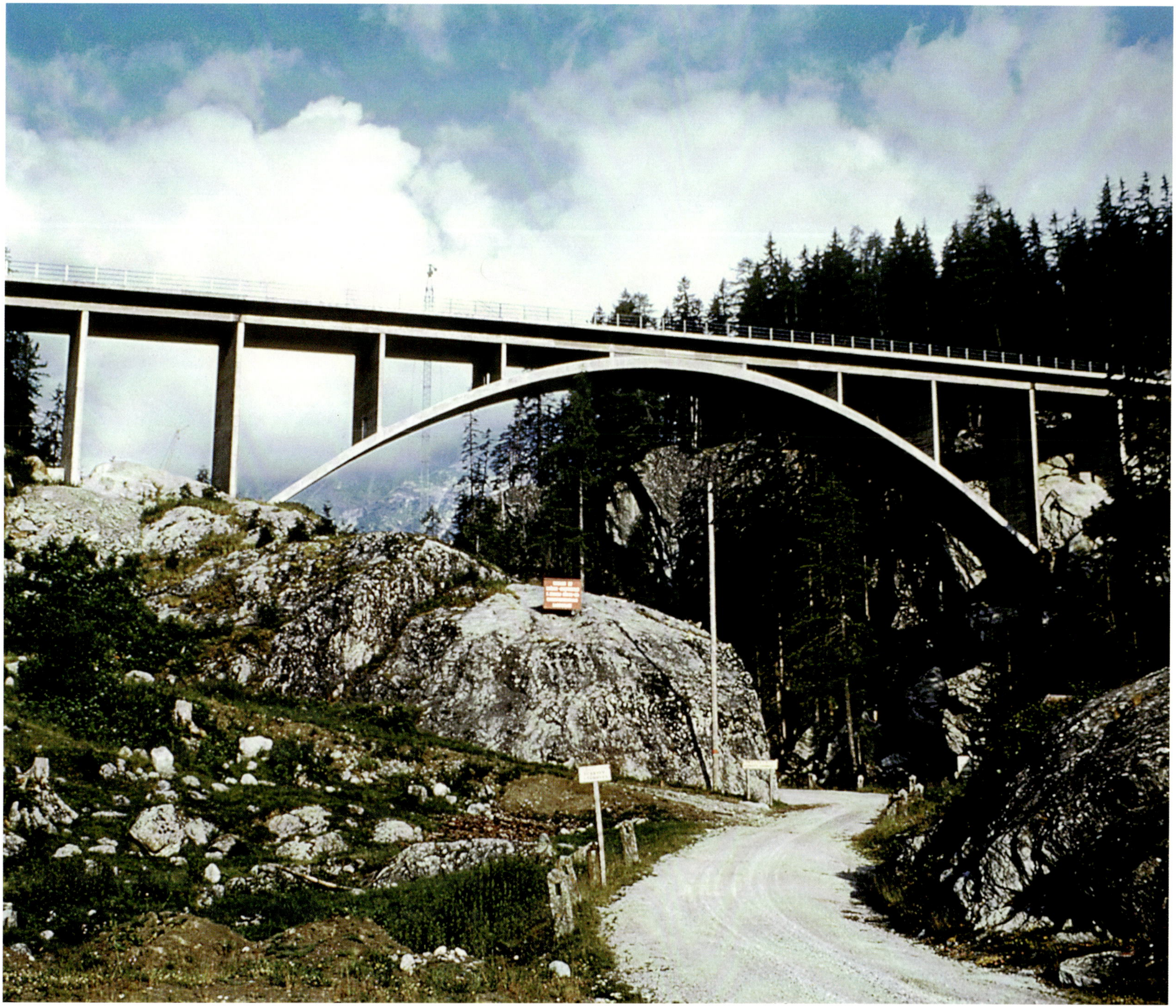

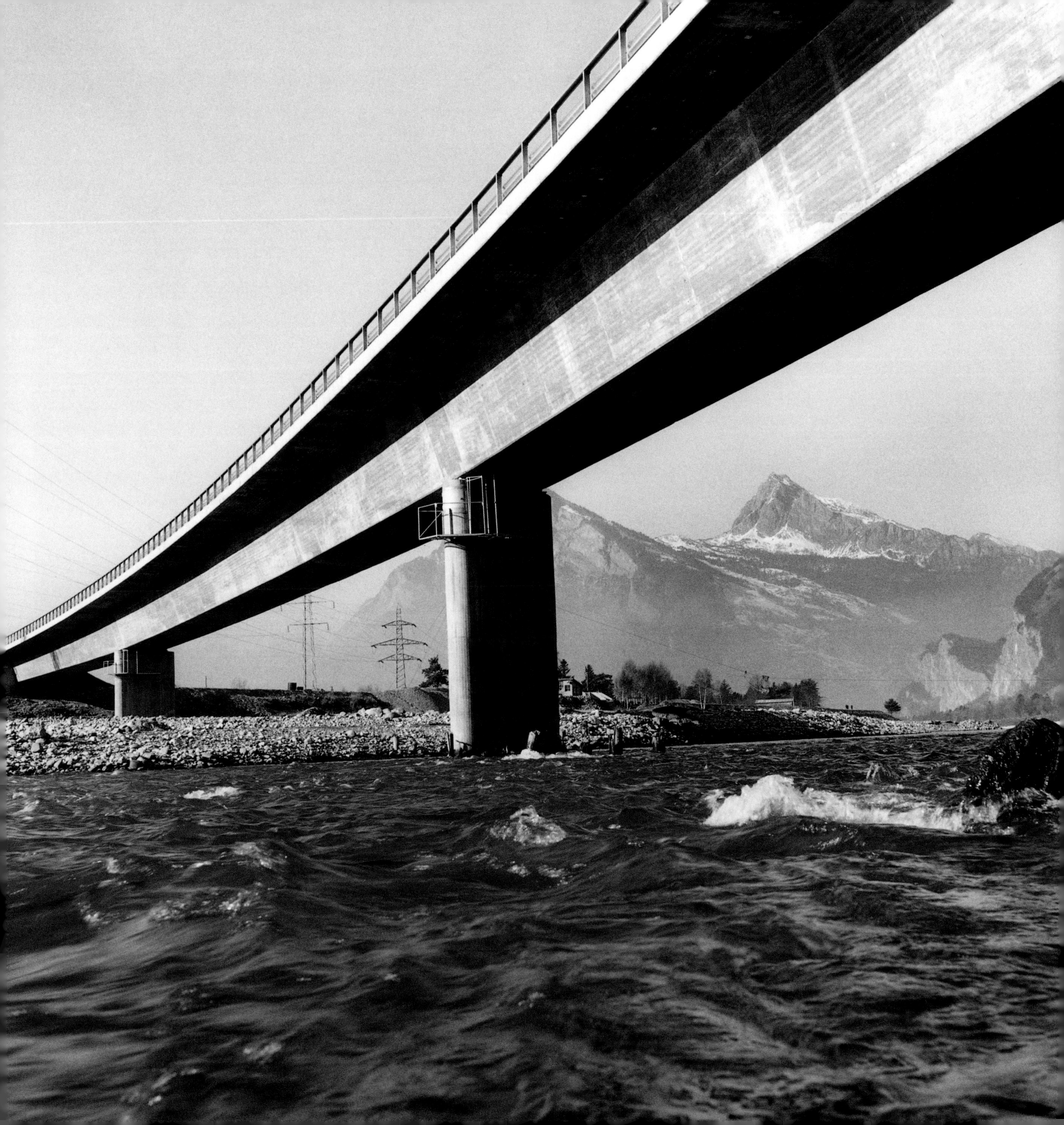

Figure 8
Rhine Bridge at Bad Ragaz, built in 1962, with a prestressed-concrete continuous beam that spans 82 meters.

Figure 9
Tamins Bridge near Reichenau on the cover of the January-February 1984 issue of *American Scientist.*

Figure 10
Valserrhein Bridge, Uors, built in 1962, with arch span of 86 meters.

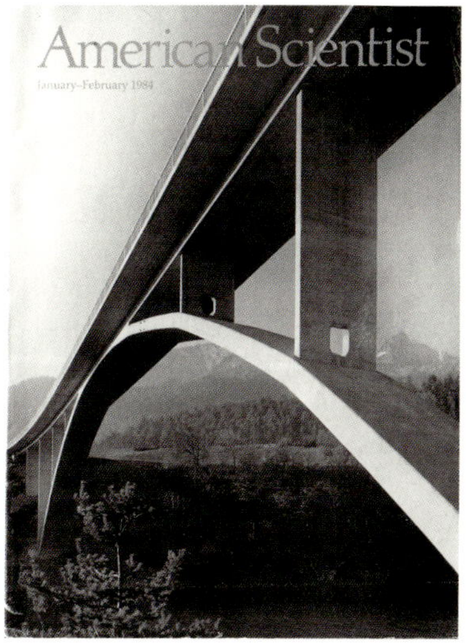

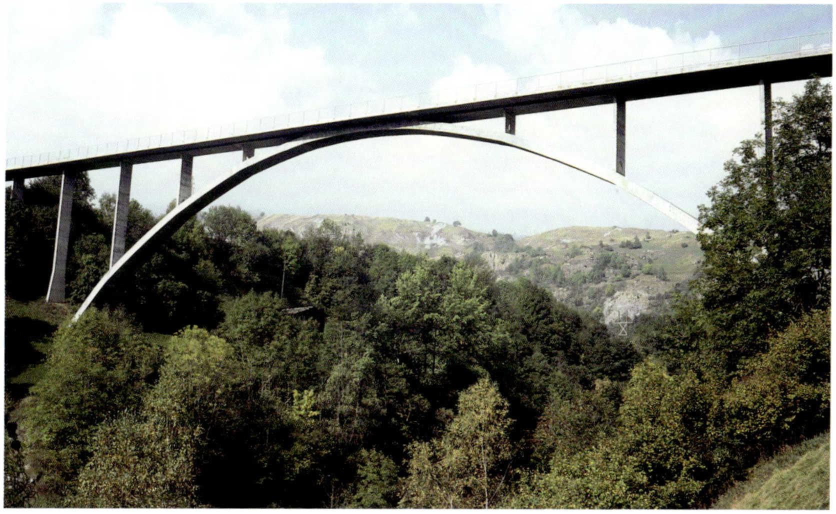

After a year in Paris working on the UNESCO structure of Pier Luigi Nervi and in Bern in a design office, Menn opened his own consulting engineering firm in Chur in 1957, and for the first few years, he designed numerous bridges in the Graubünden. Three are of special significance because they illustrate his early mastery of arch design and the influence of Maillart's ideas. The Crestawald Bridge over the Hinterrhein, built in 1959, is a deck-stiffened arch with a span of 71.5 meters (234 feet), almost twice that of Maillart's longest bridge of the same type. The next year saw completion of two bridges over the Averserrhein, a 66-meter-span (216 feet), deck-stiffened arch at Cröt and a 66.5-meter-span (218 feet), three-hinged arch at Letziwald,[18] a hollow-box structure that closely resembles Maillart's Tavanasa Bridge. As Menn put it: "Comparisons have shown that the principle of construction used for the Tavanasa Bridge was still, fifty years later, the most advantageous one." He described the decision for the form at Cröt by noting that "with respect to a light scaffolding, a deck-stiffened arch proved to be economical, a system also preferred by Maillart for the same reason."[19]

During the first four years of his independent practice, Menn designed a number of smaller beam bridges that were built in the Graubünden and gave him detailed experience with prestressed concrete. During this time, he started the design of an 82-meter-span, continuous prestressed-concrete beam bridge over the Rhine at Bad Ragaz (fig. 8), one of the longest spanning bridges of this type in Switzerland when it was built in 1962. Menn was exploring new possibilities, but for longer spans where the roadway was high enough above the water, he still preferred arches. Menn soon had the chance to fulfill a youthful goal to design a 100-meter-span arch across the Rhine. His reputation by 1962 was strong enough in his canton to secure a commission to project such a span at Tamins near Reichenau, downstream from where the Vorderrhein joins the Hinterrhein, and the wider river made that goal practical. This was a fully original variation on Maillart's innovation. It is no accident that a photo of the Tamins–Reichenau Bridge appeared in 1984 on the cover of *American Scientist* (fig. 9) to announce an article on structural art and also on the cover of the published version of Menn's retirement lecture in 1992.[20] It is a masterpiece of concrete arch-bridge design, which had followed his earlier 86-meter arch span at Uors over the Valserrhein in 1962 (fig. 10).

Menn responded to two major changes in construction following World War II: rapid increase in labor costs and the wider use of prestressing. In his 100-meter-span bridge, Menn spaced the vertical members more widely apart than did Maillart, but his structure still acted as a deck-stiffened arch. He could make the much longer spans in the hollow-box deck thanks to prestressing, and fewer vertical columns meant less labor-intensive forming. This 1963 arch bridge thus became more open–transparent–and simpler in appearance because it was easier to build (fig. 11). Furthermore, the single hollow box allows for a wide overhanging part of the roadway structure, which casts a shadow on the vertical walls of the box to make the deck appear even thinner. There are no false increases in the width of columns at the abutments. It is one thin, narrow, open, and integrated form.

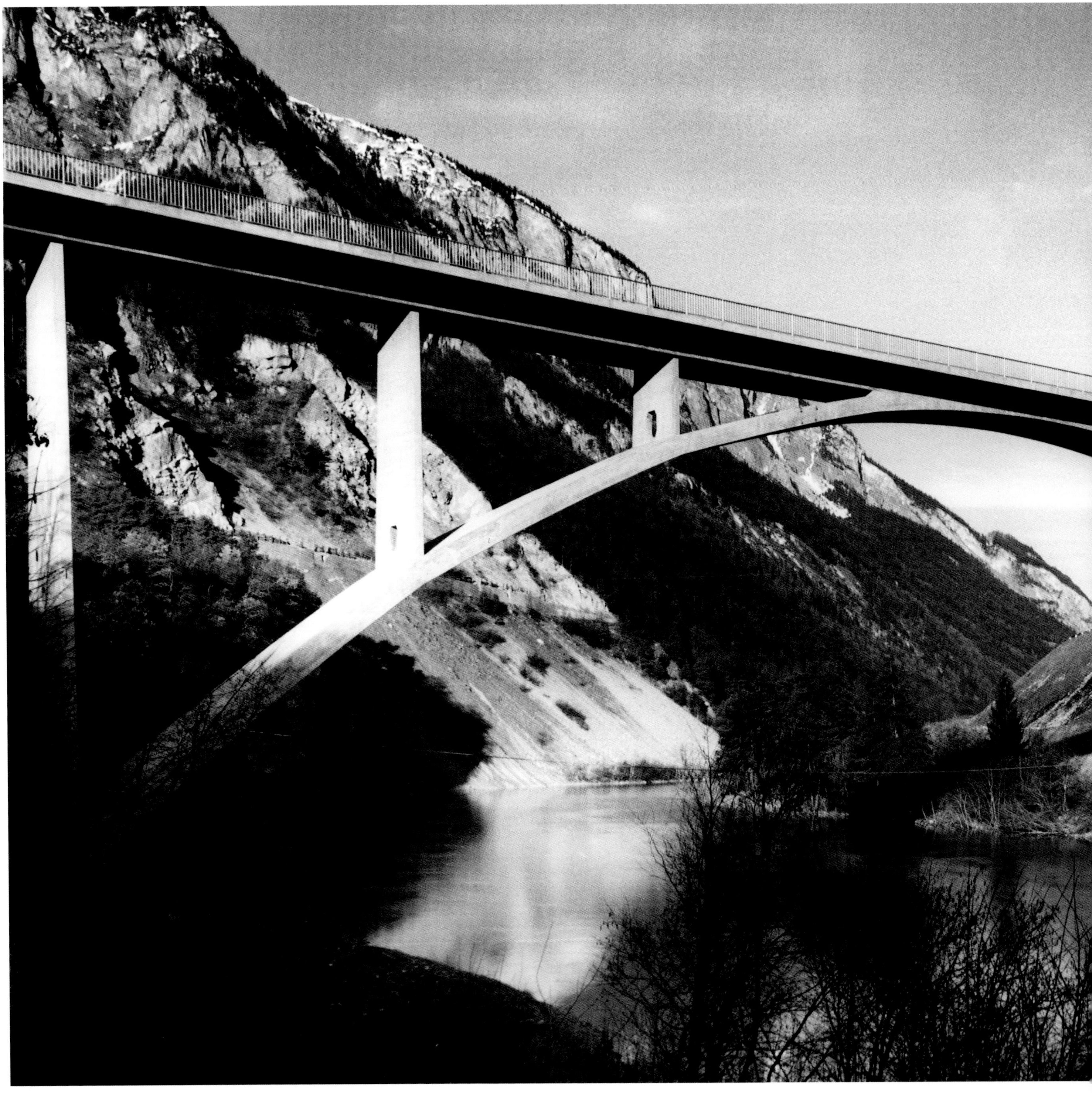

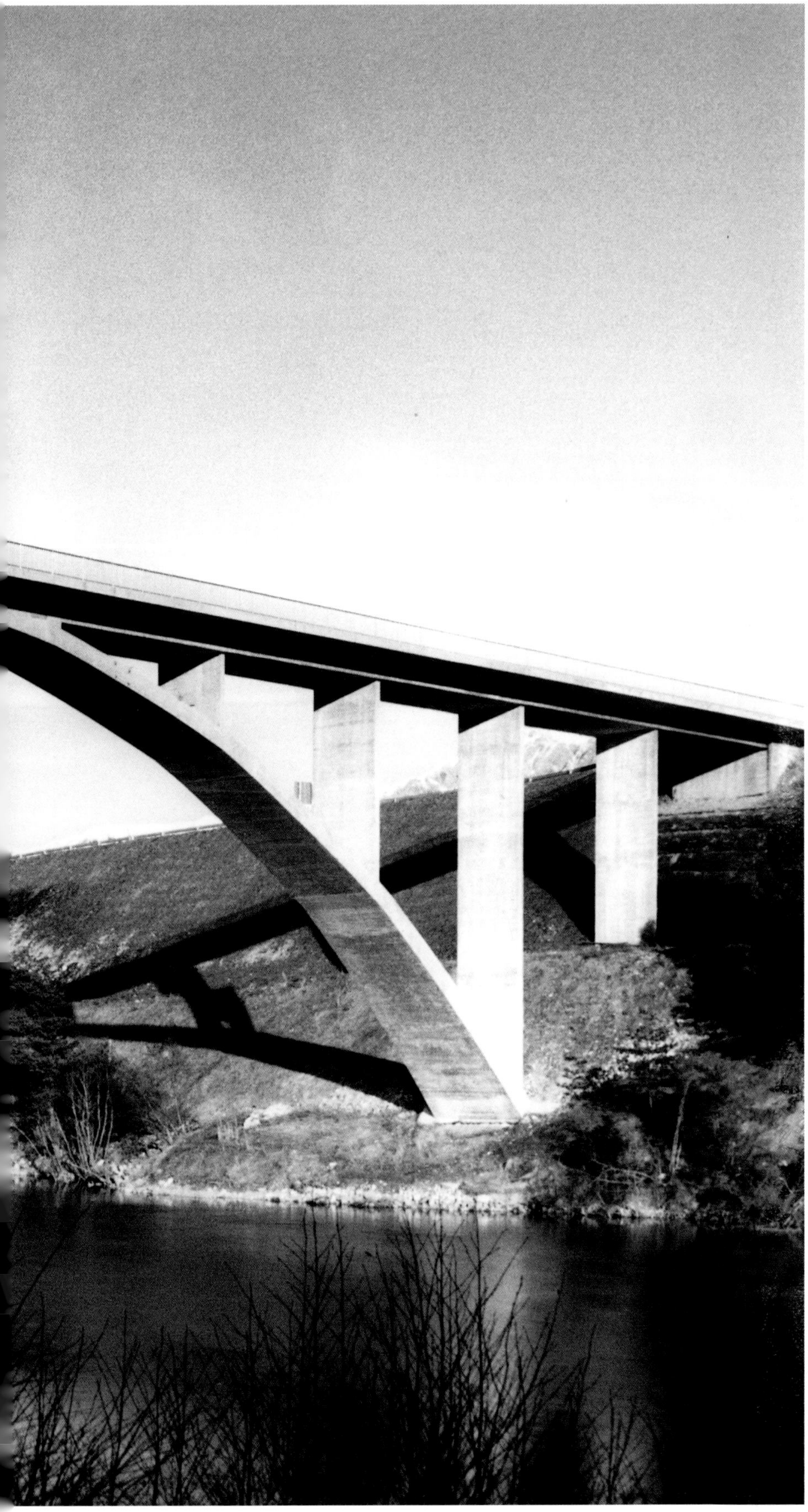

Figure 11

Tamins Bridge near Reichenau over the Rhine, completed in 1963, with arch span of 100 meters.

Figure 12
Viamala Bridge over the Rhine, built in 1967, with arch span of 96 meters.

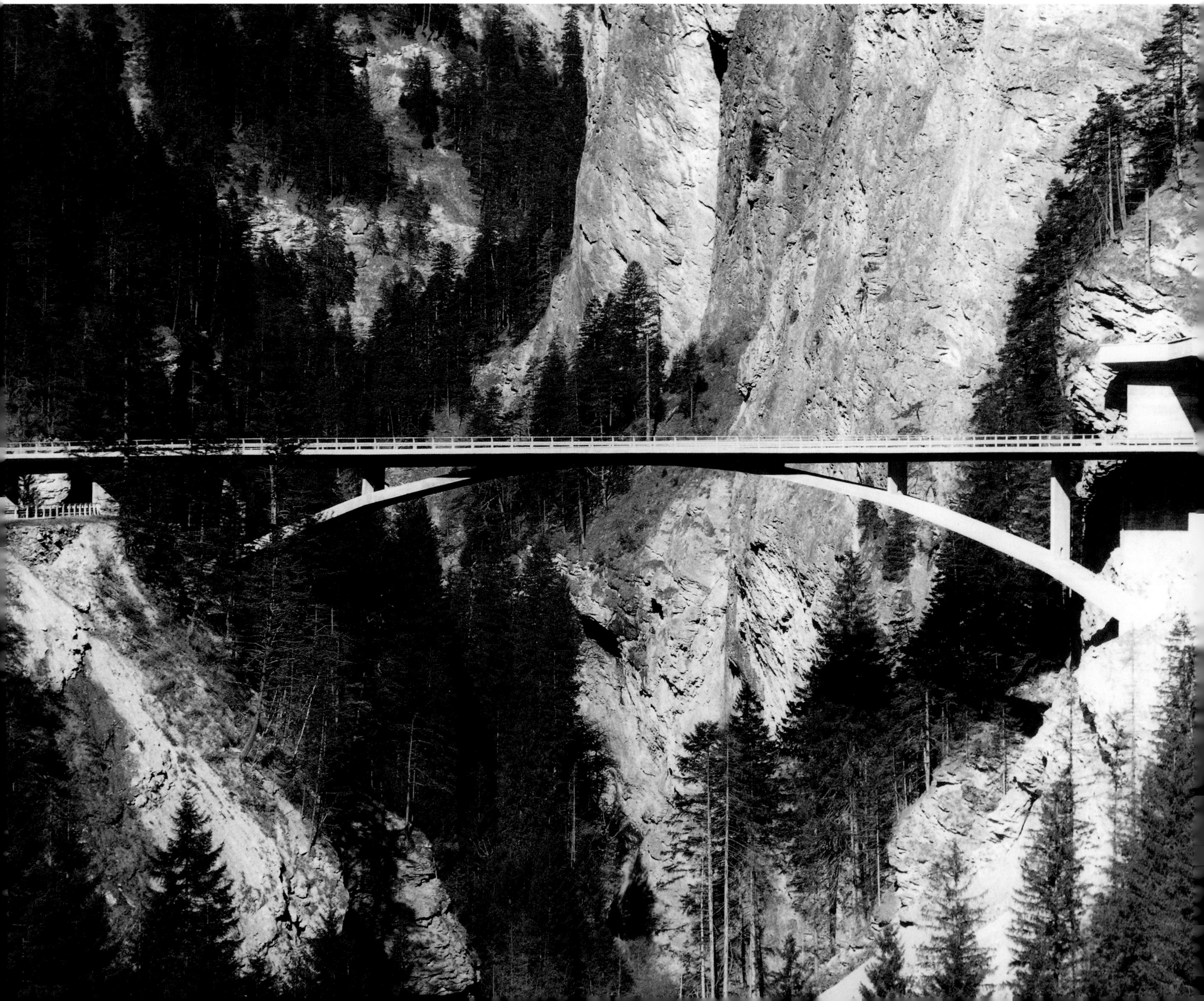

Figure 13

Double silhouette of the Nanin (1967) and Cascella (1968) Bridges, Mesocco Valley, with arch spans of 112 meters and 96 meters, respectively.

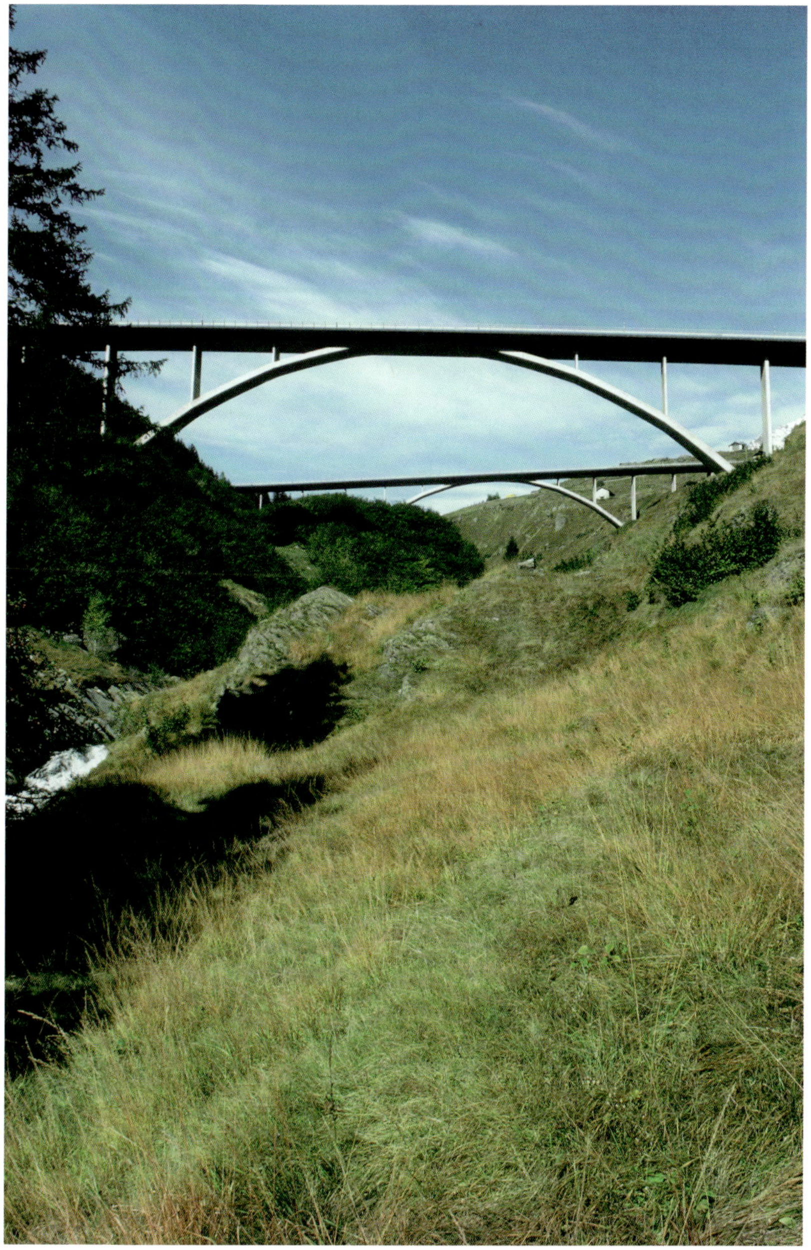

Menn continued this mature form in 1967 with his 96-meter-span bridge at the Viamala (fig. 12) with an arch of lower rise that crosses this precipitous canyon of the Rhine River. The same year saw the completion of the first of two major works in the Mesocco Valley, the Nanin arch with the even longer span of 112 meters. In 1968 the second of the Mesocco arches, the Cascella span of 96 meters, created a remarkable double silhouette when viewed from downstream (fig. 13).

As the 1960s climaxed for Menn with his mature arch bridges, he began to develop new ideas. He had solved the problem of relatively long-span, deck-stiffened arches thanks to prestressing, but now the revolutionary new material—prestressed concrete—began to open up a new vision of design: he saw that prestressing could actually replace the arch itself. The deck of his arch bridge could become the main supporting member without the arch and, moreover, without the construction difficulties of the scaffold needed to support the arch during construction and the curved formwork required to shape the arch.

In 1969 a third Mesocco Valley bridge, the Salvanei Bridge, dramatically demonstrated this new possibility. For this bridge, Menn spanned 60 meters with a horizontally curved roadway deck reminiscent of Maillart's Schwandbach design (see chap. 2, fig. 37). This bridge, however, dispensed with the arch form and carried the roadway solely by a curved hollow-box beam that was prestressed (fig. 14). The effect is a structure of exceptional lightness, a stripped-down form of transparency that resembles the work of the Graubünden sculptor Alberto Giacometti, who came from the Italian-speaking part of the canton, near the location of the Mesocco Bridges.

Following in the steps of his own professor, Pierre Lardy, Menn began a new stage in his career when in 1971 the ETH named him Professor for Concrete Structures. This new direction meant that by law he would have to give up his design business and could thereafter be only a consultant. Thus, from 1971 until his retirement in 1992, he focused on developing his ideas about bridge design for teaching, the primary result being his book *Prestressed Concrete Bridges,* published in 1986 and revised in 1990. The constriction on his design practice, however, actually resulted in a great advantage for his design career. During this next period only a few bridges would be constructed for which he was the designer, but each of these represents a major event in the history of twentieth-century bridges. He no longer would need a large number of smaller projects to sustain his activity. Rather, he could focus on a few structures of

Figure 14
Salvanei Bridge, Mesocco Valley, built in 1969, with prestressed hollow-box beam spanning 60 meters.

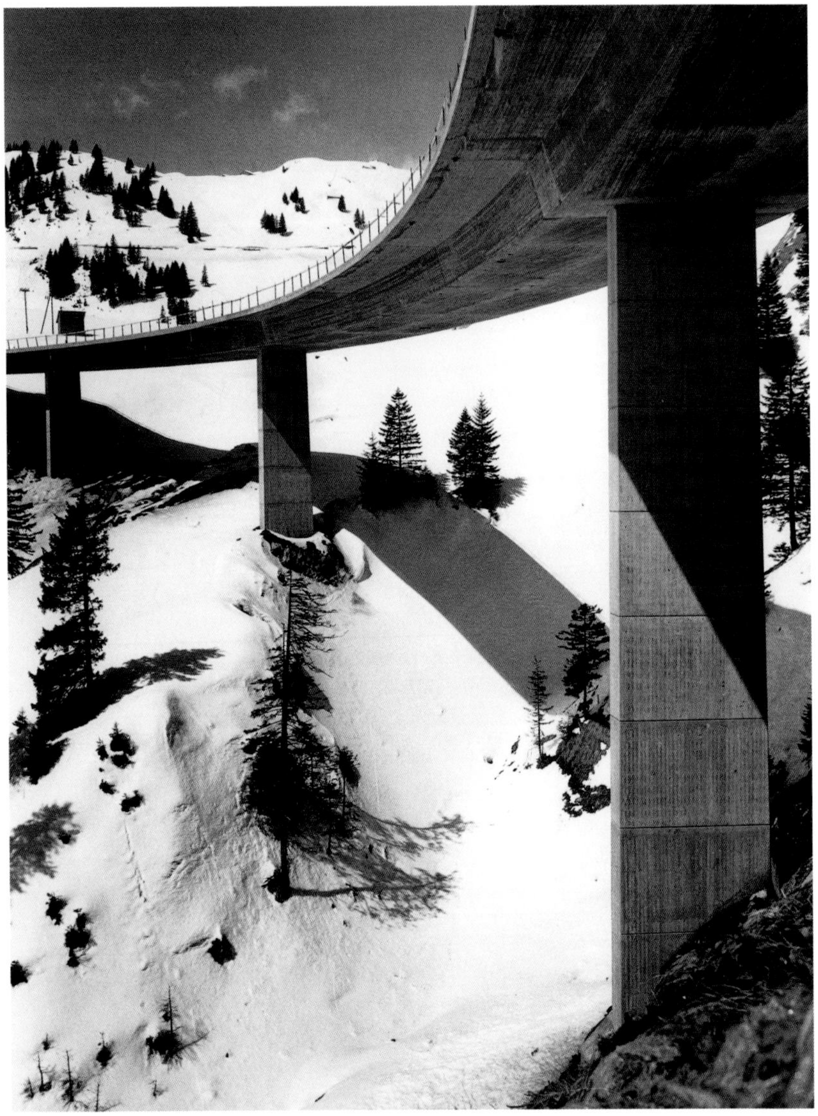

great international significance: The Felsenau Bridge (1974), the Ganter Bridge (1980), the Biaschina Bridge (1983), the Chandoline work (1989), the 1999 Sunniberg design, and the bridge over the Charles River in Boston (2002).

Many of Menn's bridges resulted from engineering design competitions that he frequently won.[21] Such events have in part been responsible for the high quality of bridge design in Switzerland. The process for the Felsenau Bridge is an example. On July 14, 1970, the cantonal office for turnpike construction in Bern published for the Felsenau Crossing of the Aare River a twelve-page program for an invitational design competition. Following rules established by the Swiss Society of Engineers and Architects, the canton invited seven of Switzerland's best-known engineers to compete. Some formed joint ventures and one remained alone. The program also listed the seven members of the jury: three engineering professors, three engineers in public works agencies, and one architect. Each of the professors had substantial experience with bridge design, and two had won such competitions.[22] The program listed the sum of 560,000 Swiss francs ($280,000 in 1971) as prize money for a structure estimated to cost about 30 million Swiss francs. Of that prize money, 50,000 francs were to be paid to each competitor and the remaining 210,000 francs allocated according to ranking by the jury. In fact, the winner received 100,000 francs, and the competitor ranked seventh earned 60,000 francs. Usually, the first two or three highest-ranking participants would receive enough to pay their costs. In addition to outlining numerous details referring to the site and to good practice for deck drainage, expansion joints, and foundations, the program specified the judging criteria: overall concept, performance, aesthetics, cost, construction procedure and time, and interruption of regular services. The competition was to begin on August 1, 1970, and all entries were to be delivered by February 28, 1971. The designs were to be based on a construction that would start early in 1972 and be completed by October 1974 at the latest.

The jury met three times between March 4 and June 30, 1971, when their final report appeared. In addition to hiring consulting engineering firms to check the quantities and building contractors to develop cost estimates for each design, members of the jury reviewed all the calculations and intensively discussed the appearance of each project, which had been assigned a number. The jury knew each design only by number and did not know which design belonged to which designer until after the completed final report was signed. When the sealed envelopes connecting the numbers with the names were opened, the first prize was awarded to the joint venture of Emch and Berger in Bern and Christian Menn from Chur.

Figure 15
Felsenau Bridge, a segmental cantilever bridge over the Aare River, completed in 1974.

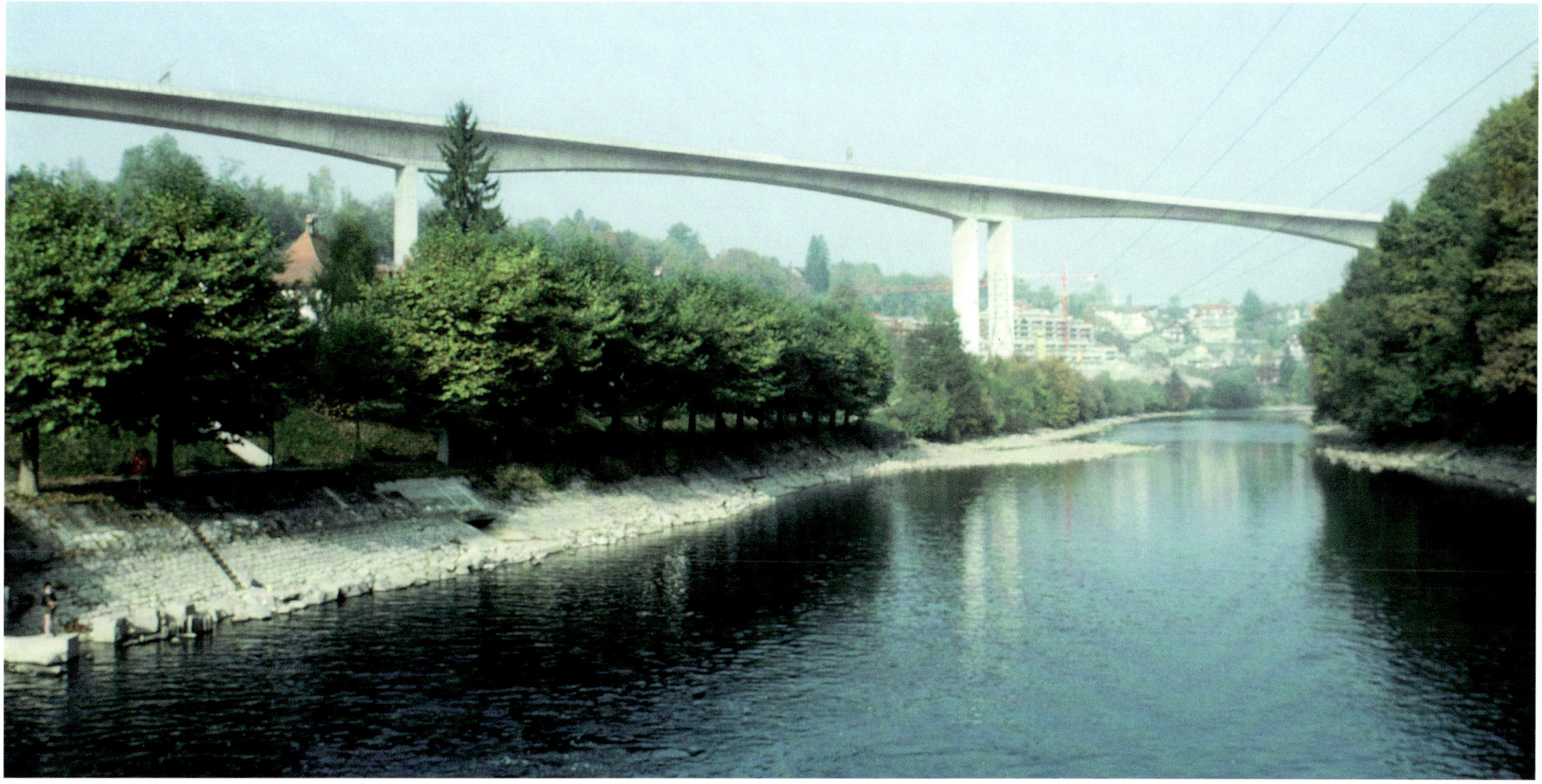

The jury report is a public document. Shortly after it was released, the canton of Bern held a public exhibition of the seven competing designs in a large room in the Post Museum in Bern. Along with each design was the jury's criticism.[23] For the winning design, the jury found the calculations very carefully done, including the torsional bending of the single hollow box curved in plan. For the aesthetics, the jury found the design "well balanced and convincing. The scale of this design fits well in the landscape. The conical hollow box is elegantly formed throughout the entire length of the bridge and gives an impressive profile with its wide cantilever roadway slab." The only criticism of the construction was in the design of one pier foundation. The cost estimate was one of the lowest. In one of the other proposals, by contrast, the jury noted that the torsional problem was not even studied and that aesthetically "this design does not fit well into the Aare Valley. The formal unity of the superstructure does not go well with the very different double columns, and the four columns in the Aare River bed are especially disturbing. The long spans in the approach structures are unnecessary." The report concluded by recommending that the first-prize design be the basis for the final construction. The canton agreed, and the design was put into final form during the next nine months so that construction could begin in early 1972. The bridge structure was completed by 1974, the approaches and wearing surface were finished in the summer of 1975, and the bridge was officially opened in September of that year.

Although the bridge competition was a joint venture, Christian Menn was the designer, since he conceived the overall form, defined the basis for calculations, and identified the concept for construction. At first glance, the Felsenau design (fig. 15) appears to be merely another good example of the segmental cantilever bridge construction pioneered by Ulrich Finsterwalder.[24] Closer study shows that Menn's design represents a unique application of that method. At least three unusual aspects characterize Menn's search for forms that satisfy all three main criteria of design: efficiency, economy, and aesthetics.

Figure 16
View of the Felsenau Bridge under construction showing the single box with partially prestressed cantilever slabs.

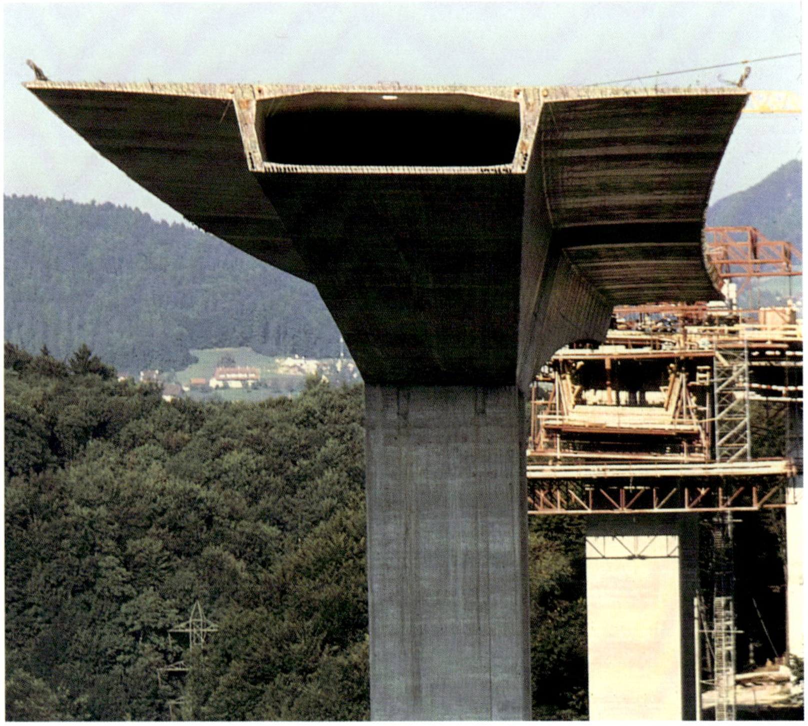

The first unusual characteristic is the curved roadway plan for which Menn provided a curved box girder supported in the main spans by two thin and narrow hollow-box walls about 48 meters (157.4 feet) high and spaced 12 meters (39.4 feet) apart. This solution provides relatively long spans of 156 meters (511.7 feet) with a girder whose maximum moments are based only on the 144-meter (472.3-foot) clear spans. Furthermore, in spite of very thin walls (2.4 x 7.5 meters; 7.8 x 24.5 feet), the longitudinal stiffness is great because the spacing is so wide (12 meters).[25] Therefore, the double-column solution is efficient. It is also economical because the wide-column spacing permits a simple scaffolding for the girder sections over the columns and a long platform from which to launch the first cantilever scaffold from both sides simultaneously. Finally, the narrow double columns permit a more open structure visually. This judgment refers to alternatives proposed during the competition, as well as to examples of such viaducts that had been built prior to the Felsenau design. Earlier viaducts usually had two columns side by side supporting a pair of box girders. Thus, the views from beneath those bridges tended to present wide, solid walls that blocked out the landscape, especially since one set of columns would merge with the next set from an angled view. Particularly with a curved bridge, such merging would have closed a striking view that Menn's solution opens up.

The second distinctive aspect arises from the profile of the roadway through a wide valley, partly wooded and partly suburban. Here Menn designed relatively long spans out of haunched single-box girders made continuous both at midspan and at supports. These haunched single boxes are efficient because they require less material than a girder of constant depth. The continuity at the midspan, while requiring extra reinforcement, avoids the rotational discontinuity that arises from concrete creep and that tends to give a bump in the riding surface, leading to extra maintenance. The haunching, however, does require an increase in field construction costs above those for a girder of constant depth: the savings gained by material efficiency would be at least partly offset by the extra cost of field labor. Nevertheless, the haunching, while not the least costly, does provide a more striking visual form than would a girder of constant depth.

The third uncommon feature of the Felsenau design is the single box for a wide roadway (fig. 16). Previously, prestressed segmental bridges had two boxes, one for each half of the bridge, producing two bridges side by side. The single box with wide cantilever slabs requires less material than would two boxes; but those transverse cantilevers could have caused deflection problems had Menn not decided to use partial prestressing. The slabs are transversely prestressed for dead load only, and additional non-prestressed steel resists the live load. In this way, the permanent loads are balanced, and there is little tendency for vertical deflection caused by creep under prestress-induced flexure. At the same time, some cracking is tolerated under full live load as it would be in a normally reinforced concrete structure. Menn had used partial prestressing in the decks of his bridges as early as 1963 with his Tamins Bridge near Reichenau (see fig. 9). The successful performance of that and other such designs has confirmed the high-quality performance and the efficiency of partially prestressed structures.

The single box compared to a double one also has the cost-saving advantage of halving the forming operations and of allowing the cantilever slabs to be built at a later time on simple scaffolding supported from the completed box girder. Finally, the single-box solution together with the narrow column walls presents a strikingly light appearance. Of equal importance is the way Menn designed the transition between the box and the columns. The box not only changes depth longitudinally, but it also changes width, becoming narrower near the supports (fig. 17). Menn achieved this narrowing by tapering the box so that its

Figure 17

Felsenau Bridge haunching at supports, showing changes in the depth and width of the box.

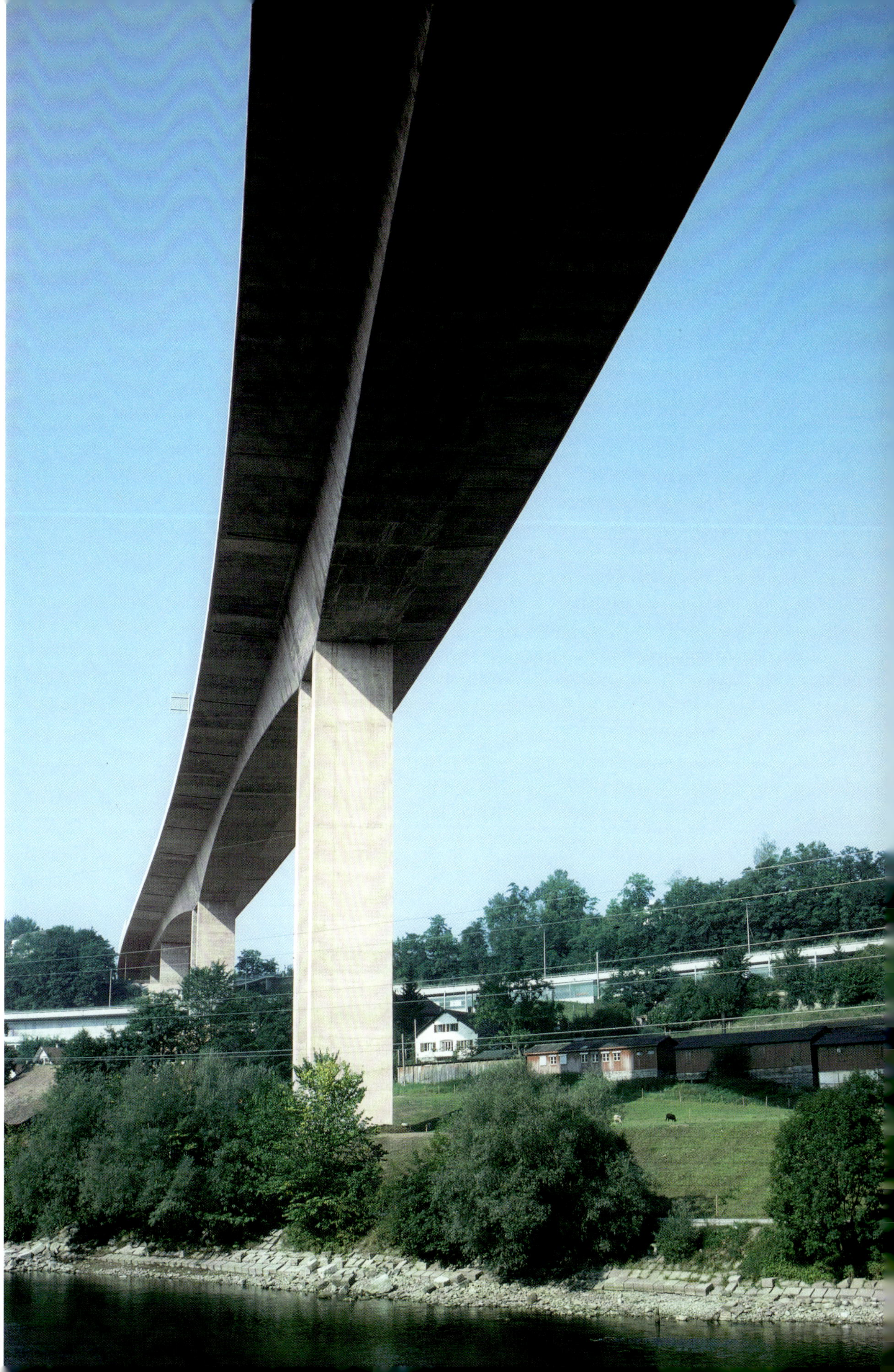

Figure 18
The maximum bending moment for Felsenau depicted geographically as one metric ton with a lever arm of 142 kilometers.

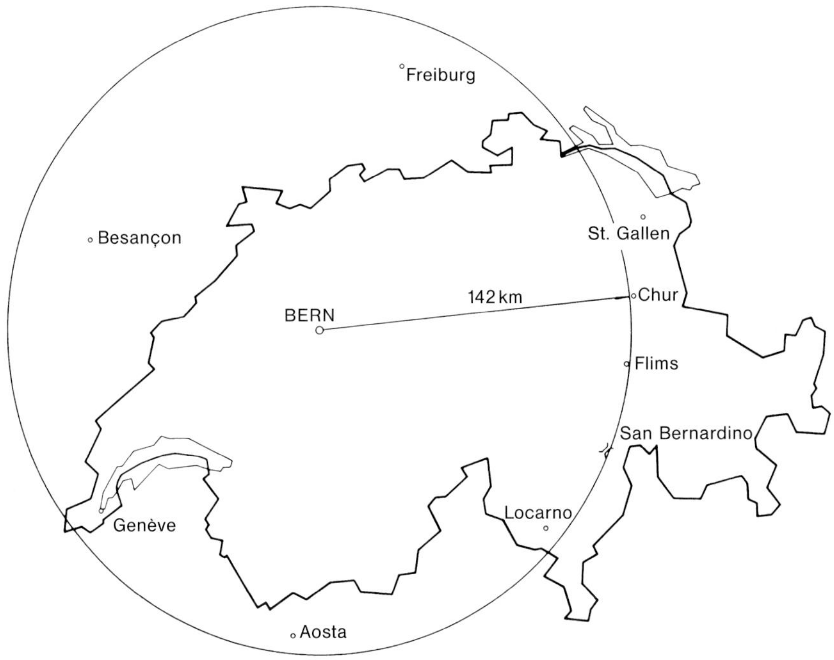

top width remained constant while its bottom width decreased as the box depth increased. Thus, he provided a wide top of 11 meters (36.1 feet) to reduce the slab cantilevers to 7.6 meters (24.9 feet), while the bottom meets the columns with a width of only 7.5 meters (24.5 feet). This smooth transition at the support is aesthetically superior to a transition in which the bottom box width is greater than the column's transverse dimension. The cost would have been somewhat less with vertical web walls, a bottom width equal to the top 11-meter (36.1-foot) width and consequently 11-meter-wide column walls, but the resulting appearance would have been substantially heavier. Since it would have been difficult to know if the overall costs would be influenced even in a minor way, Menn did not hesitate to design the more handsome form. His competition entry was judged one of the least costly. Generally speaking, as long as the concept–performance and construction procedure–is sound, minor changes in form will not appreciably change costs. When designers realize the relative freedom they have in form making, their designs can begin to take on the visual power of works like Menn's Felsenau Bridge without adding measurably increased costs. It is the structural form that creates the strong aesthetic appeal rather than arbitrarily sculptured details.

As the Felsenau was nearing completion in the spring of 1975, Menn received a telephone call from the canton engineer of the Valais, who asked him for advice about a new bridge on the Simplon Road.[26] The canton had proposed to avoid the bad foundation conditions in the Ganter Valley above Brig by building a tunnel whose costs were estimated in the range of fifty million Swiss francs. Worried about this high figure, the canton engineer asked advice from his counterparts at the federal highway department in Bern, who in turn suggested consultation with Menn. After inspecting the site and the foundation conditions, Menn developed a proposal for a bridge that he calculated would cost about half of the tunnel estimate. He drew up preliminary plans and had a model made. Normally on such a major project, there would have been a design competition, but Franz Steiner, the minister of transportation of the canton of the Valais, was impressed with Menn's ideas. Worried that a jury might select another design, he commissioned Menn to proceed with final plans.

At Felsenau, Menn had purposely created long spans (the longest then in Switzerland) to avoid piers in the waterway and to achieve a lighter-looking structure in the heavily traveled Aare Valley. In the Ganter Valley, his starting point was also the support conditions. He began by locating the tallest pier at the base of the solid rock as well as at the end of the roadway curve on the north side, the next pier high up at the existing road level, and the other main-span pier at the end of the straight roadway section in order to span as much as possible of the lower part of the critical slope. This slope on the south side is continuously moving northward. Menn arranged these piers in such a way that the bridge's main spans would be symmetrical.

With these preliminary ideas, Menn began to think about the superstructure in detail. At Felsenau he had been struck by the immensity of the maximum bending moment of 142,000 ton-meters (1 million Kip feet). Typical of his imagination, he had seen that moment as a weight of one metric ton with a lever arm of 142 kilometers (90 miles), about the distance between Bern and his home in Chur (fig. 18). Although there is no scientific reason why numbers larger than 142,000 ton-meters could not be accommodated by the design ideas at Felsenau, Menn felt that the larger moments arising from a span of 174 meters (570.7 feet) required a fresh approach. His solution was to bring the cantilevers (87 meters; 285.4 feet) on that main span down to about the same dimensions as the shorter spans by providing an intermediate support through cable stays. Menn followed his Felsenau precedent of using a single hollow box. The Ganter, being only a two-lane roadway, led him to make the box width equal to the full bridge deck width of 10 meters (32.8 feet).

He used the same computer-based analysis as at Felsenau for the curved hollow-box behavior, and as in the earlier bridge, he introduced diaphragms to prevent excessive torsional distortion of the thin box section. For the high pillars, two separated thin walls as at Felsenau would have required much heavier sections because of the greatly increased height leading to higher lateral moments. Therefore, Menn used single hollow boxes that, for the pillar (150 meters high; 492 feet), has flanges that increase in longitudinal width from 6.5 meters (21.3 feet) at the top to 10 meters (32.8 feet) at the base. Like those of the Felsenau Bridge, the roadway girder and the pillars were designed as rigid frames so that in the main spans there are no joints or bearings to separate the structure from the creeping valley except at the bases of the south pillars.

The comparisons between Felsenau and Ganter show how Menn modified his earlier experience for a newer and larger work. To see his ideas more clearly, it is essential to analyze the Ganter Bridge itself. As before, we consider separately the plan, profile, and section, and for each, the questions of efficiency, economy, and aesthetics. In plan, the roadway curves away from the north bank, runs a straight section over the valley, and curves back to meet the existing road going up the south bank to the Simplon Pass. Menn's first criterion for positioning the supports was based upon foundation conditions, but he adjusted them with two other standards in mind: symmetry and uniformity. Table 2 shows the spans finally chosen. This table shows how the main span symmetry and the choice of span lengths led to two uniformly long cantilever subdivisions of 87 meters (285.4 feet) and 40 meters (131.2 feet). Menn envisioned that the 35-meter (114.8-foot) spans with 10-meter (32.8-foot) cantilevers would be cast in place on scaffolding because they are built close to the existing grade, whereas all other parts would be built by the cast-in-place segmental cantilever method. The choice of single hollow-box pillars of about the same width as the two-lane roadway leads to visually strong supports that, compared to the great valley width, still give an overall appearance of openness on account of the wide spans. The simple hollow-box pillars, therefore, are efficient forms for resisting the heavy lateral moments, mostly from design wind speeds of 150 kilometers per hour (94 mph). They are economically positioned to permit substantial deck-girder form reuse, and are shaped and spaced to present a striking and open total structure.[27]

The Ganter Bridge shows its new form most clearly in profile, owing to Menn's decision to avoid deep girders at the supports (fig. 19). Rather, he has introduced cable stays that provide a total structural depth of 16 meters at the supports while

Table 2

Spans	35	50	127	174	127	80	50	35
Construction Subdivisions	35	10+40	40+87	87+87	87+40	40+40	40+10	35

reducing the cantilevers from 87 meters (285.4 feet) to 36 meters (118 feet). Because the stays extend well into the curved side spans (127 meters; 416.6 feet), they must be encased in concrete walls to permit them to follow the plan curvature. These walls then become prestressed by the stays, a new procedure that has three advantages: the cables are protected from corrosion; there is no fatigue danger because the cables being bonded to concrete walls experience only about one-fifth the range of stress change as cables in a normal stay bridge; and greater overall safety results from the substantial girder cross section, which includes the concrete walls above the box. These advantages in structural efficiency combine with the substantial construction advantage of reducing all the cantilever scaffoldings to one standard form: from a depth of 2.5 meters (8.2 feet) at midspan to a depth of 5 meters (16.4 feet) at support. The construction of the concrete walls, however, was expensive and could hardly have been justified without the rigorous standardization in cantilever scaffolding.

Finally, the aesthetic questions were of primary concern. The real origin of the profile did not come from ideas on efficiency and economy, but rather from Menn's simple observation that "the extreme relationship between the extraordinarily powerful pillars and the very narrow bridge girder led to an unusual design (*Ausbildung*) for the superstructure."[28] Here the German word *Ausbildung* connotes the building-up of a form during the conceptual stage rather than the vague associations of the English word *design*. Thus, Menn's major motivation was aesthetic; the ordinary cantilever form, while technically feasible, would have led to a visually weak horizontal profile compared to the powerful vertical elements required. This aesthetic sensitivity led Menn to explore the problem more deeply and to seek a solution that would then be as efficient and economical as possible.

Within these disciplines, Menn had to find a design that would satisfy the constraints of an unstable slope on the south side, high wind forces above the valley, and the curving roadway between the valley sides. Such constraints can stimulate the structural artist to search for new designs within the already existing state-of-the art. In this case, Menn was employing prestressing and cable stays, two major innovations of the second half of the twentieth century, to develop his highly original form.

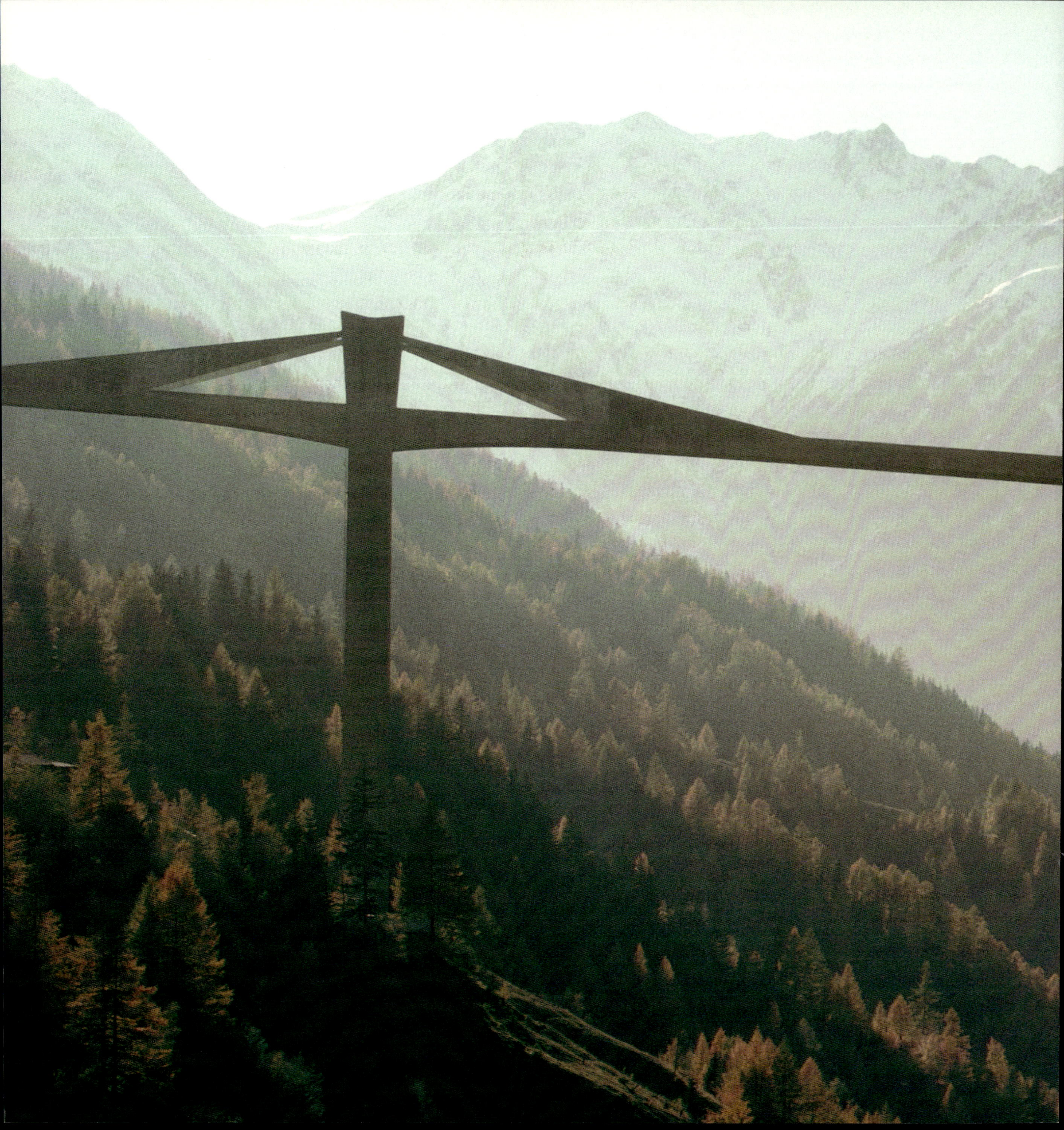

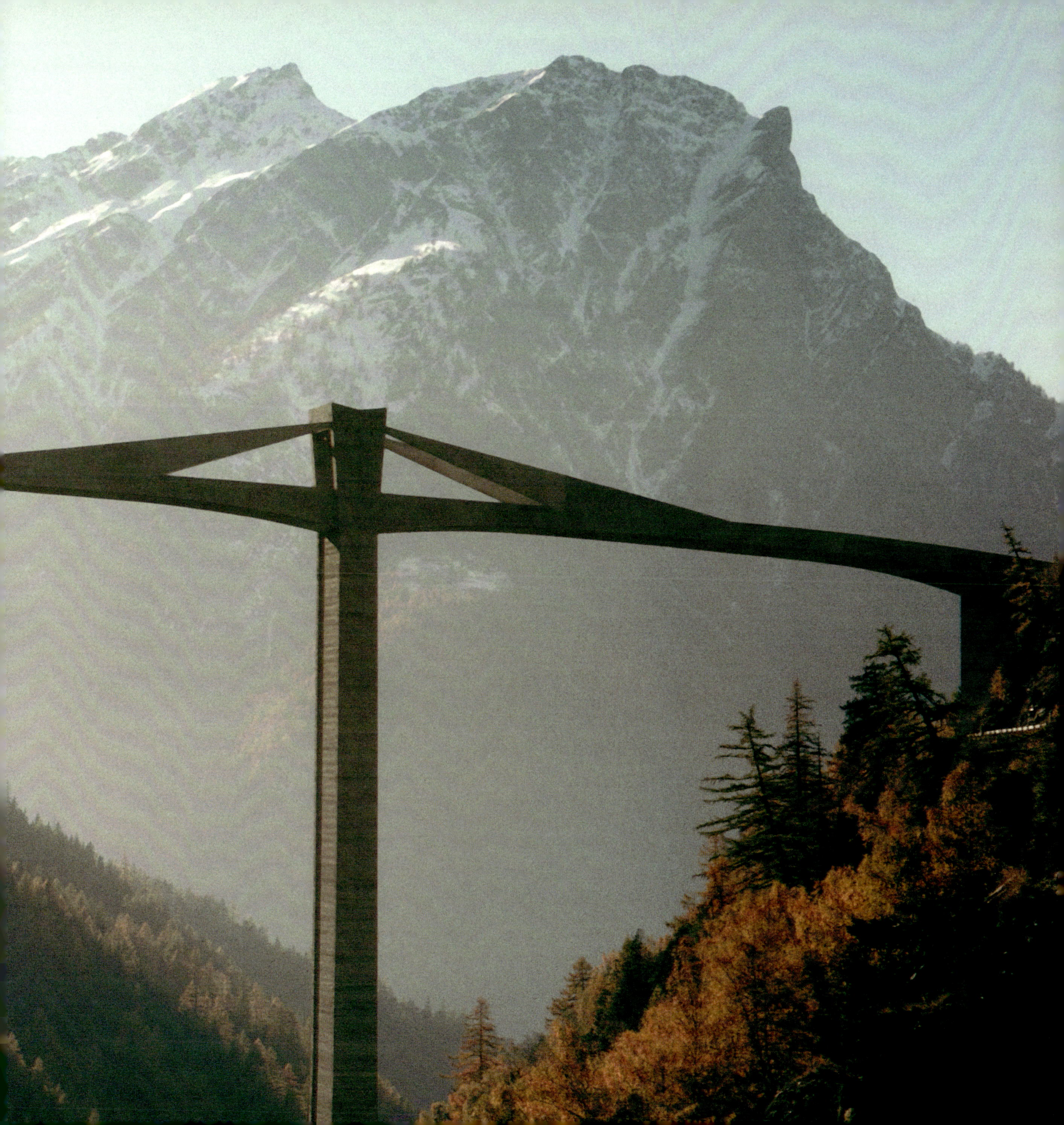

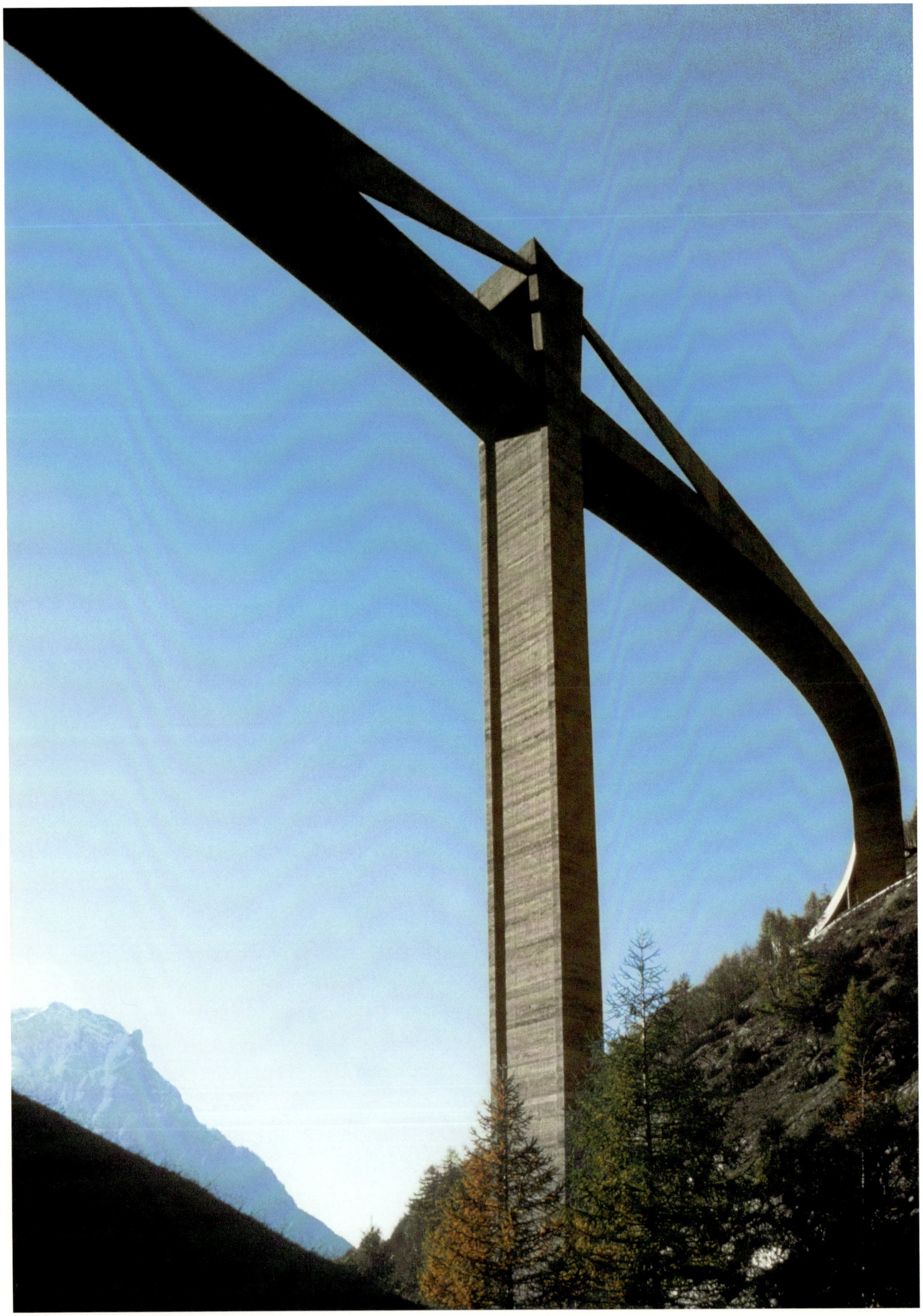

Figure 19 (previous page)

Ganter Bridge, complete in 1980, showing the profile where Menn avoided deep griders at the supports.

Figure 20

Ganter Bridge, showing the tall pier and curved roadway.

Figure 21

Biaschina Bridge, Ticino Valley, completed in 1983, with a span of 160 meters.

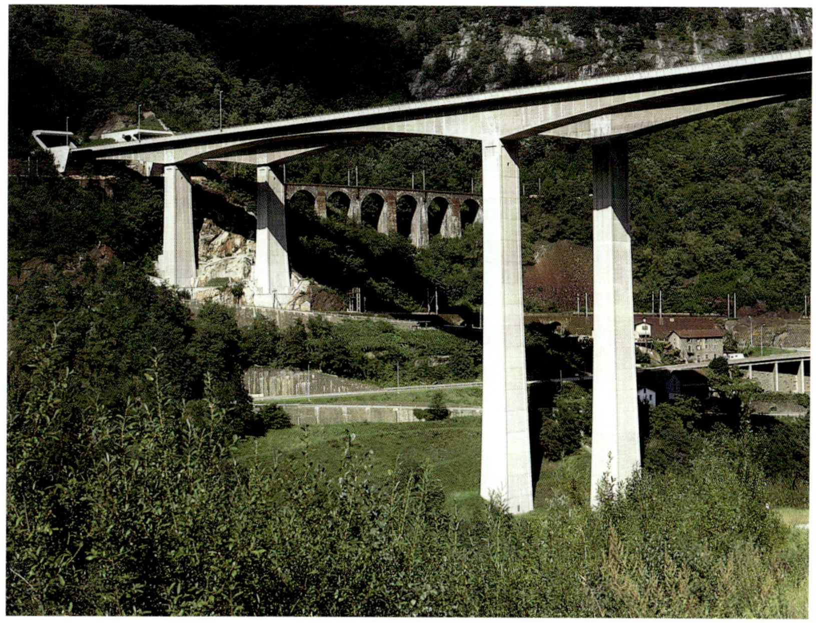

Having decided to make the horizontal profile visually stronger, Menn began by carrying the pillar up over the girder to make a pylon reaching about 10 meters (32.8 feet) above the roadway to serve as a support for cable stays (fig. 20). These stays then permitted the maximum cantilever depth to be only 5 meters (16.4 feet). Although the curved side spans led to encasing the stays in concrete walls, such walls were not technically needed on the main span; but one has only to imagine the visual consequences of leaving the light cables exposed on that span of 174 meters (570.7 feet) to realize the aesthetic improvement achieved by encasing them everywhere. Full encasement also entails additional strong technical advantages. Thus, even in the best design, it is impossible to isolate any one decision with visual consequences as purely aesthetic or as purely technical. In the best designs, the aesthetic is the primary motivation although it must never violate the technical.

Aesthetically, the cross-sectional view was the most difficult to predict. The form evolved from a perspective that emphasized the profile; and in its completed stage, it shows an impressive combination of articulation and lightness. As one drives over the bridge, however, it appears heavy because of the scale of the high walls compared to that of the narrow roadway. In addition, the pylons themselves appear to be somewhat heavy; this is especially evident when comparing the horizontal girder to the relative thinness of the verticals. This problem of two scales—one in profile and one in section—is reminiscent of the Salginatobel Bridge. In Maillart's most famous work, the view from the deck is dominated by heavy parapet walls, and the strongly foreshortened views from either approach emphasize a power and massiveness of the three-hinged, hollow-box form that is fundamentally different from the lightness and articulation of its well-known profile.

The Ganter structure replaced the Felsenau as Switzerland's longest spanning bridge. Based on square meters of roadway surface, the bridge was more costly than the Felsenau because of its greater spans, the higher columns, and the far more difficult foundation conditions, particularly on the south bank. There is some evidence, however, that it was a relatively inexpensive solution. The final cost of 23,500,000 Swiss francs is less than half the estimate for the originally proposed tunnel solution. In addition, during the bidding stage, another designer prepared a competing design that the general contractor, Zublin & Cie., priced along with Menn's design. Zublin found Menn's design to be at least two million francs cheaper and thus submitted a formal bid only on it.

In addition to performance as a measure of efficiency and competition as a measure of economy, the aesthetics of the Ganter Bridge stirred substantial interest. The form arose primarily from aesthetic ideas, which came from an engineer's imagination. No architects or other aesthetic consultants were involved in any way on Menn's designs. As such, the Ganter Bridge stands in a long tradition of modern bridges beginning with Telford's iron arches, a tradition of structures designed to combine technical and aesthetic ideas.

Following the Ganter Bridge, Menn studied two other major Swiss bridge designs that would lead to his greatest Swiss masterpiece, the Sunniberg Bridge (1999) over the Landquart Valley, not far from the Salginatobel Bridge. Both of these bridges—the Biaschina and the Chandoline—resulted from design competitions for which Menn was consultant for the conceptual design; the final form of each came from his imagination even though a design office carried out the final dimensioning and produced the drawings. The first of these was the Biaschina Bridge, built in 1982 in the Ticino Valley south of the Gotthard Pass. Menn's conceptual design received first prize in a design competition.[29] Here the roadway is nearly 100 meters above the valley floor, and Menn projected a main span of 160 meters with a profile similar to that at Felsenau. The latter has haunches 8 meters (25.2 feet) deep at the supports and a section 3 meters (9.84 feet) deep at midspan, whereas the Biaschina structure has comparable dimensions of 9.51 meters (31.02 feet)

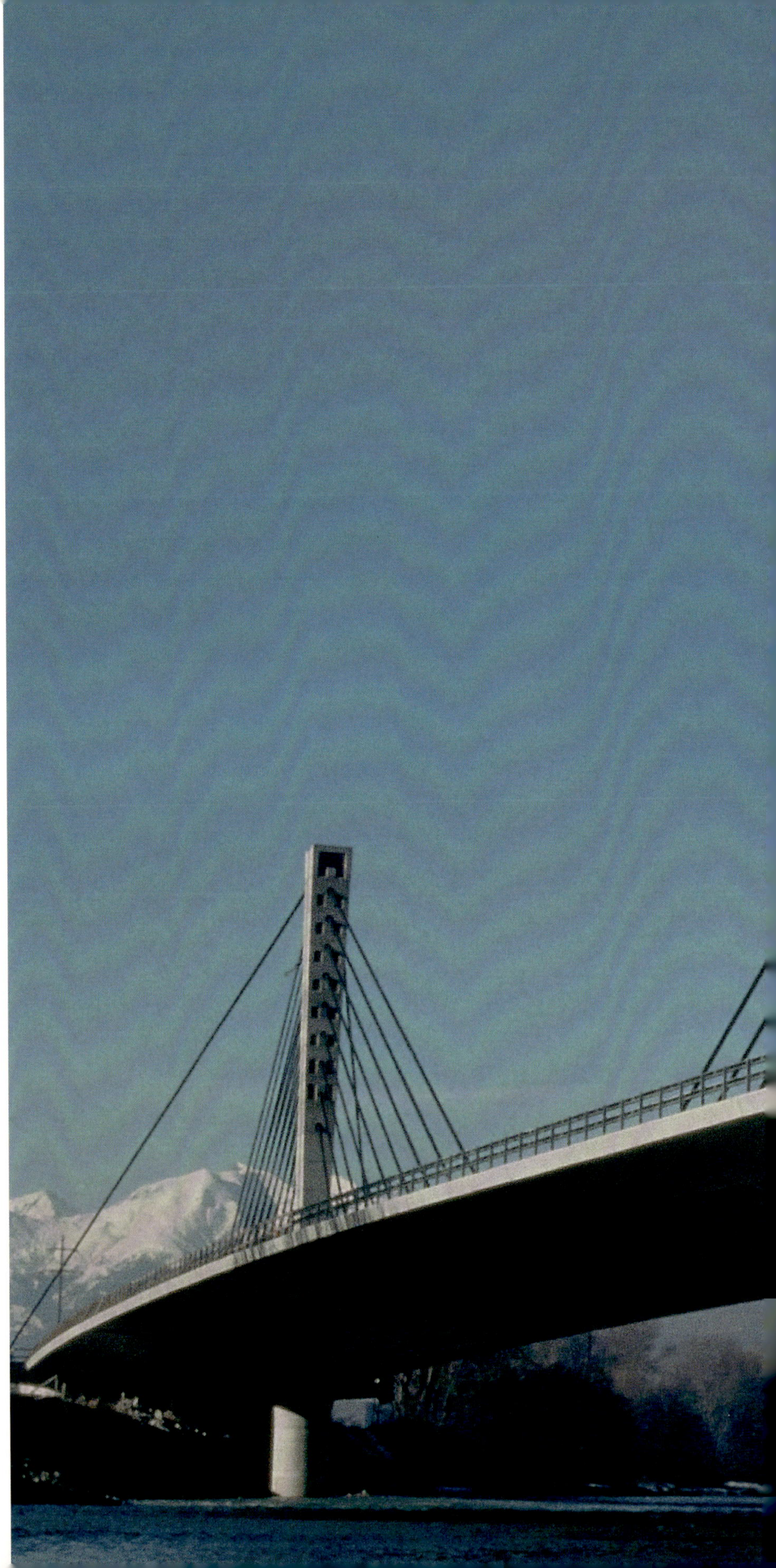

Figure 22
Chandoline Bridge over the Rhône River, with a span of 140 meters, completed in 1989.

and 3 meters (9.84 feet). The columns, however, are radically different largely because of the much greater height at Biaschina (88.7 meters instead of 48 meters; 291 feet instead of 157.5 feet).[30] The towers are similar to those at Ganter except that Menn gave the Biaschina ones a more expressive vertical shaping (fig. 21). The 150-meter tower at Ganter has a constant width equal to the 10-meter roadway, and in profile it increases from 6.5 meters just beneath the girder to only 10 meters at the base. By contrast, Biaschina increases in width from 7 meters beneath the girder to 10 meters at the base, while in profile it goes from 5.5 meters to 8.5 meters—a visually stronger slope of 1.5/88 versus 1.75/150 (.0170 versus .0117).[31] The structure is composed of two separate and roughly parallel bridges of almost identical form, but unlike Felsenau, the roadway goes straight across the main spans. The major interest, apart from the shaped columns, is the design competition because, again unlike Felsenau, there was a greater variety of forms to be judged, and by comparing them, it is possible to understand the higher quality of Menn's conceptual design.

In the second bridge, at Chandoline over the Rhône River, Menn chose to study a cable-stayed bridge of a more standard form for the 140-meter main span. In profile, the bridge looks like many such works with diagonal cables fully supporting the long deck (284 meters; 931 feet). But moving around the bridge (fig. 22), we observe that all the cables lie in the central median of the roadway except for two that fan outward to be anchored near the bridge ends and outside the roadway. These extra cables carry forces that arise from the horizontal curvature of the bridge.[32] From below, we also note the open trusslike sloping walls of the hollow-box deck girder, an aesthetic element that adds visual interest and lightens the structure. This unusual design represented an experiment for Menn; the span is short for a cable-stayed bridge, and he could have designed a prestressed hollow box like Felsenau or Biaschina. Nevertheless, during this period Menn was experimenting with new ideas of form even though, apart from Ganter, they did not appear to be radical departures.[33] That changed during the decade after Chandoline with a series of projects, two of which have been completed: the Sunniberg in 1999 and the bridge over the Charles River in Boston in 2002.

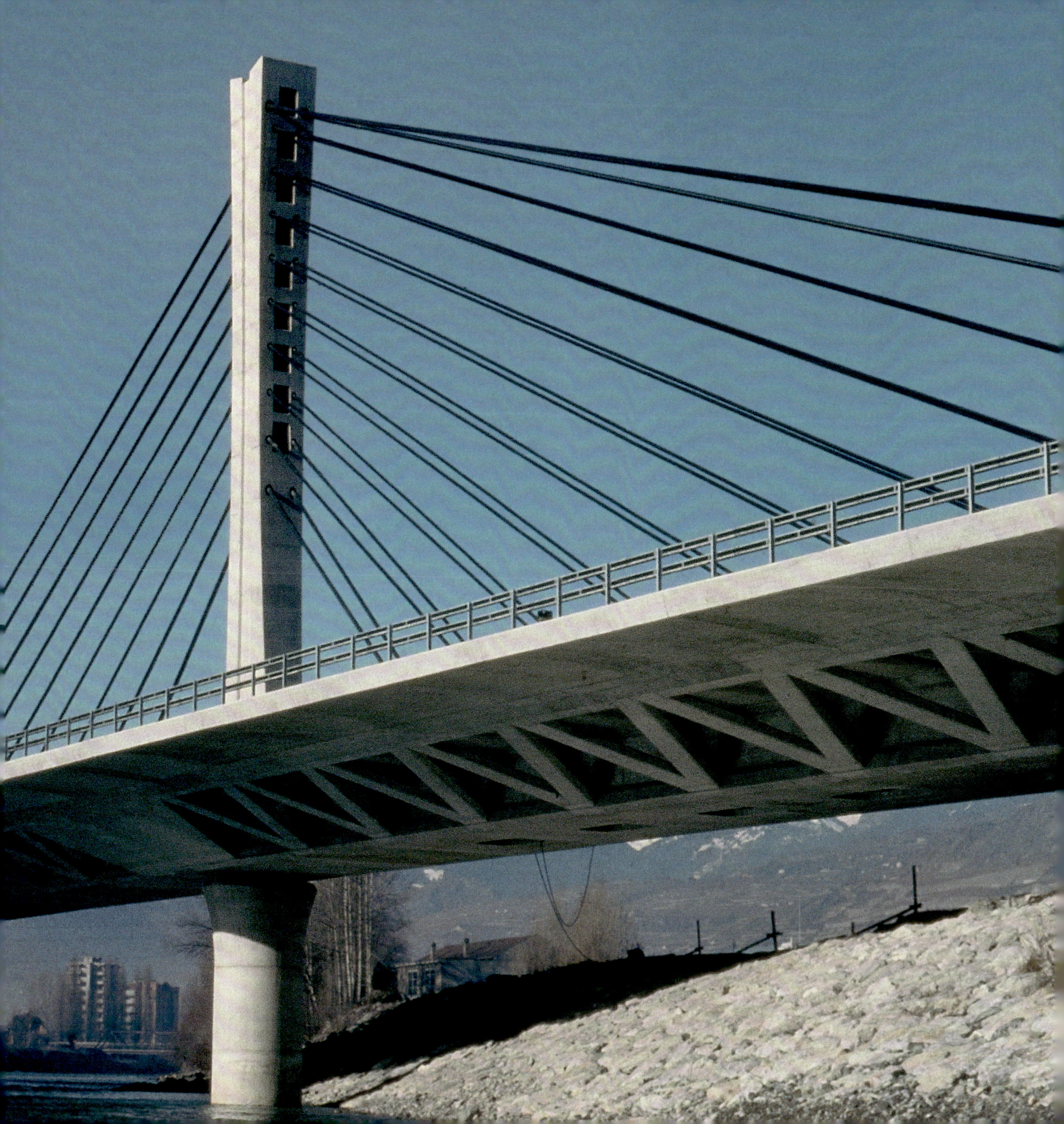

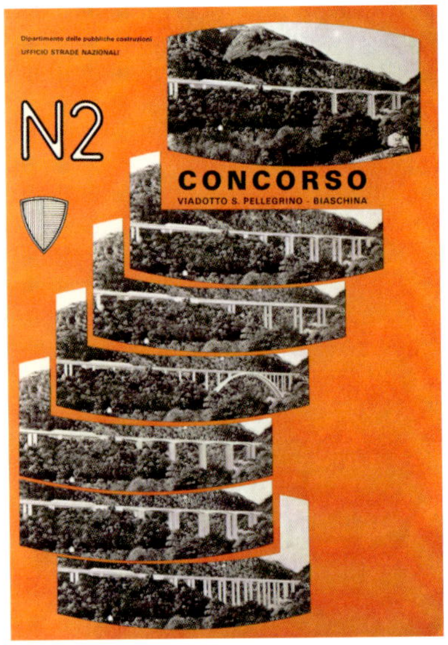

Figure 23
Cover of the jury report for the Biaschina Competition, showing seven alternative designs.

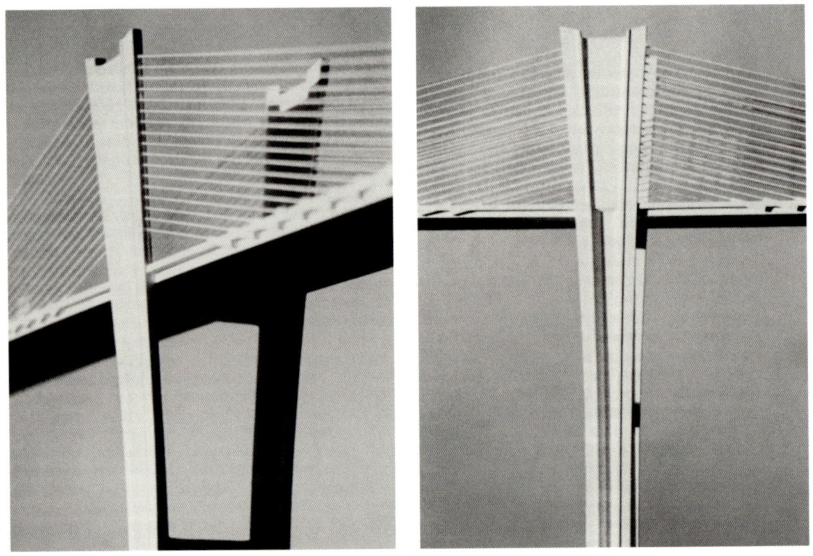

Figure 24A, B
Menn's rejected proposal for the Poya Bridge over the Sarine River.

Menn had been thinking deeply about the design problem of high viaducts over wide valleys since the Felsenau competition of 1970. His major bridges had previously been, for the most part, over deep and narrow valleys where arches seemed appropriate. These designs focused on one main span with relatively minor approach spans. The national roadway system, however, had opened up new bridge locations that required long viaducts often over rivers that were relatively small compared to the valleys that these new works had to traverse.

Menn's later bridges are all on major Swiss highways: Felsenau carries national highway N1, Biaschina carries N2, Chandoline N9, and Ganter N9. Apart from Chandoline, the other three are similarly comprised of long viaducts where the river plays only a small role in their design. This new concern led Menn to express the valley rather than the river in his design at Felsenau and in the same way at Ganter and Biaschina. A composite view of the seven competition designs for the Biaschina bridge illustrates the visual differences and, in the view of the jury, the aesthetic superiority of Menn's concept (fig. 23). The jury's report included two terms that Menn used, "transparency and generosity." This last word, *grosszügig* in German, also means noble, grand, large-scale, in bold outlines, grandiose.[34]

These same words become crucial guidelines for the Sunniberg Bridge, completed long after the Biaschina report. During this period, Menn's ideas changed slowly but, in the end, reached a dramatic climax in one of the most visually spectacular valleys of Switzerland or anywhere else. But the direct origin of the Sunniberg Bridge began earlier, near the city of Fribourg, in a bridge over the Sarine River.

In the late 1980s, the traffic in Fribourg became so congested that the federal government decided to build a bypass over the Sarine in the suburban region south of the historic city. The location chosen required a bridge 650 meters long and 75 meters high above the valley. Known as the Poya Bridge, the canton held a design competition with Menn on the jury, which judged the results in November 1989. Arches, segmental girders, and cable-stayed designs were natural solutions, but Menn was not satisfied with the competition designs and proposed a design of his own (fig. 24A, B) with a main span of 176 meters and a tower height of 100 meters, of which 25 meters would be above the roadway. A normal cable-stayed design would have towers about 45 meters above the roadway (roughly 25 percent of the main span), but to Menn that would have appeared awkward when set on such high towers. His low cable-stayed solution required special attention to the larger compression forces induced into the deck, but its appearance seemed much better because, as he put it, referring both to Chandoline and to his Poya design, "the two examples show that aesthetically impressive, economical bridges are possible under a wide range of constraints."[35] When he wrote that statement, no decision had yet been made on Poya, but soon thereafter the canton did not accept his design: it had nevertheless served to stimulate Menn's thinking in preparation for Sunniberg.

Figure 25
Sunniberg Bridge, over the Landquart River Valley, completed in 1999.

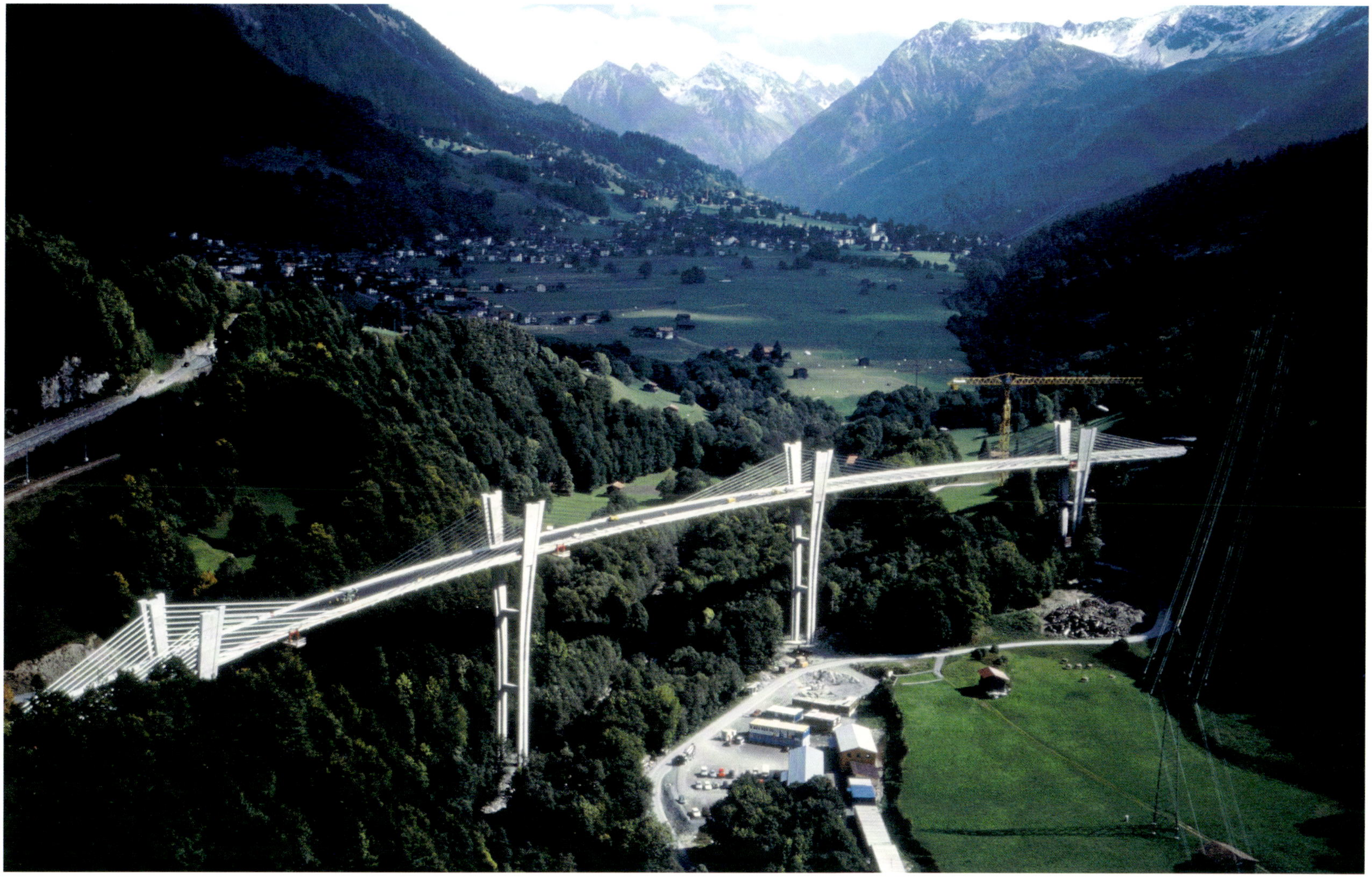

Shortly after the Poya event, the canton of the Graubünden approved a plan to bypass the famous ski-resort town of Klosters with a series of tunnels and a bridge across the valley. Here again there was a design competition with three firms invited and with Menn on the jury. Once again, Menn was not satisfied with the results and presented his Poya proposal to the highway department's consulting architect, who enthusiastically advised the department to accept it. Thus, Menn's conceptual design became the basis for the bridge, and one of the competition engineering firms, Bänziger, Bacchetta, and Partner, carried out the detailed dimensioning and produced the drawings.[36]

Menn described his conceptual design in an article in 1998, and Chelsea Honigmann translated it and worked out the numerical explanations to illustrate how Menn's ideas of aesthetics and efficiency interacted in this masterpiece.[37] Unlike the Poya design, the Sunniberg roadway (fig. 25) curves over the valley with a radius of about 500 meters so that the study of forces becomes more complicated, but the appearance is greatly enhanced. Based upon that curved plan and the height above the valley, Menn designed the four main elements (pier, pylon, cables, and deck) each to satisfy technical requirements and each to achieve separate aesthetic goals that in addition create an integrated form.

Figure 26
Sunniberg Bridge, showing the slender pier that flairs out above the roadway.

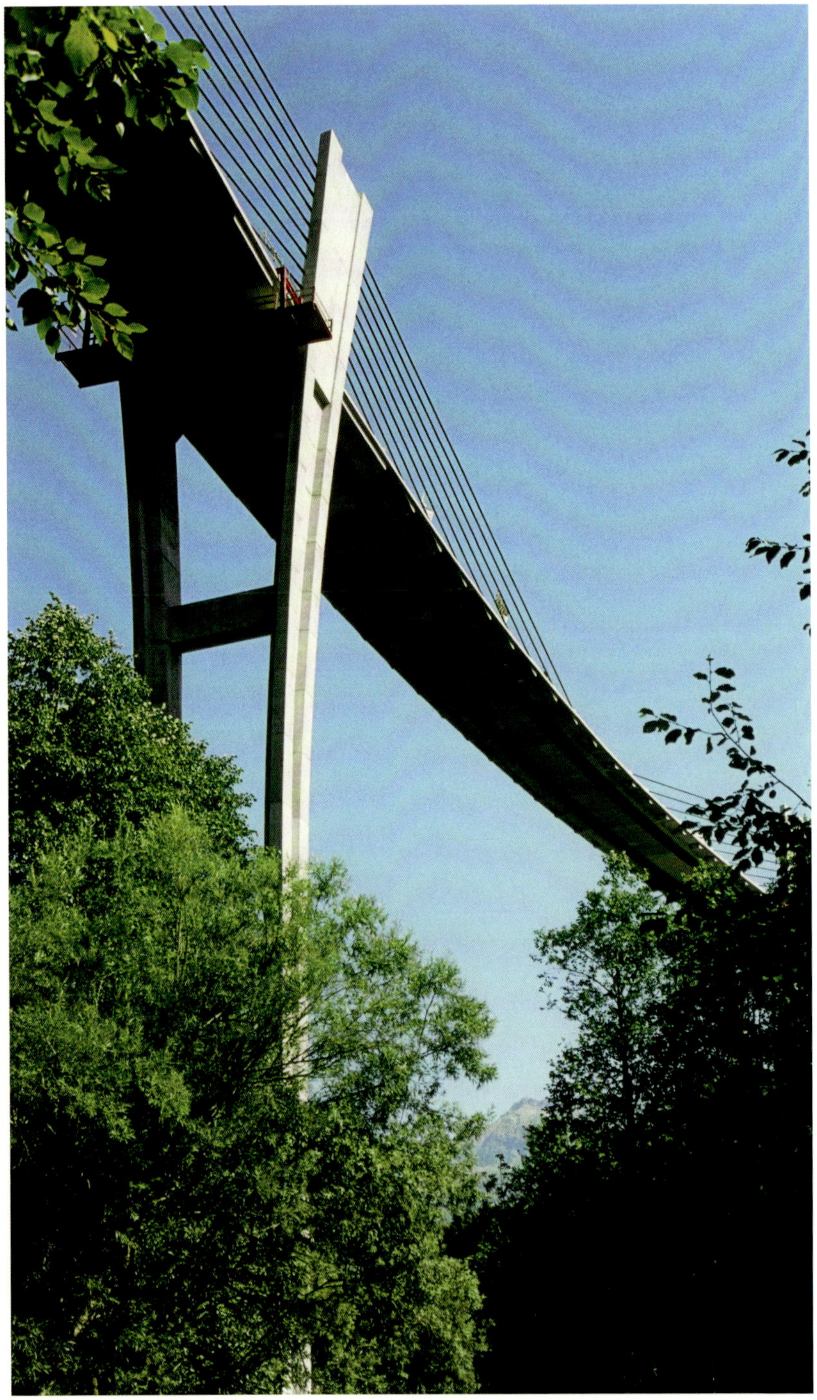

The first two elements—and visually the most prominent ones—are the pier (below the roadway) and the pylon (above the roadway). In many cable-stayed bridges, these two are separate elements, both visually and structurally, but here they are completely integrated (fig. 26). The piers' openness—transparency—contrast with the closed, solid piers at Felsenau. They represent a critique of Menn's earlier work. Moreover, the piers are shaped to express the bending moment diagram under live load on one span only. That moment is maximum at the junction of pier and deck and decreases toward the ground, as does the longitudinal spacing of the two legs of each pier. The pylon is only 14.8 meters high, slightly over 10 percent of the span length to avoid the awkward and overpowering visual effect of the normal ratio of about 25 percent, or a 35-meter-high structure above the already tall piers and the relatively short spans.

The pylons are flared in the direction (longitudinally) of the roadway (fig. 27) and are tilted outward (laterally) (fig. 28) to create a surprising visual character while solving two crucial structural problems. The longitudinal flare resists the bending moment due to live load on only one span and thus prevents that load from causing large deflections in the deck; the lateral flare allows the cables to remain straight and follow the roadway curve—the inseparable interplay of aesthetics and efficiency.

Menn designed the third element, the cables, to be parallel in the harp form so that the two cable planes would not appear visually cluttered in partial profile. But because of the low pylons, the vertical deck loads put high tension forces in the cable and correspondingly high compression forces in the deck, which forces must be carried by the thin deck.[38] To avoid buckling, Menn increased the deck thickness near the piers where compression (from all the cables) is greatest.

Finally, for the deck, Menn sought a continuous thin line throughout the entire length of the bridge. Because of its horizontal curve, he could dispense with expansion joints, always a maintenance problem, since the deck can expand or contract horizontally as an arch. Because there are many cables supporting the deck and because the stiff pylons prevent live-load deck bending, the roadway slab can be extraordinarily thin. The deck slab is 50 centimeters thick, and the edge beams are only slightly greater than one meter. Thus, the structural ratio of span to deck depth is 136. The curb, however, makes the visual depth 1.83 meters and the visual ratio 77 (fig. 29).[39]

Figure 27
Intergration of pier and pylon of the Sunniberg Bridge.

Figure 28
Roadway of the Sunniberg Bridge.

Figure 29
Sunniberg Bridge, with a structural ratio of span to deck depth of 136.

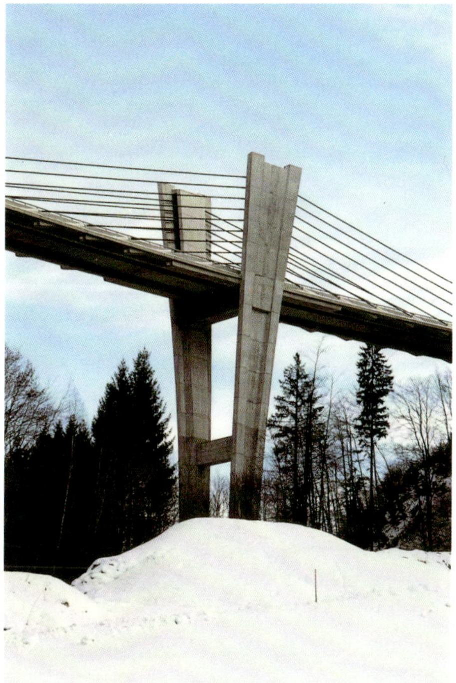

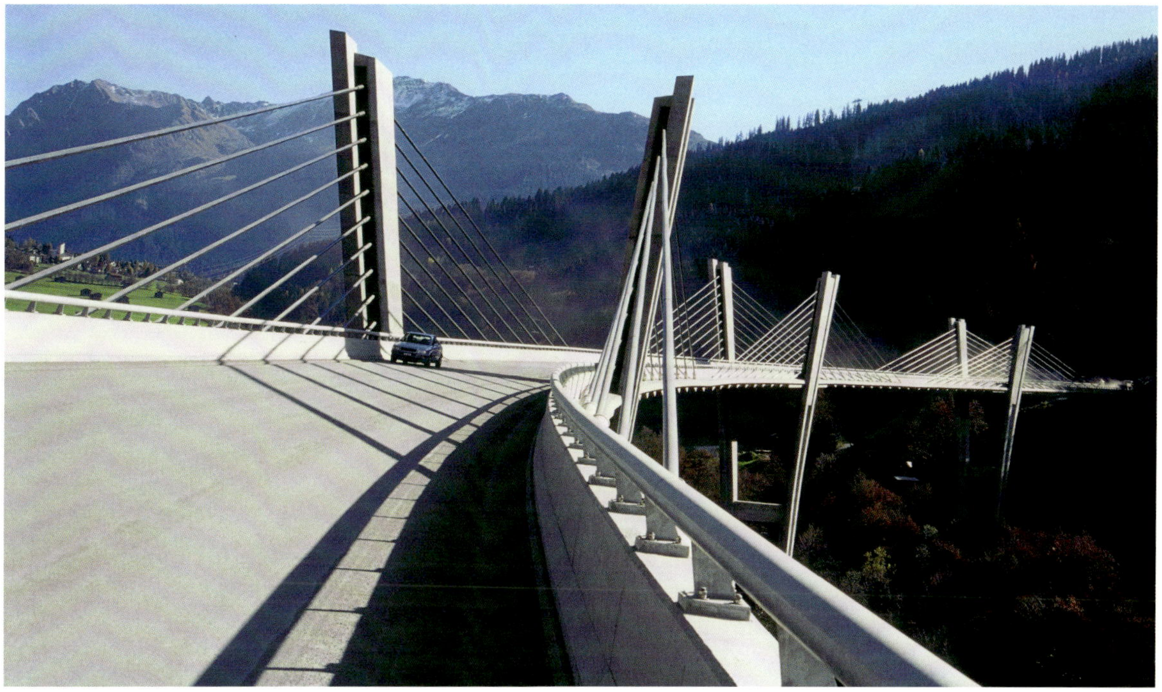

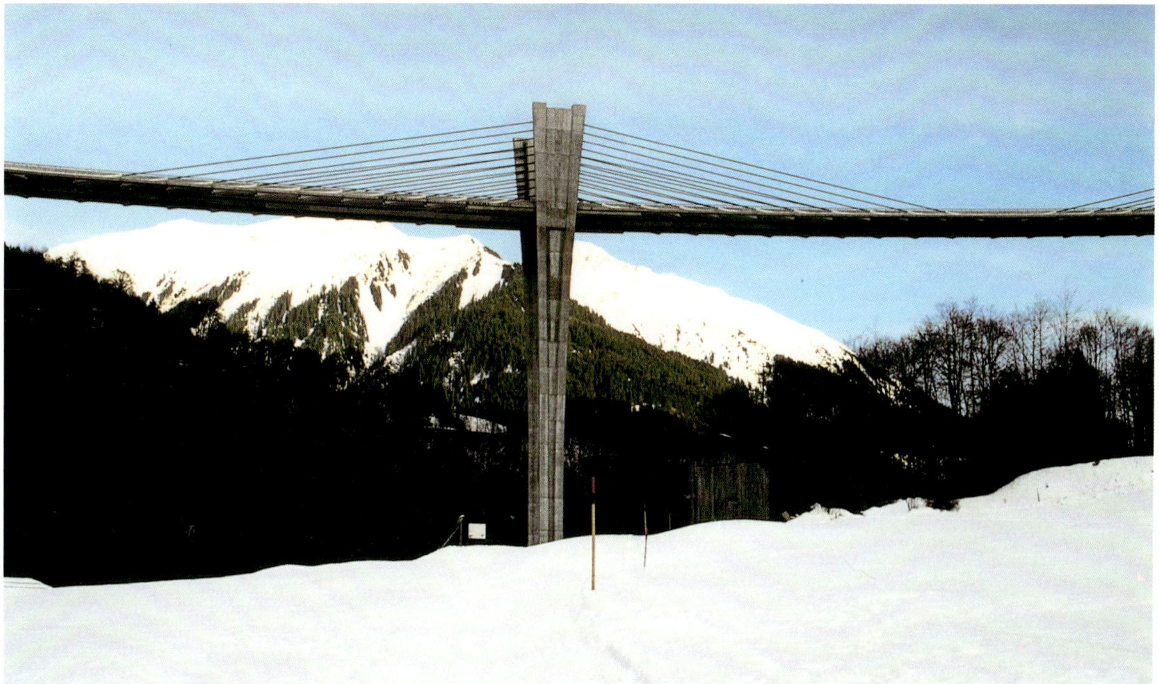

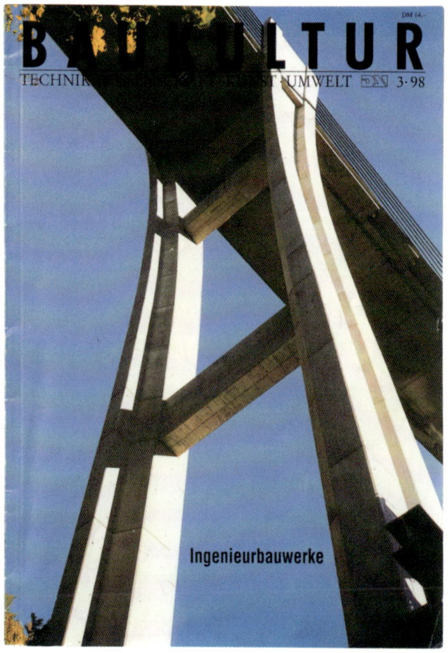

Figure 30
Sunniberg Bridge on the cover of *Baukultur*, 1998.

There are many other detailed structural problems that Menn addressed in his conceptual calculations, but the main outline sketched above shows how his thinking developed. As he expressed it: "From an engineer's point of view, it is inconceivable to accept flaws in the structural system and details or disproportionately high cost in return for architectural embellishment. The art of the structural engineer consists of balancing economy and elegance against each other on a case by case basis, to achieve the desired design objectives."[40]

The Sunniberg design is one of the landmarks of the art of the structural engineer because it has no "flaws in its structural system" and it did avoid "high cost in return for architectural embellishment." Its structural system, which required an extra cost of about 15 percent above the cheapest proposal, is definitely generous. The total cost was 20 million Swiss francs, while the total bypass project had been estimated at 506 million Swiss francs. Since the least expensive design would have been about 17.4 million Swiss francs, the expenditure of an additional 2.6 million, or about 0.5 percent more, gained for the Swiss a magnificent monument to their tradition of bridge engineering.[41] With the completion of the bridge, which has appeared on covers (fig. 30) and in advertisements, the little valley has two great works close together: the Salginatobel Bridge and the Sunniberg Bridge.

Menn's first lecture in the United States came during the 1972 Maillart Symposium at Princeton University, where he illustrated Maillart's influence on his work and showed his own emerging style.[42] He immediately took a great interest in Maillart studies at Princeton, opened many pathways for me in Switzerland, and in 1978 helped to prepare an exhibition at the Princeton University Art Museum devoted to his own bridges.[43] Four years later, he came again to Princeton to give a course in bridge design, which was part of the preparation for his textbook (1986). During that visit, he traveled and lectured in the United States, including a stop in Boston. These events led Kenneth Kruckemeyer, associate commissioner of the Massachusetts Department of Public Works, to organize a conference in 1986 devoted to the engineering aesthetics of bridges. A central focus there was the Princeton Menn exhibition augmented by an exhibition of California bridges.[44]

One result of the conference was Kruckemeyer's effort to develop an engineering design competition for a Massachusetts bridge to follow the format of a Swiss competition. Although that project did not go forward, the competition idea did succeed several years later in Maryland. A conference was organized by Earl Freedman and Fred Gottemoeller during which Menn met with Maryland Governor Schaeffer and explained his designs. A competition for a bridge over the Severn River at Annapolis was organized following Menn's advice, and he became a central figure in the jury decision in November 1989.

Menn was disappointed with the results, and, as at Poya, he made drawings of a proposal of the same low cable-stayed form. Since this competition was the first of its kind in the United States, it was not possible to accept a proposal from a jury member, and the winning design by Tom Jenkins of the Greiner Corporation was built. Several jury members worked with the winner to make changes that the jury felt necessary to achieve a more elegant design, and the bridge eventually won many awards for design excellence.[45]

During a visit in 1991 to the United States, Menn lectured at Harvard University; a high Massachusetts public works official in attendance there asked Menn to look at a bridge project currently planned for the huge Boston restructuring project known as the Central Artery/Tunnel project, or "the Big Dig." Menn immediately saw a great possibility for the Charles River Bridge, and returned to Chur where he had a model made. He returned to Boston several weeks later and convinced the authorities to accept his conceptual design (fig. 31). This unique cable-stayed bridge was completed in 2002, Menn's first bridge outside Switzerland and the first major bridge in Boston (fig. 32).[46]

Figure 31
Menn's model for a bridge to span the Charles River, Boston.

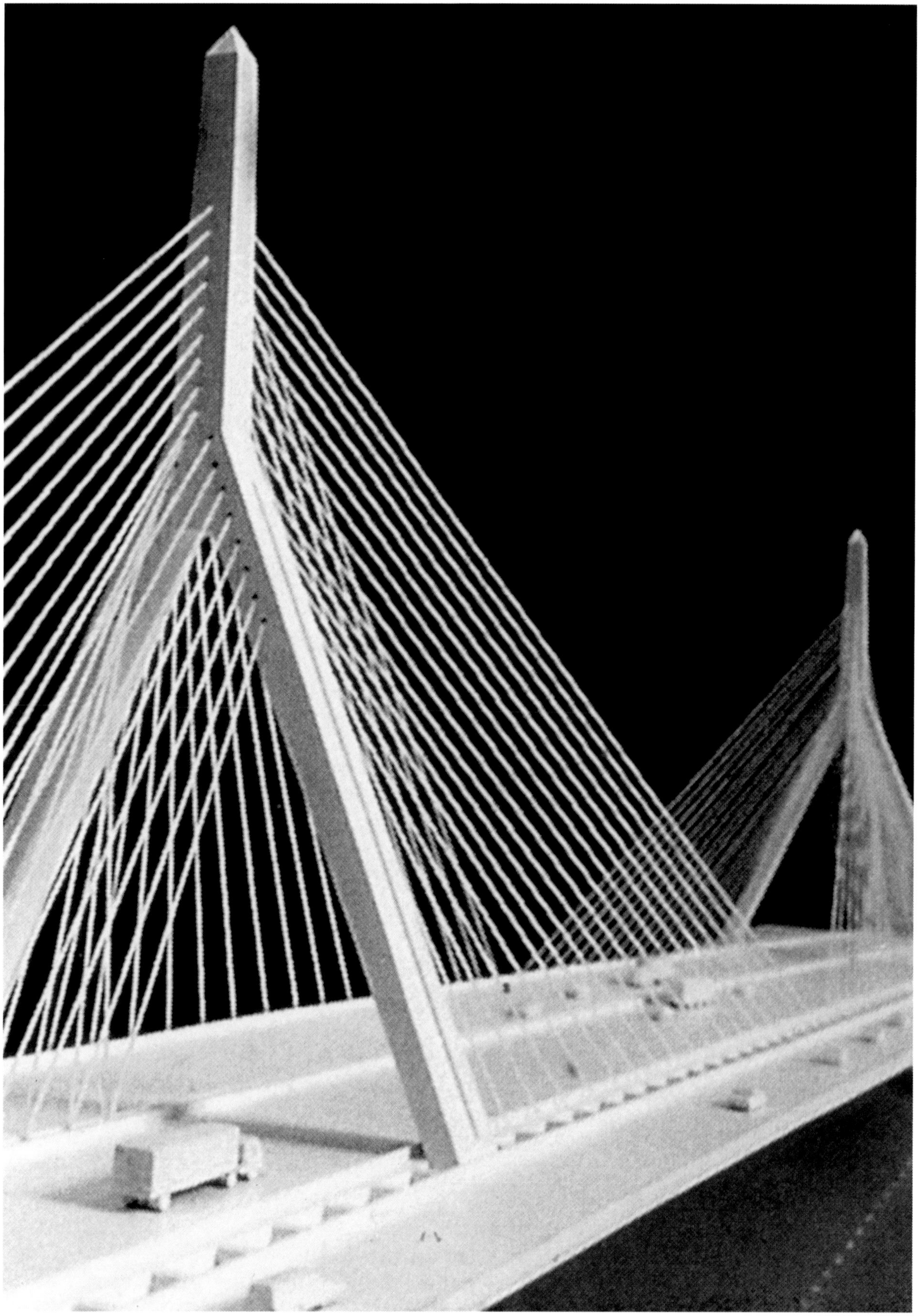

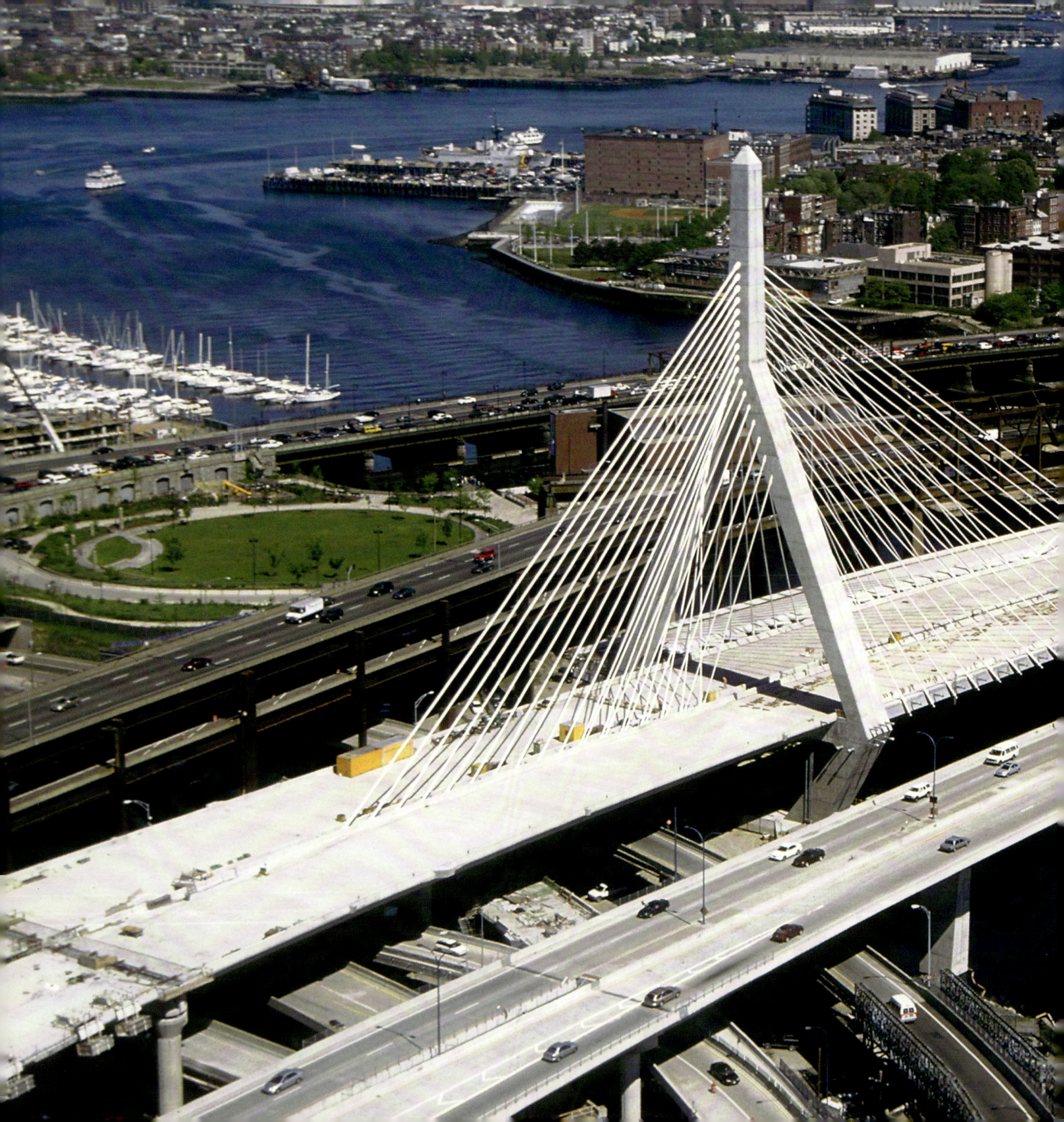

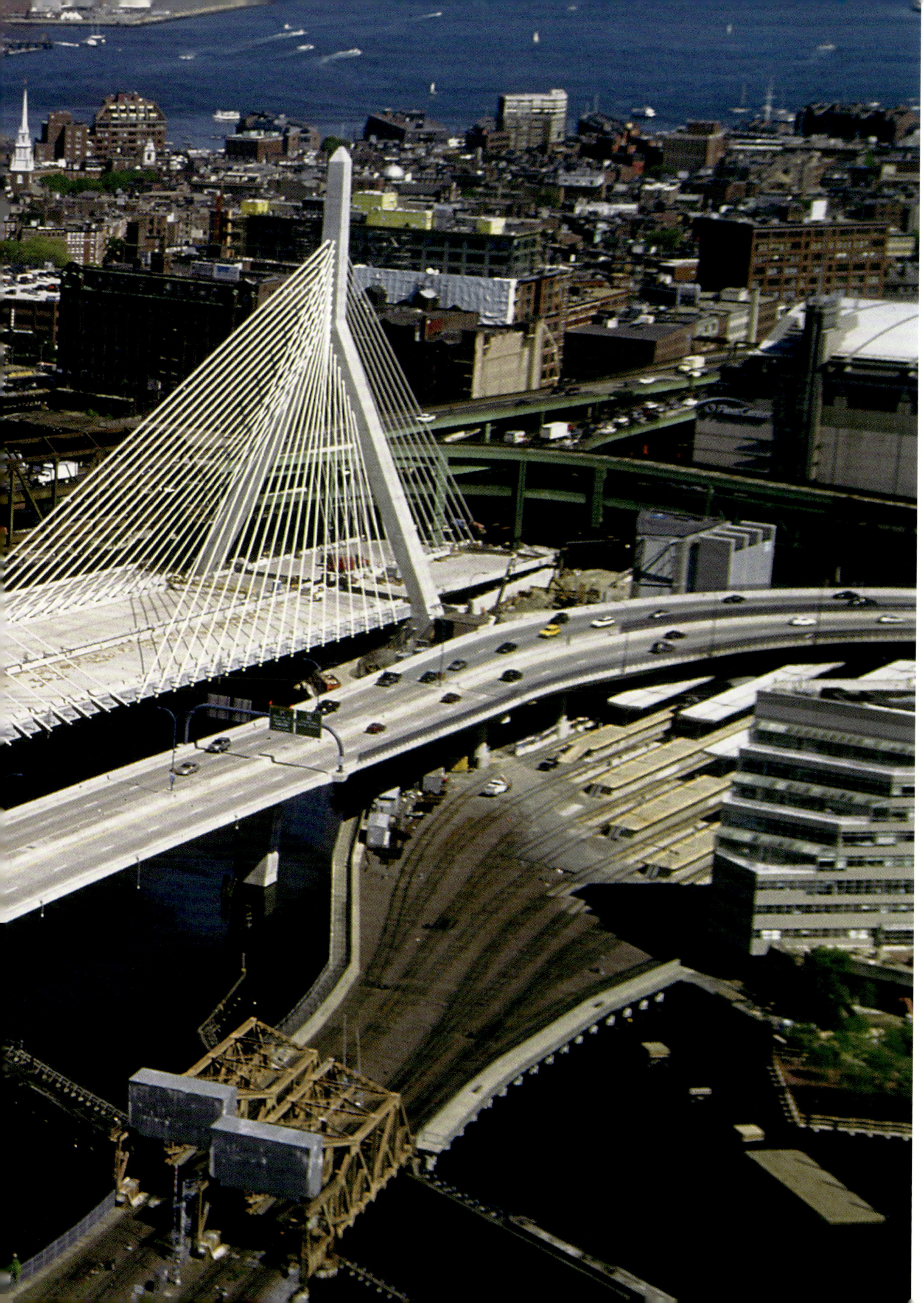

Figure 32
Leonard P. Zakim Bunker Hill Bridge over the Charles River, Boston, Menn's first bridge outside Switzerland, completed in 2002.

While this New England project went forward, Menn became involved with another Maryland competition, this one for the Woodrow Wilson Bridge replacement over the Potomac River at Alexandria, Virginia. Here Menn served as conceptual design consultant to one of the four invited competitors. One of the competition requirements stated that this long, over-water bridge had to be carried by arches to "match" the Arlington Memorial and Key Bridges in the District of Columbia. Menn sketched a design that avoided arches, and the jury gave this design first prize (unfortunately, the design was partly reworked for submission in the competition).[47]

At an October 1994 conference in Deauville, France, engineers in one session presented papers on new ideas for extremely long-span bridges.[48] The conference coincided with the completion of the Normandy Bridge close by, at the time the longest spanning cable-stayed bridge anywhere. Menn attended and felt unsatisfied by the long-span proposals, which were in part stimulated by the continuing study for a 3000-meter span bridge across the Straits of Messina. Menn, therefore, wrote a brief paper on his own proposal, which was translated for publication in the journal *Structural Engineering International.* The following description comes directly from that paper.[49]

For this extra-long span, Menn proposed a new type of structure that consists of a regular system of cable suspension from pylons that rise 360 meters above the deck and support the central 1100-meter portion of the bridge. Cable stays carry the 300-meter portions closest to the towers, while a combination of vertical suspenders and cable stays take the deck loads in the remaining 650-meter sections (fig. 33). Each pylon is centered in the roadway, making it "simpler to erect, more stable, and aesthetically satisfying" (fig. 34). Menn assures the crucial design issue of dynamic stability "by placing on either side of the deck girder a sloping cable-stayed system carried by slender pylons which are supported by the central pylon."[50] Cables connecting these sloping pylons to the central pylon provide simple construction and avoid the bending and buckling of these light angled elements.

Menn outlined the construction procedure and identified a series of design problems that needed to be studied carefully within this conceptual framework. He then explained why new materials such as carbon fiber reinforced polymers would not be needed for such spans until their costs could be substantially reduced. The much heavier steel cables are still adequate for spans of 3000 meters, but once longer spans are required, the entire structural system of cables and deck will need to be made of synthetic materials. Menn concluded the paper with these new thoughts:

Simple extrapolations of rather conventional concepts usually do not lead to great progress in technological development. This is true as well for extremely long spans in bridge structures. Because today lightweight high-strength materials are not yet inexpensive and cannot yet be produced in sufficient quantity, economical and practical designs for extremely long spans require unconventional structural systems. The proposed structural concept, with its relatively simple construction technique, is certainly suited for this design problem. It is suitable, not only for great spans, but also for narrow bridges and shorter spans and especially also for long-span pipeline bridges.[51]

The highly unusual form of the preceding proposal is typical of Menn's urge to continue exploring new forms even without a specific design commission. In so-called retirement he has shown a restless energy toward design, and yet his works are all expressive of the highest goals in bridge design: the creation of motionless objects of stunning elegance. The guidelines he formulated during the last years of his teaching career are consistent with those goals. To understand Menn's ideas and how they are realized in his works, we need to return to the seminal chapter 2 in his 1990 book with the revealing title "Economy and Aesthetics." There is no other major essay on aesthetics in structural engineering that uses such a title. An intellectual contribution of Menn's, it makes clear that the ideals of cost and appearance are inseparable and that no bridge can be designated as a good bridge if either of these ideals is missing.

Menn places his discussion of aesthetics, therefore, within the context of efficiency (minimum materials) and economy (competitive costs), and defines two basic issues: abstract structural form, and structure and environment.[52] The former considers the bridge as a thing in itself, while the latter views the bridge in its specific setting. The two are not mutually exclusive, and the designer should consider both. He further stresses the centrality in design of evaluating the bridge visually from all possible positions, usually by means of three-dimensional models: either large-scale physical ones or those using computer graphics.

For abstract structural form, Menn identifies three primary characteristics: efficiency, harmony, and artistic shaping. For efficiency, he states positively that "from Roman times to the present day, bridges that have achieved renown for their elegance have almost without exception been remarkable for their efficient use of materials."[53] He follows this evaluation with a crucial definition of structural art. "The visual expression of efficient structural function is thus a fundamental criterion of elegance in bridge design. It is one of the primary distinguishing factors between structural engineering art and architecture."[54]

Figure 33
Design for a 3000-meter-span bridge.

Figure 34
Model for a 3000-meter-span bridge.

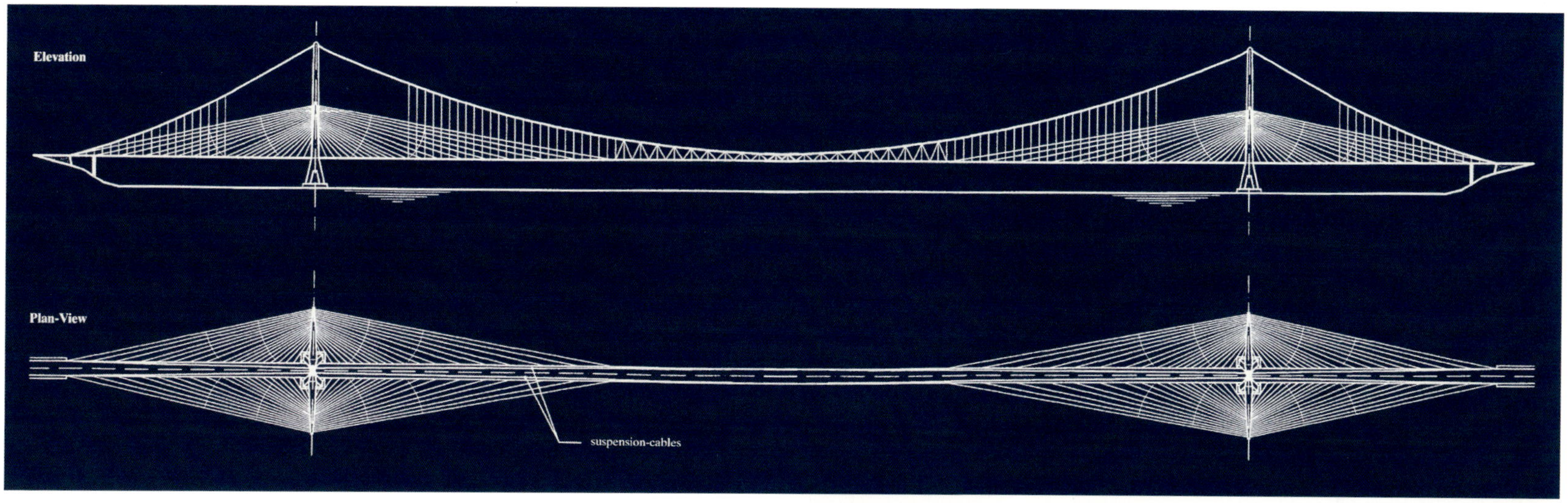

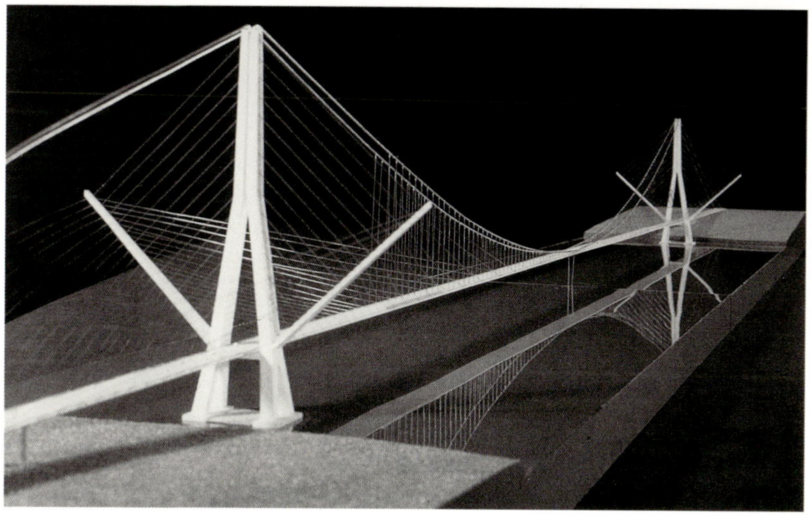

He then explains his two major ideals for efficiency and aesthetics: transparency and slenderness. The former relates primarily to the columns, which often are solid, heavy forms that block views, and he suggests good ratios of deck width to column width. He rejects hammerhead columns and multiple column bents because they drastically reduce the transparency of long bridges. This emphasis on columns is critical because they are often so visually prominent. In like manner, the superstructure can be prominent but less so in the sense of blocking a view and more so for its longitudinal bulk. Here Menn identifies thinness as an ideal and emphasizes the apparent visual slenderness he defines as "the ratio of the perceived uninterrupted length of the superstructure to the perceived superstructure dimension perpendicular to length."[55] This is the deck thinness that Menn achieved at Sunniberg, the arch thickness Maillart designed at Schwandbach, the deck thickness Ammann sought, and the edge thinness Isler expressed at Sicli and Heimberg.

Following the expression of efficiency by transparency and slenderness, Menn describes the ideal of harmony through the visual means of symmetry, order, and regularity:

Our perception of harmony in bridges, like our perception of efficiency, has originated from familiarity with naturally occurring structural forms which have grown using minimum energy and materials. These forms invariably possess symmetry; the visual impression of balance and stability they create is always matched by a state of stable equilibrium and low stress. A symmetrical structural system forms the basis of almost all of the bridges that have been acclaimed for their elegance, from Roman times to the present day.[56]

This statement must be understood as calling attention to the natural world as one from which the designer draws deep analogies but never superficial forms. Isler's shells, like Menn's bridges, do not visually mimic natural forms. They are disciplined by natural laws, not the undisciplined copying of natural forms. The designer is in full control of making the forms, which, as Menn goes on to argue, also requires (in addition to symmetry) the order that orients and arranges bridge elements to avoid ambiguous or unstable appearances. Menn further explains the virtue of regularity through such goals as spans of equal length and constant cross section. He does not mean to make all girders of constant depth or all spans of equal length where environmental reasons override strict regularity, as, for example, at Felsenau. Each of these factors related to harmony must be understood as suggestions that apply differently to different bridge designs.

Menn's final ideal for the abstract design is artistic shaping:

The raw structural form required for safety, serviceability and economy is rarely the most elegant. It can normally be refined into an elegant form, however, through artistic shaping of the structural members. The associated additional cost is insignificant. Shaping that follows the flow of internal forces is recommended in most cases. Forms that disregard the flow of forces normally produce a chaotic effect and should thus be avoided. In the hands of a gifted designer, however, member shaping based on purely aesthetic considerations and ornamentation can produce particularly charming results.[57]

Examples of shaping to follow the flow of forces would be his Biaschina columns that splay at the base and the Sunniberg columns that splay at their tops. For the purely aesthetic shaping, he shows a photo of Maillart's Vessy Bridge and means us to see the X-walls that support the deck. He is correct to advise against such a play with form unless "in the hands of a gifted designer." Here, however, the flow of internal forces is not disregarded; rather, Maillart is controlling the forces by means of a unique form. Nevertheless, Menn's point is valid; no designer without the gift of one of these Swiss engineers ought to attempt such a play.

Finally Menn discusses briefly the relationship between structure and environment. Quite properly, this issue, while important, is secondary to his much more lengthy discussion of abstract structural form. Primarily, Menn means that the designer can greatly improve the aesthetics by shaping of the roadway alignment, as he did, for example, at Ganter with a symmetrical S-shaped plan layout. He then presents in a short paragraph some general thoughts on the landscape:

A subdued form is preferable in flat or gently rolling terrain that lacks conspicuous topographical features. Prominent obstacles such as wide rivers or deep canyons are best crossed by structures in which one span has been given particular visual emphasis. Bridges in built-up areas should normally be unobtrusive. In certain cases, however, a positive visual effect can be achieved by a structure that stands out from an urban environment.[58]

These are by no means strict rules and merely give some indications. He avoids completely the superficial ideas about bridge design that lead to false facades or contorted forms. He is still thinking within a disciplined view of structural art, which is emphasized, once again, in the chapter's final two paragraphs: "aesthetics and economy." Here he notes that careful attention to aesthetics in the sense described already "will normally result in a net change in construction cost of no more than about 2 percent."

In the final paragraph, he observes that where span lengths are increased for aesthetic reasons to make the structure more grand (*grosszügig*), "a cost increase of up to about 7 percent of the cost of the most economical solution should be allowed for this purpose."[59] Although this concludes Menn's chapter on aesthetics, he returns to the subject in the book's longest chapter, "Design and Construction of Special Bridge Types," where he includes sections on conceptual design that draw on his ideas about efficiency, economy, and elegance.[60] The entire book can be seen as culminating in that chapter because Menn avoids extensive mathematical analysis and always focuses on design ideas. He was a skilled professor for twenty-one years, but his emphasis never left the field of design, and his ideas have always remained embedded in the goals of structural art.

The greatest challenge that Menn believes he has faced, and hence the most satisfying bridge he has designed, is the bridge over the Charles River in Boston (fig. 35). Named the Leonard P. Zakim Bunker Hill Bridge, it presented him with the most difficult functional constraints he had yet faced. When Menn studied the proposed designs, he was convinced that a better solution could

Figure 35
Leonard P. Zakim Bunker Hill Bridge, over the Charles River, Boston.

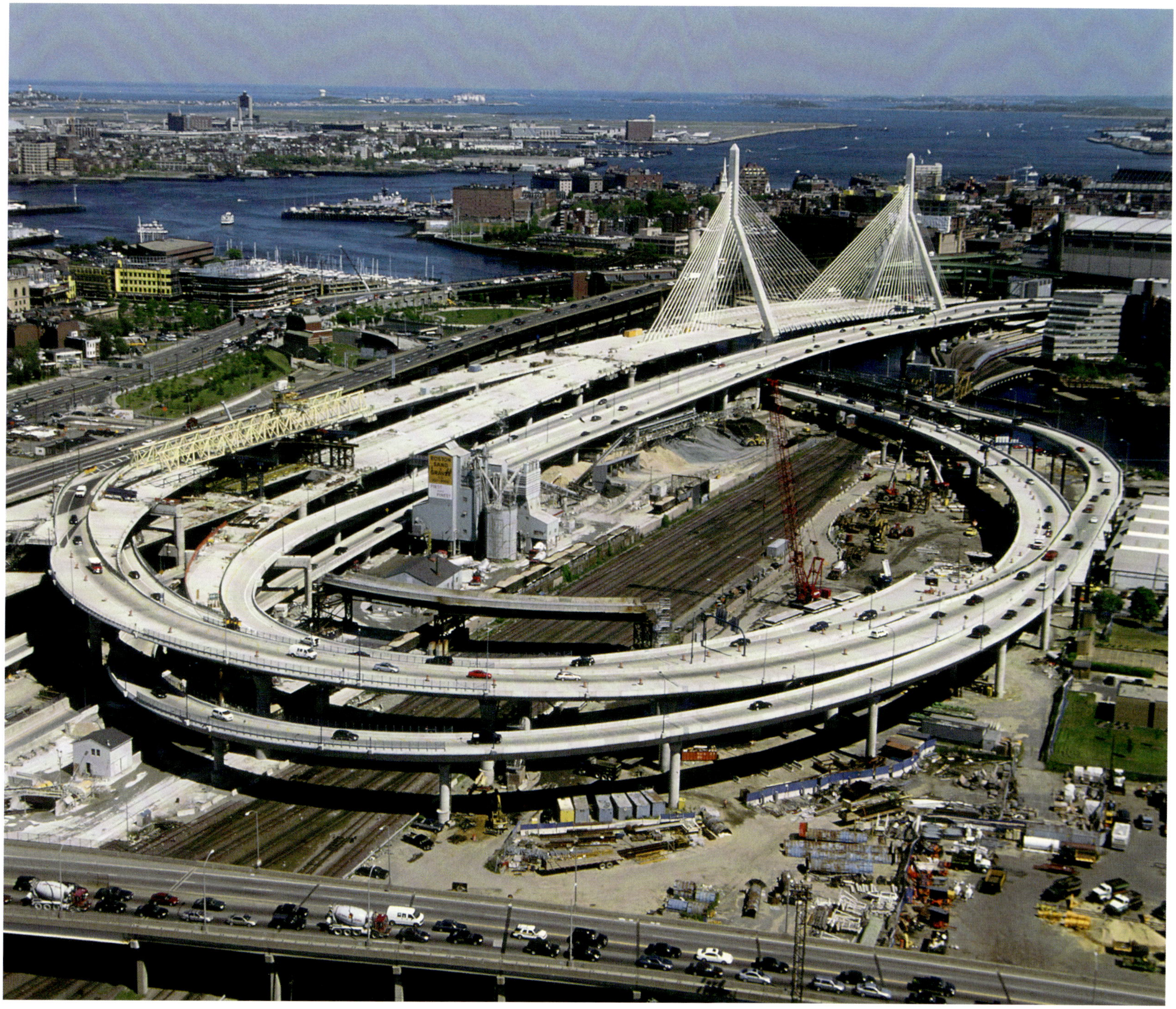

Figure 36
Leonard P. Zakim Bunker Hill Bridge, showing the unusual layout of the cable stays.

Figure 37
Leonard P. Zakim Bunker Hill Bridge at night.

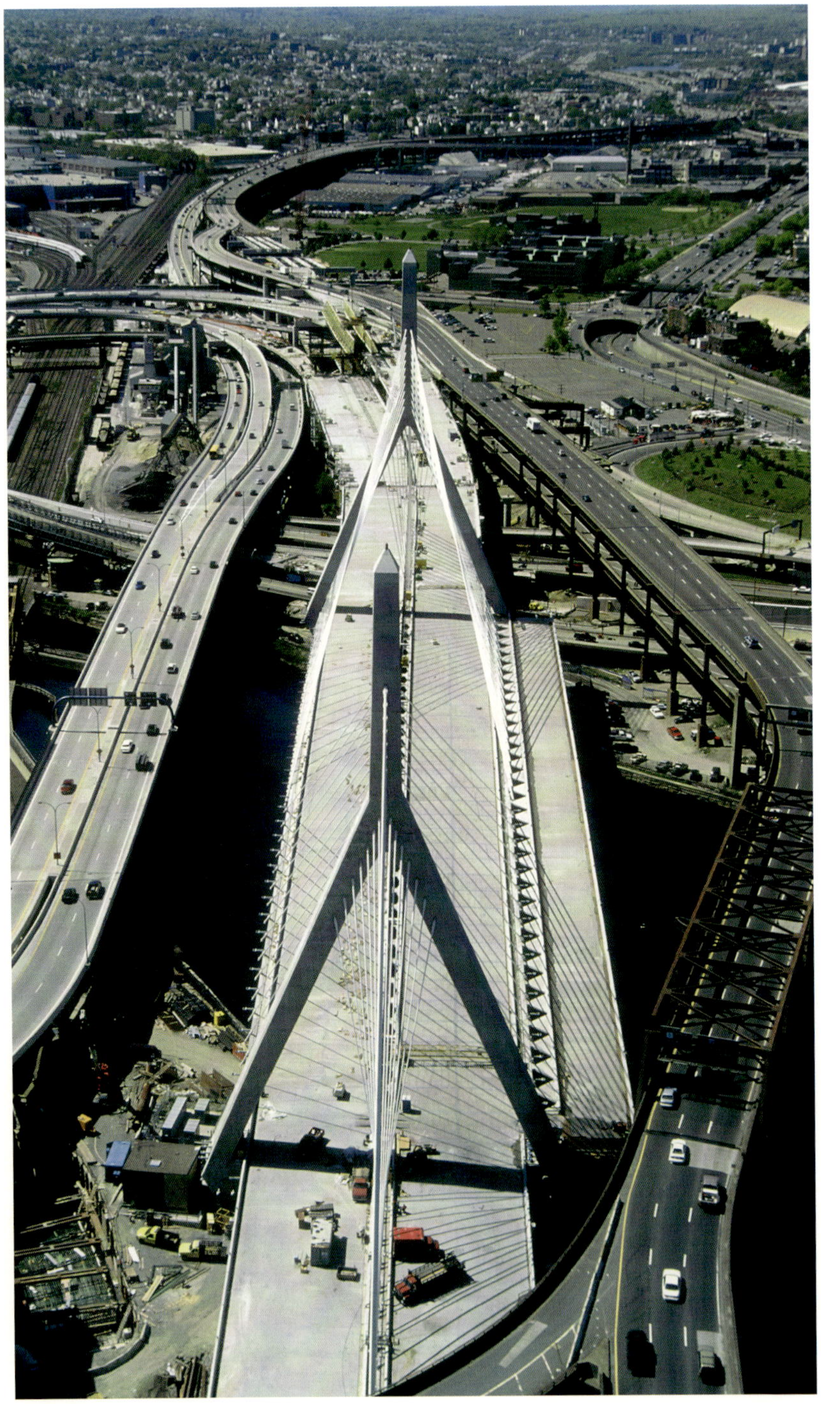

be found. The American engineers were initially skeptical about his model, but they soon recognized its merits and collaborated with him in a friendly, collegial manner. During its planning and construction, the structure was continually discussed in Boston and elsewhere because its form was so unusual. For example, the main span has two sloping planes of cables, whereas the back spans have only one plane of cables anchored along the center line of the bridge deck (fig. 36).

More and more, it has come to be seen as a great work—indeed, a realization of the often misused term "signature bridge." Menn achieved this ideal by a logical structural form which, however, was the product of his imagination. As he put it: "I think that I found a concept, which is very close to the ideal solution (that means a maximum of economy and aesthetics at the same time)." The governor of Massachusetts met Menn on the deck of his bridge in November 2000, where he presented to the Swiss engineer "an award with the proclamation that November 3, 2000 should be C. Menn Day in Massachusetts." In May 2002, more than 200,000 people massed to cross the bridge on foot.[61]

This is the only completed design of Menn's that is outside Switzerland, and it has added a new interest to the skyline of Boston (fig. 37). Christian Menn came much later to America than did his predecessor, Othmar Ammann, but he arrived with the same background in education; and as with Ammann, Menn has become an educator through his designs. Of the four designers, he is the only one to be a full-time professor as well as structural engineer, but his American bridge signals a new challenge to educators in the United States. Works like the Salginatobel Bridge, the George Washington Bridge, the Sicli Building, and the Leonard P. Zakim Bunker Hill Bridge all stand for the potentials for an improved environment through structural art. The age of technology makes possible the art of the engineer who can fuse the efficiency of form, economy of constraints, and elegance of expression into symbols of conservation, accountability, and aesthetics.

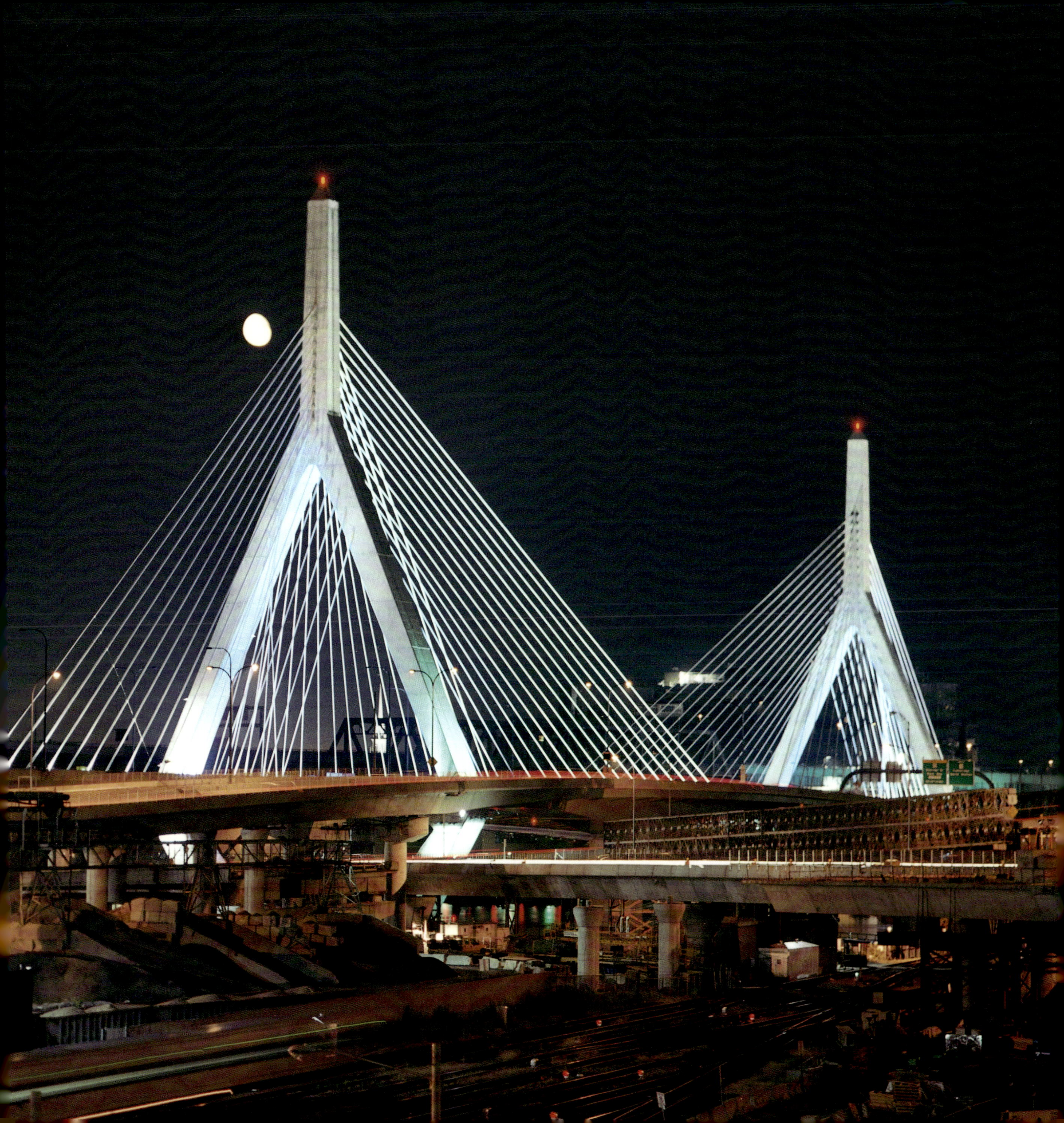

Notes

1. C. Menn, "Aesthetics in Bridge Design," *Bulletin of the International Association for Shell Structures* (August 1985): 53.

2. Ibid.

3. C. Menn, *Prestressed Concrete Bridges*, 2d ed., ed. and trans. P. Gauvreau (Berlin, 1990), 49. Although the first edition was titled "Stahlbeton" meaning "reinforced," all the design examples were in prestressed concrete, so the change to "prestressed" gave a clearer meaning to the English edition.

4. Ibid., 55.

5. Menn, "Aesthetics in Bridge Design," 49.

6. Ibid.

7. D. P. Billington, "The Bridges of Christian Menn," in *The Bridges of Christian Menn*, exh. cat., Princeton University Art Museum (Princeton, 1978), 7.

8. C. Menn, "Recollections of Prof. Dr. Pierre Lardy," April 22, 2002, p. 1, Princeton Maillart Archive (hereafter PMA).

9. Ibid.

10. Ibid.

11. Ibid., 2.

12. H. Isler, Notebooks for courses by Pierre Lardy, Book 6, Massive Brücken, 1949.

13. Ibid., Book 8, Baustatik und Massivbau Ausgewählte Kapitel, 1950.

14. Menn, "Recollections," 2.

15. Ibid.

16. Ibid.

17. Ibid.

18. *Christian Menn, Brückenbauer*, ed. T. Vogel and P. Marti (Basel, 1997). This book, which also served as a catalogue for an exhibition in Switzerland, contains listings of Menn's works with dates and spans from which I have taken relevant information for this chapter. Menn's website gives for a few of his works slightly different dates, which I have used in this text.

19. C. Menn, "New Bridges," in *The Maillart Papers from the Second National Conference on Civil Engineering: History, Heritage, and the Humanities*, Princeton University, Department of Civil and Geological Engineering, October 1972, ed. D. P. Billington et al. (Princeton, 1973), 105.

20. D. P. Billington, "Bridges and the New Art of Structural Engineering," *American Scientist* 72 (Jan.–Feb. 1984): 22–31; and the offprint, C. Menn, "Brückenbau–Weg und Ziel," *Schweizer Ingenieur und Architekt* 26 (1993), cover.

21. This summary is taken from D. P. Billington, "Swiss Bridge Design Spans Time and Distance," *Civil Engineering* 51, no. 11 (Nov. 1981): 42–44.

22. "Felsenaubruecke: Program für Projektwettbewerb," Bern, July 14, 1970, 12 pages.

23. "Felsenaubruecke: Schlussbericht der Jury," Bern, June 1971, 23 pages.

24. U. Finsterwalder, "Free-Span Prestressed Concrete Bridges," *American Concrete Institute Journal, Proceedings* 49, no. 3 (Nov. 1952): 225–32 and "Prestressed Concrete Bridge Construction," *American Concrete Institute Journal, Proceedings* 62, no. 9 (Sept. 1965): 1037–46.

25. C. Menn, "Projektgrundlagen und Entwurf," in *Felsenaubrücke: Projekt + Ausführung* (Chur, 1975), 16–20; and H. Rigendinger and W. Maag, "Die Projektierung des Überbaues," in *Felsenaubrücke: Projekt + Ausführung*, 32–55.

26. Billington, "Swiss Bridge Design Spans Time and Distance," 44–46. See also "Prestressed Stays Stiffen Spans," *Engineering News-Record*, December 18, 1980, 68–69 and the cover.

27. C. Menn, H. Rigendinger, and W. Maag, "The Bridge over the Ganter Valley for the New Simplon Road in Switzerland," *Cemento* 52, no. 6 (June 1982): 509–35.

28. C. Menn, "Ganterbruecke," paper presented at Bridge Engineering Conference, St. Louis, Sept. 25–29, 1978; revised German version, August 7, 1979; quote from pages 10–11.

29. "Concorso Viadotto S. Pellegrino-Biaschina: Concorrenti e Giuria, Bellinzona," September 1977, 24 pages, PMA. See also "Swiss Highway Soars over Rugged Terrain," *Engineering News-Record*, January 27, 1983, 29–31.

30. "Viadotto della Biaschina," in *Prestressed Concrete of Switzerland, Fédération Internationale de Précontrainte, Congress* (Wildegg, 1982), 58–61. An oversight in this publication and the one cited in the preceding note is the omission of Christian Menn's name as the originator of the conceptual design.

31. U. Guzzi and B. Meier, "Viadotto della Biaschina," *Schweizer Ingenieur und Architekt* 38 (1979): 739–45

32. "Mémoire Technique," competition report for the Menn design, Sion, 23 pages plus 5 drawings, undated; but the competition was in 1985; published in *Pont sur le Rhône Chandoline*.

33. C. Menn, "An Approach to Bridge Design," *Engineering Structures* 13 (April 1991): 106–9.

34. *Concorso*, 1983, cover and p. 20; report of a design competition in the PMA.

35. Menn, "An Approach to Bridge Design," 110–12.

36. H. Figi, C. Menn, D. J. Bänziger, and A. Bacchetta, "Sunniberg Bridge, Klosters, Switzerland," *Structural Engineering International* 7 (Feb. 1997): 6–8.

37. C. Menn, "Generelle Berechnung," *Schweizer Ingenieur und Architekt* 44 (1998) and C. Honigmann, "Conceptual Design of Bridges with a Focus on Christian Menn," M.S. thesis, Princeton University, Department of Civil and Environmental Engineering, June 2001, 43–67.

38. The cable and deck forces are high when the pylons are low because $T = V/\sin ø$, and for small angles $ø$, the $\sin ø$ becomes small.

39. Menn ("Generelle Berechnung") described the bridge as follows: "The structural system of the Sunniberg Bridge corresponds more or less to a relatively slender cantilever system with hinges at midspan. The cantilever moments due to dead load are balanced. The cantilever moments due to live load on only one span are distributed at the pier approximately as follows: 20% into the adjacent span and 80% into the pier."

40. Menn, "An Approach to Bridge Design," 112.

41. T. Trüb, "Aspekt der Örtlichen Bauleitung," *Schweizer Ingenieur und Architekt* 44 (1998): 20, and "Sunnibergbrücke," Baudepartment Graubünden, ca. 1997. For the increase of 15 percent, see C. Menn, "Baukultur im Brückenbau," *Baukultur* 3 (1998): 26. As Menn more recently described it: "The Klosters bypass is extremely expensive. On grounds of environmental considerations 400 Mio CHF will be spent for additional tunnels. On the other hand [an extra] 2 Mio CHF have been granted reluctantly for an extraordinary bridge."

42. See note 19 above.

43. *The Bridges of Christian Menn*, 15 pages; see note 7 above.

44. *Bridge Design: Aesthetics and Developing Technologies*, ed. A. Fleet Bacow and K. E. Kruckemeyer (Boston, 1986), 171 pages.

45. D. P. Billington, "Better Bridges through Engineering Design Competitions," *Proceedings of the World Congress on Structures, San Francisco, July 1998* (San Francisco, 1998), 8 pages.

46. C. J. Pennington, *The Charles River Bridge and Twenty-first Century Signature Design* (senior thesis, Princeton University, 1998), 56–60.

47. "Woodrow Wilson Bridge: Design Competition, Report of the Jury," 2001, 20 pages. Copies of Menn's design drawings are in the PMA.

48. "Theme 4, Very Long Span Bridges," in *Cable-Stayed and Suspension Bridges, Proceedings of the Conference of the International Association of Bridge and Structural Engineers* (Deauville, 1994), vol. 1, 489–594.

49. C. Menn and D. P. Billington, "Breaking Barriers of Scale: A Concept for Extremely Long Span Bridges," *Structural Engineering International* 1 (1995): 48–50. I translated this article into English, but the ideas are entirely those of Christian Menn.

50. Ibid., 49.

51. Ibid., 50.

52. Menn, *Prestressed Concrete Bridges*, 58.

53. Ibid, 58.

54. Ibid.

55. Ibid., 61.

56. Ibid., 62.

57. Ibid.

58. Ibid., 63–64.

59. Ibid., 64.

60. Ibid., 293–438.

61. Letter from Christian Menn to D. P. Billington, July 17, 2002, p. 2.

SELECTED MAJOR WORKS

Robert Maillart*

Date	Place	Crossing/Covering	Type
1. 1899	Zurich (Stauffacher Strasse)	Sihl River	Unreinforced, three-hinged arch
2. 1901	Zuoz	Inn River	Three-hinged, hollow-box arch
3. 1904	Billwil	Thur River	Three-hinged, hollow-box arch
4. 1905	Tavanasa	Rhine River	Three-hinged, hollow-box arch
5. 1912	Rorschach	Filtration plant roof	Flat slab
6. 1923	Wägital	Flienglibach	Deck-stiffened arch
7. 1924	Chiasso	Roof shed	Roof-stiffened inverted arches
8. 1925	Châtelard	Eau-Noire	Continuous hollow-box beam
9. 1925	Donath	Valtschielbach	Deck-stiffened arch
10. 1930	Schiers	Salginatobel	Three-hinged, hollow-box arch
11. 1932	Schwarzenburg	Rossgraben	Three-hinged, hollow-box arch
12. 1933	Hinterfultigen	Schwandbach	Deck-stiffened arch
13. 1933	Felsegg	Thur River	Three-hinged, hollow-box arch
14. 1934	Wülfingen-Winterthur	Töss River	Deck-stiffened arch
15. 1936	Vessy	Arve River	Three-hinged, hollow-box arch
16. 1938	Zurich	Cement Hall roof	Thin-shell
17. 1940	Lachen	Railroad	Three-hinged, hollow-box arch

*For a complete listing of Maillart's works, see D. P. Billington, *Robert Maillart: Builder, Designer, and Artist* (Cambridge, 1997), 300–315.

Othmar Ammann*

Date	Name**	Crossing	Type
1. 1931	George Washington Bridge	Hudson River	Suspension
2. 1931	Bayonne Bridge	Kill van Kull	Spandrel-braced, trussed-arch
3. 1936	Triborough Bridge	East River	Suspension
4. 1936	Little Hell Gate Bridge	Hell Gate	Plate-girder arch
5. 1939	Bronx-Whitestone Bridge	East River	Suspension
6. 1949	Harlem River Pedestrian Bridge	Harlem River	Plate-girder
7. 1961	Throgs Neck Bridge	East River	Suspension
8. 1962	George Washington Bridge	Hudson River	Second deck
9. 1964	Verrazano Narrows Bridge	The Narrows	Suspension

*The firm of Ammann & Whitney, formed in 1946, designed many works for which Ammann, as senior partner, was in general charge. These have yet to be fully catalogued, and none, except those listed above, are large-scale steel bridges designed by Ammann himself.

**All bridges are within New York City except the first two, which connect New York City to New Jersey.

Heinz Isler*

Date	Place	Covering: Roofs	Type
1. 1958	Thun	Eschmann Factory	Pneumatic**
2. 1960	Wangen bei Olten	COOP Distribution Center	Pneumatic
3. 1962	Solothurn	Wyss Garden Center	Geometric
4. 1965	Lyssach	Isler office building	Watertight flat roof
5. 1965	Recherswil	Kilcher factory	Flow
6. 1965	Paris	Clause Garden Center	Flow**
7. 1967	Lommiswil	Parish church	Hyperbolic parabaloid
8. 1968	Deitingen	BP service station	Hanging membrane reversed
9. 1969	Geneva	Sicli Factory and office	Hanging membrane reversed
10. 1971	Camorino	Bürgi Garden Center	Flow
11. 1976	Port de Pully	Sculpture of Angel Duarte	Hyperbolic
12. 1977	Grötzingen (Germany)	Outdoor theater	Hanging membrane reversed
13. 1979	Heimberg	Swimming pool	Hanging membrane reversed**
14. 1979	Heimberg	Tennis courts	Hanging membrane reversed**
15. 1981	Brugg	Swimming pool	Hanging membrane reversed
16. 1987	Dübendorf	Aircraft museum	Hanging membrane reversed
17. 1991	Norwich (England)	Swimming pool	Hanging membrane reversed

*For a complete listing of Isler's works, see J. Chilton, *Heinz Isler* (London, 2000), 165–66. See also *Heinz Isler Schalen*, ed. E. Ramm and E. Schunck, 3d ed. (Zurich, 2002), 97–101.

**Isler has designed a large number of similar thin-shell roofs of these types for which those listed here are typical.

Christian Menn*

Date	Place	Crossing	Type
1. 1958	Crestawald	Hinterrhein	Deck-stiffened arch
2. 1959	Letziwald	Averserrhein	Three-hinged, hollow-box arch
3. 1959	Cröt	Averserrhein	Deck-stiffened arch
4. 1962	Bad Ragaz	Rhine River	Prestressed-concrete girder
5. 1962	Uors	Valserrhein	Deck-stiffened arch
6. 1963	Tamins bei Reichenau	Rhine River	Deck-stiffened arch
7. 1967	Viamala	Rhine River	Deck-stiffened arch
8. 1967	Nanin	Moesa River	Deck-stiffened arch
9. 1968	Cascella	Moesa River	Deck-stiffened arch
10. 1969	Salvanei	Moesa Valley	Prestressed-concrete, hollow-box girder
11. 1974	Felsenau	Aare River	Prestressed-concrete, hollow-box girder
12. 1980	Simplon Road	Ganter Valley	Cable-stayed, prestressed-concrete, hollow-box girder
13. 1983	Biaschina	Ticino River	Prestressed-concrete, hollow-box girder
14. 1989	Chandoline	Rhône River	Cable-stayed
15. 1999	Sunniberg	Landquart River	Cable-stayed
16. 2002	Boston	Charles River	Cable-stayed

*For a complete listing of Menn's works, see *Christian Menn: Bruckenbauer*, ed. T. Vogel and P. Marti (Basel, 1997), 42–49.

GLOSSARY OF TERMS

Abutment: The end support for a bridge where the last spanning element rests on foundations and beyond which the roadway rests on ground.

Anchorages: Devices at either end of a prestressed concrete beam that hold the prestressing steel tension elements (bars or wires). Also, the massive concrete blocks at the end of a suspension bridge that resist the inward pull of the bridge cables.

Beam: A horizontal member, usually of constant depth, that carries a vertical load by bending.

Bending moments: Internal forces on a structural element that cause the element to bend and thus create internal compression on one side of the element and internal tension on the other.

Cable-stayed bridge: A form defined by a deck supported by diagonal stays that emanate from a tower.

Cantilever: A structural element with fixed support at one end and free of any support at the other end. *Fixed support* means that the element cannot rotate or translate at that point.

Column capital: A structural element that connects a column to a horizontal element (usually a slab).

Column: A vertical member that carries a vertical load primarily by axial (vertical) forces in compression.

Concrete: Artificial stone made by mixing crushed stone or gravel, sand, cement, and water and allowing the cement to react chemically with the water to bind the mix into a solid.

Continuous beam: A beam that extends unbroken over at least three supports.

Cross wall: A wall that is vertical (or nearly so) and oriented in the direction perpendicular (or nearly so) to the bridge axis.

Crown (of an arch): The high point of an arch, usually at the midspan. In stone arches the keystone is usually at the crown.

Deck beam: A horizontal element supporting the roadway slab and usually oriented in the direction of the bridge axis.

Deck: On a bridge, the horizontal structure consisting of a roadway slab (or deck slab), deck beams, and spandrel walls (when they are above the roadway).

Deck-stiffened arch: An arch connected by vertical elements to a deck beam (or spandrel beam) that stiffens the arch in such a way that the arch carries almost no bending moments and can therefore be thin.

Diaphragm: The usually vertical slab that stiffens a bridge locally or against twisting.

Elliptic paraboloid: A geometric shape defined by two sets of parabolas, both convex with respect to the ground.

Flanges: The horizontal plates at the top and bottom of plate girders.

Formwork: The structure (usually wooden) into which fluid concrete is cast and allowed to harden.

Girder: A primary beam that usually carries loads directly to a column. A *plate girder* is a beam fabricated out of steel plates having horizontal flanges top and bottom connected by a vertical web.

Graphic statics: A method of analyzing a structure by visual diagrams rather than by algebraic formulas.

Haunched beam: A beam without a constant depth that usually has a straight horizontal top edge and a bottom edge that becomes lower as it approaches a support point. This lower part (the additional depth added to an otherwise constant depth of beam) is called the *haunch*.

Hinge: In concrete structures, a point where the element is purposely reduced in size to such an extent that it is free or nearly free to rotate at that point. In some cases, an actual steel hinge is embedded in the concrete; in other cases, the hinge may be made by creating a plane of weakness in the concrete.

Hollow box: A structural element (beam, column, or arch) made up of a combination of vertical and horizontal walls that together carry the loads. The walls are usually thin and leave a substantial open space within. In a hollow-box arch, the top wall can be the roadway, and the bottom wall can be an arched slab, while the side walls connect the top and bottom elements.

Hyperbolic paraboloid: A geometric shape defined by two sets of parabolas, one convex and the other concave with respect to the ground.

Joist: A secondary beam used to carry loads from a floor or roadway slab to a larger beam; often called a girder.

Longitudinal wall: A wall that is vertical (or nearly so) and oriented in the direction of the bridge axis.

Parapet: The longitudinal element along either side of a bridge deck.

Pier: The support of a bridge below the roadway.

Pneumatic form: A concrete roof shape defined by inflating a bladder that gives a form defined by pressure loading, i.e., loading perpendicular everywhere to the surface.

Prestressed concrete: A method of construction whereby compression forces are applied to concrete elements (usually by steel bars or wires under high tension, anchored at either end of the element) with the goal of counteracting the tension that would otherwise occur due to loads.

Pylon: The tower above a bridge deck that usually carries cable stays.

Quarter span: The points on an arch that are halfway between midspan and support points measured horizontally.

Reinforced concrete: Steel bars embedded in concrete and designed to take the tension that occurs on account of loads.

Segmental bridge: A bridge, usually constructed as cantilevers, built by prestressing together a series of segments longitudinally.

Slab: A concrete element whose thickness is small in relation to its length and width.

Span: The length of a horizontal element measured horizontally between its two end supports.

Spandrels (walls): Side walls on a bridge. In some cases, they are above the roadway and serve as parapets or guardrails; and in other cases, they are below and serve to fill the spaces between roadway slab and arch.

Springing line: The imaginary line drawn between the two end-support points of an arch.

Stiffened suspension bridge: A suspension bridge with a stiff horizontal deck.

Suspension bridge: Usually defined as a bridge where the deck load is carried vertically (or mostly vertically) by suspenders up to cables that transfer the loads by nearly horizontal forces to towers and then over them to side spans ending in heavy concrete anchorages.

Thin shell concrete structure: A curved concrete surface whose thickness is small in relation to its length and width.

Three-hinged arch: A curved structure with hinges usually at each abutment and at the crown.

Truss: A structure composed of a usually horizontal top element and a usually horizontal bottom element, the two connected by vertical and diagonal elements to create a beamlike form.

DRAWING AND PHOTOGRAPHY CREDITS

DRAWINGS

William Cooch, Princeton University, Class of 2003

Chap. 1: fig. 6 (after W. Ritter, *Der Brückenbau in den Vereinigten Staaten Amerikas* [Zurich, 1895]: fig. 52B); Chap. 2: figs. 3–6, 14, 20A–D, 25 (after Mark Reed), 26, 46a–d; Chap. 4: figs. 2–4; Chap. 5: fig 31 (after *Heinz Isler as Structural Artist* [Princeton, 1980], fig. 32); Chap. 6: fig. 2 (after C. Menn, in Bulletin of the International Association for Shell Structures [August 1985]: 53).

Evans and Agans

Chap. 2: figs. 38, 39, 43, 44.

Clark Fernon

Chap. 2: figs. 18, 19, 31, 32, 33.

Jill Moraca

Maps p. 10, 11.

Mark Reed

Chap. 2: figs. 7, 15, 54.

William F. Shellman

Chap. 2: figs. 11, 12.

PHOTOGRAPHS

Courtesy of Margot Ammann-Durrer

Chap. 3: figs. 1, 18.

Courtesy of the Baugeschichtliches Archiv, Zurich

Chap. 2: fig. 10.

David P. Billington

Chap. 2: figs. 51, 52; Chap. 5: figs. 2–4, 14, 15, 17B, 27, 28, 38, 40; Chap. 6: figs. 11, 21.

Courtesy of Madame M.-C. Blumer-Maillart

Cover: illus. 4, 7, 17; Chap. 2: figs. 1, 17, 22, 24, 27, 37, 47, 53, 55.

Courtesy of Mrs. Dorothy Candela

Chap. 5: figs. 9, 10.

Case Western Reserve University, Department of Special Collections

Chap. 3: fig. 20.

Frank Döring

Cover: illus. 5; Chap. 2: fig. 36; Chap. 5: figs. 1, 11, 17A, 25, 36.

Image Archive of the ETH Library, Zurich

Chap. 1: figs. 1–3; Chap. 2: figs. 2, 8, 9, 16, 34, 35; Chap. 3: fig. 23; Chap. 4: figs. 1, 24; Chap. 6: fig. 1.

Courtesy of the Institute Eduardo Torroja

Chap. 4: figs. 22, 23.

Courtesy of Heinz Isler

Cover: illus. 2, 12, 15; Chap. 5: figs. 5–8, 12, 13, 16, 18, 19–24, 26, 29, 30, 32–35, 37, 39.

Library of Congress

Chap. 3: fig. 10.

Mancia/Bodmer FBM Studio, Zurich

Chap. 2: figs. 13, 21, 23, 29, 30, 40, 45, 49.

Robert Mark

Chap. 2: fig. 28.

Courtesy of Christian Menn

Cover: illus. 8, 10, 14; Frontispiece; Chap. 6: figs. 7, 8, 10–12 (Fotoateuer Vonow, Chur, Switzlerand), 13, 14 (Foto L. Crameri, Graubunden, Switzerland), 15–17, 19, 20, 22, 25, 26, 28, 31, 33, 34.

Courtesy of the Massachusetts Turnpike Authority, Boston

Chap. 6: figs. 32, 35, 36.

Courtesy of the Metropolitan Transportation Authority, New York

Chap. 3: fig. 28 (Photo © MTA Bridges and Tunnels Special Archive 1990, Richard Averill Smith, ca. 1939).

Courtesy of the Museum of the City of New York

Cover: illus. 13, 16; Chap. 3: figs. 22, 27 (The Gottscho-Schleisner Collection), 33.

Courtesy of the Museum of Modern Art, New York

Cover: illus. 11; Chap. 2: figs. 42, 48.

Courtesy of the Port Authority of New York

Chap. 3: figs. 19, 21, 26.

Courtesy of the Rosengart Collection, Lucerne

Chap. 2: fig. 50.

Emily Webster

Chap. 6: figs. 27, 29.

Bruce M. White, Caldwell, New Jersey

Cover: illus. 1, 3, 6, 9; Chap. 1: figs. 4, 5 (after W. Ritter, *Der Brückenbau in den Vereinigten Staaten Amerikas* [Zurich, 1895], fig. 7), 8 (ibid., fig. 56), 9 (after W. R. Hutton, The Washington Bridge [New York, 1889], ix), 11 (after Ritter, *Brückenbau*, fig. 60), 12 (ibid., fig. 29); Chap. 2: figs. 11, 12, 18, 19, 31, 32, 41, 43, 44, 56 (after M. Ros, Bericht 99, Eidgenössische Materialprüfungsanstalt an der Eidgenössische Technische Hochschule, Zurich, addendum 2 [August 1, 1940], 21–60); Chap. 3: figs. 2 (after O. Ammann, in Transactions of the ASCE 82 [1918]: fig. 2), 3 (ibid., fig. 3), 4 (ibid., fig. 4), 5 (ibid., fig. 5), 6 (ibid., fig. 6), 7 (ibid., fig. 7), 8 (ibid., fig. 8), 9 (ibid., fig. 9), 11, 12, 13 (after W. Ritter, *Anwendungen der Graphischen Statik . . .*, pt. 4, Der Bogen [Zurich, 1906], fig. 40), 14 (ibid., fig. 45), 15 (after Ammann, in Transactions 82, fig. 55), 16 (ibid., fig. 54), 17 (after Ammann, in Transactions of the ASCE 97 [1933]: figs. 34–36), 24 (after Port of New York Authority Progress Report 1930), 25, 29, 31, 32, 34; Chap. 4: figs. 5, 6 (after L. C. Urquhart and C. E. O'Rourke, *Design of Concrete Structures* [New York, 1940], 4th ed., fig. 93), 7–11, 12 (ibid., fig. 117), 13–15, 16 (ibid., fig. 161), 17 (ibid., fig. 178), 18–21; Chap. 6: figs. 3–6, 9, 23, 24 (after C. Menn, in Engineering Structures 13 [April 1991]: 111, fig. 12), 30, 37.

Wayman Williams

Chap. 1: fig. 10; Chap. 2: figs. 30, 57; Chap. 3: fig. 30.

C.M. Woodward, *A History of the St. Louis Bridge* (St. Louis, 1881)

Chap. 1: fig. 7.

INDEX

F

G

H

I

J

K

L

M